GOVERNMENTAL MANAGEMENT OF CHEMICAL RISK
Regulatory Processes for Environmental Health

Rae Zimmerman

LEWIS PUBLISHERS

Library of Congress Cataloging-in-Publication Data

Zimmerman, Rae.
    Governmental management of chemical risk/by Rae Zimmerman.
      p.   cm.
    Includes bibliographical references.
    1.   Hazardous substances—Law and legislation—United States.
  2.   Risk Management—Government policy—United States.   I. Title.
T55.3.H3Z56   1990
363.17′5—dc20   89–13325
    ISBN 0–87371–143–2

LEWIS PUBLISHERS, INC.

PRINTED IN THE UNITED STATES OF AMERICA

We gratefully acknowledge receipt of permission to use material from the works cited below.

D. L. Bazelon. "Risk and Responsibility," *Science* 205:277–280 (1979), p. 278 (p. 253 in this volume). Copyright 1979 by the American Association for the Advancement of Science.

D. L. Bazelon. "The Judiciary: What Role in Health Improvement?" *Science* 211:792–793 (1981), p. 792 (p. 94 in this volume). Copyright 1981 by the American Association for the Advancement of Science.

M. A. Berry. "A Method for Examining Policy Implementation: A Study of Decisionmaking for the National Air Quality Standards, 1964–1984," PhD dissertation (University of North Carolina, 1984), four figures (5.14–5.17, pp. 239–242, in this volume).

S. Breyer. *Regulation and Its Reform* (Cambridge, MA: Harvard University Press, 1982), pp. 3, 163 (pp. 236, 81 in this volume). Reprinted with permission. Copyright 1982 by the President and Fellows of Harvard College. All rights reserved.

C. H. Foreman, Jr. "Congress and Social Regulation in the Reagan Era," in *The Reagan Regulatory Strategy: An Assessment,* George C. Eads and Michael Fix, Eds. (Washington, DC: The Urban Institute Press, 1984), p. 189 (p. 172 in this volume).

T. Greenwood. *Knowledge and Discretion in Governmental Regulation* (New York: Praeger Publishers, 1984), p. 4 (p. 210 in this volume). Copyright 1984 by Praeger Publishers.

M. Kraft and R. Kraut. "Citizen Participation and Hazardous Waste Policy Implementation," in *Dimensions of Hazardous Waste Politics and Policy,* C. E. Davis and J. P. Lester, Eds. (Westport, CT: Greenwood Press, 1988), p. 65 (p. 54 in this volume). Copyright 1988 by the Policy Studies Organization.

A. A. MacIntyre. "The Multiple Sources of Statutory Ambiguity: Tracing the Legislative Origins to Administrative Discretion," in *Administrative Discretion and Public Policy Implementation,* D. H. Shumavon and H. K. Hibbeln, Eds. (New York: Praeger Publishers, 1986), pp. 67–88 (pp. 46, 201, 225–226 in this volume). Copyright 1986 by Praeger Publishers.

R. A. Merrill. "Risk-Benefit Decisionmaking by the Food and Drug Administration," *George Washington Law Review* 45:994–1012 (1977), p. 1001 (p. 231 in this volume). Copyright 1977 by the *George Washington Law Review*.

D. F. Morgan and J. A. Rohr. "Traditional Responses to American Administrative Discretion," in *Administrative Discretion and Public Policy Implementation,* D. H. Shumavon and H. K. Hibbeln, Eds. (New York: Praeger Publishers, 1986) p. 221 (p. 275 in this volume). Copyright 1986 by Praeger Publishers.

D. M. O'Brien. "Administrative Discretion, Judicial Review, and Regulatory Politics," in *Administrative Discretion and Public Policy Implementation,* D. H. Shumavon and H. K. Hibbeln, Eds. (New York: Praeger Publishers, 1986), pp. 41–42 (pp. 257, 277 in this volume). Copyright 1986 by Praeger Publishers.

D. M. O'Brien. *What Process is Due? Courts and Science—Policy Disputes* (New York, NY: Russell Sage Foundation, 1988), pp. xii, 3, 62 (pp. 108, 272–273 in this volume). Copyright 1988 by The Russell Sage Foundation.

W. Rowe. *Anatomy of Risk* (New York: John Wiley & Sons, Inc., 1976), p. 24 (p. 26 n. 4 in this volume). Copyright 1977 by John Wiley & Sons, Inc. All rights reserved.

D. H. Shumavon and H. K. Hibbeln. "Administrative Discretion: Problems and Prospects," in *Administrative Discretion and Public Policy Implementation,* D. H. Shumavon and H. K. Hibbeln, Eds. (New York: Praeger Publishers, 1986), pp. 1–2 (p. 17 in this volume). Copyright 1986 by Praeger Publishers.

Preface for the Series

Given the complex and ever-expanding body of information in toxicology and environmental health, the purpose of the Toxicology and Environmental Health Series is to present a genuine synthesis of information that not only will offer rational organization to rapidly evolving developments but also will provide significant insight into and evaluation of critical issues. In addition to its emphasis on assessing and assimilating the technical aspects of the field, the series will offer leadership in the area of environmental health policy, including international perspectives. Thus, the intention of this series is not only to provide a careful and articulate review of critical areas in toxicology and environmental health but to influence the directions of this field as well.

The Editorial Board will oversee and shape the series, while individual works will be peer-reviewed by appropriate experts in the field.

Edward J. Calabrese
University of Massachusetts
Amherst, Massachusetts

Preface

Human health risks associated with chemicals in the environment periodically claim the public's attention in a variety of ways. The latest revival of this issue has been as an outgrowth of the environmental protection movement of two decades ago. Many supporters of the earlier environmental protection movement were quickly drawn into the environmental health risk issue. It is a relatively new concern in the history of regulation and now takes its place as one of the major components of social regulation.

Government is a major player in the control of chemical risks. This book explores the ways government adapts its management system to this new policy area. It addresses the question of whether the risk issue has altered the organization and decisionmaking processes of government—in particular, whether a new risk bureaucracy has been spawned or whether the customs and traditions of normal government continue to dominate. Those who manage policy in government are initially faced with the decision of how to approach a new problem. A common reaction is to start with those management tools and analytical techniques that are at hand. When these no longer work, new initiatives are explored to solve the problems. Understanding how government has reacted to the risk issue in this manner sheds light on its future capacity to meet the new challenges posed by chemical risk through readjustments in the management system in view of these challenges.

In the course of focusing on how government manages the policy formulation and policy implementation processes with traditional approaches and new initiatives, this book views governmental action in terms of the intensity and direction of actions by the branches of the federal government, their own organizational tendencies toward bureaucracy and discretion, and their sensitivity to the outcomes of their own behavior. The first step in this process is to understand the backdrop or context of "normal" government, its norms and traditions, with which governmental actions in risk management are compared. Such a comparison has to be placed in a historical context. Thus, this exploration into the governmental management of risk is in part a history, though a recent one.

The complexity of an issue such as chemical risk often encourages an exploration of only a few of the pieces of a much larger picture. This book attempts to address the larger picture by placing the management of chemical risk in the context of a wide range of governmental activity. In that regard its major emphasis is on federal government processes. It is federal governmental management and design of its programs that have set the stage for state and local government action by often defining, guiding, or at least motivating these governments. Within that context, the book begins by covering two forms of management—regulation by directive and incentive-based forms of risk control.

This is followed by an analysis of the way Congress, the executive agencies, and the judiciary have drawn on traditional mechanisms as well as new initiatives to manage risks within their domain of responsibilities.

Given the extent of coverage of the federal government's involvement in the risk issue, the book is intended to serve as an empirical work and as a text for students and scholars of governmental processes for the management of public policy, especially a policy that relies heavily on scientific information. Its coverage of the substance of the risk issue—its laws, regulations, and management structure—is also aimed at those engaged in the technical aspects of managing chemical risks who seek the social and political basis for and implications of their work. In addition to addressing these two audiences separately, the book tries to provide a bridge between science and policy implementation. Its publication in a series whose major thrust is environmental science is a reflection of one of the objectives of the book: to bring the management of public policy in the risk area closer to the science of environmental risk.

This book represents the culmination of a number of years of my work in the area of governmental risk management. It began as an outgrowth of my work on government institutions responsible for environmental regulation and planning. Subsequently, under sponsorship from the National Science Foundation, I explored government's management of a dozen or so cases of chemical contamination of the environment and the alleged health risks ensuing from these episodes. The outcome of these cases in the regulatory setting first alerted me to the role of the management of public policies toward chemical risks because of the periodic reappearance of the problems or their lack of closure. This work was also the foundation for a course that I developed in 1983 on risk management in environmental health and protection that has been offered as an annual summer institute at New York University's Robert F. Wagner Graduate School of Public Service and that was initially sponsored by a New York University Curricular Challenge Fund grant.

Chemical risks, both actual and perceived, will be with us for a long time. *Governmental Management of Chemical Risk* is intended to help both government officials and students of governmental processes to see the ramifications of governmental actions in the future and provide direction for these actions.

Acknowledgments

This book has been in the making for quite some time. Accordingly, the acknowledgments extend to many individuals who have influenced the work both directly and indirectly.

I would like to thank a number of people who reviewed portions of the manuscript. Donald W. Stever, Jr., who has been a Professor of Law at Pace University Law School and is the author of a major work on the legal aspects of chemical risks, scrutinized the details of the legal foundations for risk management and of Chapter 6 (on the judiciary). Michael Baram, Professor of Health Law at Boston University's School of Medicine and Public Health, provided important insights into common law. Howard Kunreuther, Professor at the Wharton School and a noted scholar in the area of financial liability questions pertaining to technological risk, provided invaluable commentary on the discussion of insurance in Chapter 3. Michael A. Berry, Deputy Director of the U.S. EPA's Environmental Criteria and Assessment Office, generously allowed me the use of four figures on air quality decisionmaking from his dissertation at the University of North Carolina at Chapel Hill. Gail Marcus, formerly of the Congressional Research Service and currently with the Nuclear Regulatory Commission, also commented extensively on an early draft of the manuscript.

A number of my colleagues at New York University's Robert F. Wagner Graduate School of Public Service provided invaluable inputs into my work as well. In general, the exchange of ideas in public management was made possible by a community of interests at the Wagner School. This enabled me to apply public management to the discipline of environmental science and planning, which I had previously approached from quite a different perspective. Dick Netzer, Professor of Economics and Public Administration, reviewed the section on taxation and other financial mechanisms used in risk management. His precision and attention to detail alerted me to a variety of subtleties in applying financial mechanisms to risk issues. Matthew Drennan, Professor of Economics, gave me several invaluable pointers on economic indicators as measures of the backdrop of economic conditions for risk management. Roy Sparrow, Associate Professor of Public Administration and Director of the Center for Management, reviewed large portions of the book on governmental management, bringing his extensive knowledge of the broad public management literature to bear on the theme of the book. Professors Elizabeth Durbin and Elena Padilla provided useful commentary on an earlier draft, highlighting the bases of literature in economics and health management, respectively, that also deal with risk problems from a different perspective. In spite of these extensive reviews, the final responsibility for this work, of course, rests with its author.

I also owe thanks to the various agencies that over the years have funded portions of the research on which the book is based, namely the National Science Foundation, the U.S. Environmental Protection Agency, the New Jersey Institute of Technology Hazardous Substances Management Research Center, and the New York University Challenge Fund. In the course of this research, several research assistants contributed long hours and effort in the collection of some of the data on which the book's findings are based. In particular, Penny Miller, Mark Neggers, and Tina Libenson provided important support in this regard.

I would also like to thank Brian Lewis of Lewis Publishers, who provided the encouragement and initiative for me to embark on this work. I am also indebted to the editor, Robin Berry, who worked patiently with me to polish the language.

Finally, I want to thank my family—my husband, Michael, and my two children, Gabriel and Alexa—who patiently stood by me through the entire effort. It is always hard to live with a writer with a work in progress! In particular, Michael provided computer skills that kept the effort running smoothly in spite of the inevitable obstacles.

Rae Zimmerman is Associate Professor of Planning at New York University's Robert F. Wagner Graduate School of Public Service, where she teaches graduate courses in environmental planning and management, epidemiology, urban infrastructure, and risk management. She received a BA in chemistry from the University of California (Berkeley), a Master's degree in city planning from the University of Pennsylvania, and a PhD from Columbia University, with a concentration on environmental planning.

Throughout her career, Dr. Zimmerman has worked on a varied range of public management problems, most notably through research projects funded by a variety of federal, state, and local agencies. She has published numerous articles and book chapters based on this research, which is in the areas of regulation and environmental risk management, public perception of risk, the origins and management of industrial accidents, the anatomy of hazardous waste controversies, and environmental management studies focusing on institutional analysis.

Dr. Zimmerman has served on numerous advisory committees and panels for federal, state, and local governments, providing reviews and critiques of safe drinking water policy, the application of risk assessment methodologies for planning and regulation, and the selection and development of research centers for hazardous waste management.

Contents

INTRODUCTION

PART I
Legislative and Incentive-Based Systems for Managing Chemical Risk

PART II
Risk Management in the Federal System

CONCLUSION

Figures

Tables

GOVERNMENTAL MANAGEMENT OF CHEMICAL RISK

Regulatory Processes for Environmental Health

INTRODUCTION

CHAPTER 1

Risk Management
in the Governmental Context

RISK MANAGEMENT AND GOVERNMENT PROCESSES

Chemical risks have claimed the public's attention for over two decades, and the management of these risks poses a new challenge for government. When public issues such as chemical risks arise, government responds with a combination of traditional mechanisms and new initiatives. It is this interplay between tradition and initiative that shapes the way any new policy is implemented and managed. This book uses the chemical risk issue as a way of examining how one level of government in the United States—the federal government—manages a new policy area, namely chemical risks.

The risks that chemicals in the environment pose to human health are relatively new, important, and unique to the government for a number of reasons.[1] First, a wide variety of potential sources of chemicals and associated risks exists in the environment. These sources may encompass many different activities, materials, and substances. Chemicals that are potential targets of governmental action number in the millions, and recognition of the size of the problem has occurred only within the last century. Second, many individuals and organizations are potentially affected by these risks. Third, there is no single constituency concerned with chemical risks. The number of organizations and interest groups that could influence the way government manages these risks is very large, and their composition is heterogeneous, encompassing at any given time different combinations of general environmental groups, civic organizations, workers, businesses, trade associations, and property owners. Fourth, there are continual disputes over the goals for chemical risk policies and the means to achieve them. Both the goals and the means depend heavily on how willing people are to accept risks. Criteria for risk acceptance are far from well understood, and debates about "how clean is clean" and who should bear the costs and the benefits occur in almost every discussion involving chemical risks. For any given goal, the means are often in dispute with regard to the role of science, the quality of science, the types and levels of standards to accomplish the mission, and how the different branches of government should interact. The basis for choosing many of the means is uncertain, particularly regarding what their impacts are and whether they will work. The resultant governmental management system is very complex, and the rate and direction in which it is moving is often controversial.

A major purpose of this inquiry is to understand the risk management infrastructure that has evolved within government against the backdrop of traditional government processes. The investigation focuses on how the federal government has managed human health risks from chemicals in the environment during the 1970s and 1980s, and emphasizes the formulation, management, and administration of laws and standards by government.[2] The following questions provide direction to this inquiry:

1. Has risk management placed new demands on the structure and processes of government, and if so, have these demands generated new initiatives?
2. Alternatively, is risk management a part or extension of actions that are commonly regarded as protecting the health and welfare of society,[3] and is it similar to other forms of regulation and control to which government responds with its traditional mechanisms and processes?

The theme of tradition and new initiative in government parallels certain theories that attempt to explain other social phenomena. It is analogous to Kuhn's (1970) division of scientific theory into normal science and new scientific paradigms. Kuhn was particularly interested in the early signs of the emergence of new paradigms in terms of the reaction of the existing system of norms to new directions. He also observed incentives for and direction of change. New approaches, according to Kuhn, can be introduced into normal science, in part through the occurrence of anomalies or contradictions and a breakdown of conventional standards; at some point, a new paradigm arises.

On the other hand, Perrow's (1984) study of accidents involving industrial technologies implies that there are conditions under which government may not be able to change circumstances or introduce new ways of doing things. Perrow implies that what is normal at any given time in society or government is open to interpretation. He interprets many accidents that have resulted in severe consequences to human health and safety as being normal or inevitable in a technologically complex system where it is impossible to anticipate all of the interactions and factors that led to them. Once one recognizes or accepts this argument, technologies can presumably be sorted according to the inevitability of accidents and how much intervention to prevent accidents from occurring can be reasonably expected or pursued.

This book explores some of the signs of new governmental initiatives along with government's traditional reactions to chemical risks. Whether initiatives will be drastic and lead to fundamental changes in government is perhaps yet to be seen; answering that question first requires a patient examination of the steps along each path.

The theme of how government manages risk issues has relevance beyond environmental health risks. Knowledge of how relatively new policy areas such as environmental health risk management become rooted in ongoing governmental processes deepens our understanding of how government responds to new issues. The approach to existing and future risk policies can depend on how government has responded to similar issues in the past. If traditional governmental processes have dominated governmental management of risk, then these traditional mechanisms are likely to guide the way government manages new risk issues in the future. If, on the other hand, the risk issue has required new initiatives by government, tailored specifically to it and not transferable to other governmental needs, then narrowing the focus of attention to the specifics of new environmental health risk issues might be called for in managing them in the future.

Few studies in the literature, either on risk management or regulation, discuss how governmental processes involving the design and operation of organizations and programs are integrated in risk management. Risk has been viewed in a social context as a function of cultural differences (Douglas and Wildavsky, 1982) and as being driven by social-psychological motives (Fischhoff et al., 1981). The economic literature is a very rich source for concepts of risk management as they relate to economic risk and its connection with uncertainty. This literature was pioneered by the work of Knight (1921). Numerous policy and management case studies that focus on individual chemicals or episodes have been conducted in response to the rise in public concern over risk issues. Case studies of individual chemicals include vinyl chloride (Doniger, 1978; Badaracco, 1985) and dioxin/2,4,5-T (Hay, 1982).

The episodic literature includes studies of Love Canal and numerous other hazardous waste site cases; spills in transit; sudden chemical releases such as those at Bhopal and Chernobyl; or accidents, such as at Three Mile Island, that posed the potential for releases. Most of these case studies have not been conducted on a comparative basis, though there are a few exceptions, such as an international comparative study of risk management conducted by Brickman, Jasanoff, and Ilgen (1985) and comparative studies of workplace standards (Occupational Safety and Health Administration [OSHA]) and air quality stan- dards (Environmental Protection Agency [EPA]) (Greenwood, 1984; Broder and Morall, 1983; McGarity, 1979b).

Still other studies address the risk issue within single governmental processes (e.g., legislatively, administratively, judicially). Examples of individual executive agency studies are covered in Wilson (1980), and a study of OSHA decisionmaking was conducted by McCaffrey (1982). Kraft (1986), Johnson (1985), and others have focused on Congress's role as risk manager; Vig (1984) and O'Brien (1988) portray judicial activity in risk management. These administrative studies do not simultaneously address the three branches of the federal government and how they interact to affect and shape the risk issue.

THE CONCEPT AND PROCESS OF RISK MANAGEMENT

Before the investigation of risk management in government is undertaken, the terms risk and risk management are discussed as they pertain to health risks from chemicals in the environment.

Risk

Risk is the chance that a hazard or threat will be realized.[4] Risks covered in this book pertain to the probability of human health hazards occurring from environmental exposure to chemicals originating primarily from human activity.

While there is generally little difference among scholars on the definition of risk,[5] there is less agreement on how risks are categorized. The categories are an important foundation for risk management, since risk managers react to and design their programs around them. Risk categories are alternatively based on risk estimation techniques, risk attributes (such as risk level), or relationships of risk to human behavior.

Sage and White (1980) referencing Chauncey Starr's (1969) categorization, list four kinds of risk: real risk (when anticipated circumstances actually develop), statistical risk, predicted risk (usually estimated from historical information or other models when available), and perceived risk. Vlek and Stallen (1980, 276–277) present other definitions, including "the probability of loss, the size of (credible) loss, expected loss, and the variance of the probability distribution over the utilities of all possible consequences." Social psychologists contribute still another set of categories for risk that are dimensions of perceived risk, such as controllable, voluntary, common, and catastrophic risk (Slovic, Fischhoff, and Lichtenstein, 1980). Finally, the passage of environmental health risk statutes over the past two decades has introduced an even larger number of categories.

First, different terms are used to describe risk. The legislation characterizes risks as being adequate, imminent, substantial, reasonable vs unreasonable, posing grave danger, at a zero level, significant vs de minimis, and lying within an ample or adequate margin of safety (Zimmerman, 1985a; Skaff, 1979; Ricci and Cox, 1987). Chapter 2 of this book deals extensively with this terminology. Second, even within a given category, such as substantial risk, judicial and administrative interpretations can give different meanings to the same characteristic describing risk. (See Chapter 6.) Finally, the boundaries of concern for risk can be quite broad. For example, Stever summarizes EPA's definition of risk under the Toxic Substances Control Act (TSCA) as covering such diverse areas as "human health effects (single instances or patterns), environmental effects and emergency incidents of environmental contamination" (Stever, 1988, 2–35). In the course of examining risk management by government, risk concepts used and their corresponding circumstances will be identified.

Risk Management

Risk management is a process broadly aimed at achieving risk reduction, avoidance, or aversion. It involves balancing the health and economic interests of stakeholders as well as weighing scientific evidence against political judgment. While it can be used to denote the technical engineering solutions to health risk problems, its meaning, as used in this book, is confined to the legal, financial, and administrative mechanisms to develop and manage technical solutions.

The concept of risk management developed throughout this book draws meaning from the broader context of management as applied to organizations, which has had a long and controversial history.[6] Drucker (1974, 5) has characterized management as the organ of institutions, thereby emphasizing its role as essential to the functioning and survival of organizations. Parsons (1960) defined the function of an organizational management system as the means an organization uses to mediate between itself and its external environment or conditions.[7] Management has also been defined more narrowly as the administration of the internal operations of an organization (Parsons, 1960, 63–64; Scott, 1981, 97). Stoner, in a typical textbook on management in organizations, has summed up many of these definitions from a more applied point of view:

> Management is the process of planning, organizing, leading, and controlling the efforts of organization members and of using all other organizational resources to achieve stated organizational goals. (Stoner, 1981, 8)

Appendix 1.1 contains a more extensive discussion of the theories of management that provide a foundation for risk management.

While risk management and management in general share a number of concepts, the meaning of risk management is in some ways broader and in other ways more specialized than the meaning of management. This section explores how risk management has been used by those involved in risk issues in government. It identifies (1) characteristics of risk management developed by those close to the process, (2) how the risk management concept has emerged and evolved over the past two decades to address risk problems, and finally, (3) some current developments in the meaning and use of the concept. The next section then explicitly applies the management perspective to the risk management concept as a framework for the study of risk management in the federal government that follows.

Characteristics of Risk Management

Characteristics of risk management as it is commonly used can be grouped according to two dimensions: (1) the scope of decisions that are made and the decisionmaking strategies used (for example, regulatory or nonregulatory approaches), and (2) the processes and functions that risk management encompasses.

Scope of decisions and decisionmaking strategies. The term risk management has been used for more than a decade to signify a form of management applied to decisions about environmental health and safety risks. In the financial industry, a different meaning of the term risk management predates this use. As used in finance, risk management pertains to the management of financial resources to avoid losses (for example, through the use of insurance).[8]

The term risk management is also used by both governmental and nongovernmental decisionmakers to signify a strategy to reduce, avert, or avoid risks.[9] In this context, many approaches are incorporated under the overall umbrella of risk management, including regulatory and nonregulatory strategies. Covello and Mumpower (1985, 108) summarize risk management activities of the current decade (called "societal risk management") as four strategies, similar to those used earlier by Baram (1982):[10] insurance, the common law, direct governmental intervention, and private self-regulation. A chemical industry publication emphasized private sector strategies as a part of risk management "which resides jointly with the manufacturer, distributor, and public (governmental regulatory agencies are surrogates for the public)" (Smith, 1983, 9).

The process and functions of risk management. Throughout the 1970s, government and industry wrestled with the problem of how scientific information could be used as the basis for health policy decisions, especially where chemicals were a source of the risks. Risk assessment played a central role in that debate as a tool for organizing and analyzing scientific information. Finally, in 1983, a committee of the National Academy of Sciences was formed to investigate how risk assessment and risk management related to one another in government. In the context of that debate, that committee characterized governmental risk management as follows:

> Risk management is the process of weighing policy alternatives and selecting the most appropriate regulatory action, integrating the results of risk assessment with engineering data and with social, economic, and political concerns to reach a decision.
>
> [It is] the process of evaluating alternative regulatory actions and selecting among them. Risk management, which is carried out by regulatory agencies under various legislative mandates, is an agency decision-making process that entails consideration of political, social, economic, and engineering information with risk-related information to develop, analyze, and compare regulatory options and to select the appropriate regulatory response to a potential chronic health hazard. The selection process necessarily requires the use of value judgments on such issues as the acceptability of risk and the reasonableness of the costs of control. (National Research Council, 1983, 18–19)

Definitions of risk management differ according to which functions are included in the process. In particular, a major debate occurs with respect to how risk assessment functions and risk management functions are separated (i.e., whether the technical risk analysis component is incorporated within the risk management function or separate from it).

Many have argued that this separation must occur (National Research Council, 1983; U.S. EPA, December 1984c; Lowrance, 1976). They argue that risk management should be confined to decisionmaking and policy determinations alone, and that the technical results of risk analysis are separate and distinct inputs into the management process. At the other extreme, some have pointed out that judgments occur at every stage in risk assessment, from problem identification (Vig, 1984, 61) to judging the design of animal experiments and extrapolation to humans (Whittemore, 1983, 32–33), and the judgments can only be made in the context of risk management. Latin (1988) argues that social policy has to be integrated with risk assessment at the outset because of the inherent unreliability and uncertainty in even the best scientific evidence. Freudenberg (1988) presents similar arguments in support of the integration of risk assessment and management.

Lave (1986) has elaborated on this more inclusive approach to the risk management process. According to Lave, risk management consists of the steps of risk identification, risk assessment, implementation of a risk management strategy (either regulatory or nonregulatory), and finally, monitoring to evaluate the performance of the management strategy chosen. There are interactions and feedback loops throughout the process, reflecting its dynamic character in light of changing information and values. Starr gives a similarly broad construction of risk management: "Effective risk management requires an optimal combination of risk assessment, technical feasibility, human intervention, and political support" (Starr, 1985, 102).

There is less debate over the components of the risk management and risk assessment processes and in what order the components should occur than there is over where one process ends and the other begins. These components, which derive from systems theory and decision analysis, can be summarized as follows:

- identification of hazards
- estimations of the magnitude of those hazards and probability of undesirable events or situations occurring due to the hazards
- estimations of the consequences or effects of the risks from the hazards being realized, using toxicological, epidemiological, and medical information as well as exposure analysis
- formulation of alternative courses of action to avert, avoid, or reduce risk
- evaluation of alternatives in light of social, economic, and health-based criteria
- determination of risk acceptability as a function of perceptions, attitudes, beliefs, and behavior
- selection and implementation of an alternative on the basis of both technical evaluations and acceptability
- monitoring and feedback or readjustment

These steps both influence and are influenced by governmental legal and regulatory systems and nonregulatory alternatives in risk management. They occur within administrative settings in governmental and private organizations. The chapters that follow explore these legal and administrative aspects of risk management in detail within the context of the federal government.

Emergence of Risk Management Over Two Decades

It is now more important than ever to assess how government engages in the process of risk management. Government managers require these insights to deal with the very different perspectives and frequent changes in outlook that characterize the concept of risk. These different viewpoints appear to be spawning very heated debates about how risks are managed. The frequent volleying of opinion that characterizes the risk issue and its historical antecedents is briefly recounted below.

The issue of how to manage human health risks rode the tide of the environmental movement of the early 1970s and shortly thereafter emerged as an issue distinct from environmental protection. Before the 1970s, environmentalism primarily emphasized resource conservation, aesthetics, and public health problems (focusing on biological risks such as bacterial and viral diseases, not on chemicals). A significant consequence of this trend was the shifting of burden-of-proof responsibilities in litigation involving chemical health hazards from parties claiming injury to those allegedly responsible for the injury. Manufacturers and users of chemicals now shared the burden of demonstrating that their products and activities were not risky (Havender, 1982, 49–51).

During the 1980s, an important backlash occurred in response to the popularity of environmental protection and environmental health issues. This took the form of a reduction in resources allocated for the control of environmental risk and government attempts to cut funding for numerous environmental programs.[11] Even prior to these policies, earlier administrations attempted to place boundaries around the way that health risk decisions were made. These occurred as executive orders mandating explicit consideration of economic impact in the formulation of risk policies.

In spite of those policies, a persistent interest in the health aspects of chemical risks, often centering on hazardous waste problems, continued to claim the public agenda. In fact, some argue that the environmental movement became even stronger after the Reagan administration's attempt at program cuts.[12] This continued interest in environmentalism and its human health counterpart was reflected in the growing membership in and influence of environmental organizations (Zimmerman, 1987b, 246; Mitchell, 1979), a growing number of court cases and size of claims against polluters (Wenner, 1984, 182), Congressional oversight and leadership in reversing many of the executive policy decisions, and an upturn in allocation of some of the components of environmental resources by the federal government (evaluated in Chapter 5).

Finally, the middle to late 1980s saw the emergence of still another position with respect to the reach of environmental regulation. The advocates of this position stated that protection had gone too far and that too much safety engenders new, unforeseen risks, since safety systems have risks of their own (Aharoni, 1981; Huber, 1988; Wildavsky, 1988; Nichols and Zeckhauser, 1977). They claimed that one outcome of society's preoccupation with risk was the inconsistent application of standards. For example, new products were subject to a proof of safety prior to their introduction, while those already in existence required only a proof of the existence of risk as a basis for being withdrawn (Wildavsky, 1988, 34–35; Huber, 1983).

Such a cyclical trend in interest in environmental policy has been observed before, as put forth by Downs (1972). While Downs concluded that the cycling would ultimately result in a downturn of interest, the cycling of interest appears to have continued (Gale, 1987, 233). Some economists have identified cyclical trends in regulation and have linked them to a rise and fall in the business cycle (Peltzman, 1976; Shughart and Tollison, 1985). Rather than being cyclical, the environmental health risk policy issue may actually represent a polarity of opinion, whose equilibrium point may shift over time. Regardless of whether environmental interest is cyclical or polarized, controversies have existed through time over both the goals of risk policy and the means to achieve these goals. Governmental processes, whether they are traditional or new (and tailored to risk problems), are at the heart of resolving or at least managing these controversies.

Current Developments in the Risk Management Concept

Recent approaches to a theory or framework for risk management have been proposed that emphasize how problems should be formulated rather than how organizations manage the problems. Many of them emerged during the 1970s and 1980s, when much of the legislation was passed to deal with technological risk. They emphasized social institutions rather than the technical aspects of risk. Notable examples are regulatory frameworks for risk, alternatives to regulation that can be used to alter risks or their outcomes, and institutional policy analysis.

In the early 1980s, regulatory theory emerged in the form of strategies or frameworks applied to risk decisions involving chemicals. For example, Lave (1981, Chapter 2) characterized a set of risk management frameworks that combined economic and health concerns. Many of them are for nongovernment decisionmakers as well as government decisionmakers concerned with risk. These are, according to Lave:

- "market regulation" or the assumption of operation of the market mechanism, where consumers balance the risks of an activity or product against its benefits
- "no risk" (or zero risk), where products or activities involving any risk at all are automatically rejected

- "technology-based standards," where choices are made based on engineering judgment about state-of-the-art performance of techniques to reduce risks, and costs or benefits are not directly involved
- "risk-risk (direct)," where risks and benefits of only a single health effect of product consumption are quantified and balanced against one another
- "risk-risk (indirect)," where the risks and benefits of a broad range of health effects are quantified and balanced against one another
- "risk-benefit analysis," a generalized balancing of risks and benefits, not restricted to health effects
- "benefit-cost analysis," which involves a greater degree of quantification and formalization than risk-benefit analysis
- "cost-effectiveness," the maximizing of an objective at a given or fixed level of cost
- "regulatory budget," which considers only a subset of costs that are defined in a budget[13]

Similarly, incentive-based systems incorporating insurance, subsidies, taxation, and other more unusual forms of incentives began to be applied to chemical risk management. Much of this arose out of or was part of the regulatory reform literature of the late 1970s and early 1980s. Baram (1982) brought together the literature on alternatives to regulation that could be used by governmental or nongovernmental decisionmakers for risk management. Breyer (1982) has also explored regulatory alternatives. This approach is addressed in Chapter 3.

A third approach to risk management falls within the area of "institutional policy analysis." Gormley (1987) describes and critiques these theories in detail. The institutional perspective has been underscored as a major implementation process for public policy in the environmental area in general, though its linkage to risk issues has not been explicit. This approach, according to Vig and Kraft (1984, 8), addresses such questions as whether institutional resources are sufficient for agencies to implement the policies that have been set forth in legislation, given limits on the certainty of knowledge. Institutional policy analysis does not restrict itself to any given level of analysis and can encompass evaluations of organizational and political systems, single organizations, programs, and people (leadership or technical style).

A FRAMEWORK FOR RISK MANAGEMENT IN GOVERNMENT

The previous discussion emphasized how risk management has been used by those close to risk issues rather than analyzing it from the vantage point of organization and management theory. It also dealt primarily with how risk problems are broken down in order to solve them rather than how government organizes to reach those decisions and manage them. To link management to the concepts of risk management described above, two behavioral modes of operation in government are combined with three analytical perspectives stemming from the management literature.

Behavioral Modes

First, existing governmental processes can in some cases be applied to risk problems with little adaptation. Governmental processes already in place are often used when major actions are involved, such as the passing of laws and the setting up of agencies. One example is the influence of the existing Congressional committee structure on the distribution of responsibility for lawmaking involving chemicals among Congressional committees. (This is discussed in Chapter 4.)

Second, risk issues can require the design of new government processes that may or may not be transferable to other policy areas. This can occur, for example, when a program is being refined within a framework that already exists in government. Under such circumstances innovations occur at a micro-scale level. An example discussed in Chapter 3 is the resurrection of the use of federal common law under Superfund and the frequent use of the strict liability common law statute for chemical risk cases. These legal mechanisms are transferable to other policy areas. Alternatively, innovative processes might not be transferable to other areas, and may be applicable to only a few risk areas. An example is the development and application of risk analysis to provide a scientific basis for setting limits on chemical carcinogens.

Analytical Perspectives

Three analytical perspectives will be taken to characterize what happens when government manages chemical risk issues either through traditional processes or new initiatives. The first perspective pertains to the intensity and direction of activity within government, expressed as overall trends in the rate of governmental activities and the degree of emphasis placed on various activity components. The second perspective focuses on the process of government and the degree of bureaucratization that has been associated with the institutionalization of these processes. The third perspective focuses on how government uses outcomes of its own activities to respond to its own performance. (Outcomes are not evaluated normatively as measures of how government is or should be operating, but rather in terms of how they have influenced changes in government.)

These three perspectives and how they relate to the two behavioral modes (tradition and initiative) that government can take in risk management are portrayed in Table 1.1, along with some examples that are investigated in more detail in later chapters. The three perspectives are discussed below.

Table 1.1. A Framework for Governmental Risk Management and Some Examples

	Government Uses Traditional Mechanisms to Manage Chemical Risks
Intensity of Activity	Lawmaking. Agency formation.
Process	Use of the excise tax as a form of incentive-based risk management. Expansion in a conventional base of grant and loan programs.
Outcomes	Unused or underused agency authority, particularly in the areas of toxic substances control and drinking water standards, initiates new, more prescriptive legislation.
	Government Uses New Initiatives to Manage Risks
Intensity of Activity	Intense use by Congress of traditional oversight proceedings in changing agency leadership.
Process	Resurrection of federal common law. New uses of strict liability. Substantive reviews of agency technical decisions. Innovations in incentive systems, such as new forms of taxation, marketable rights, and insurance. Organizational adaptation of agencies to the use of new analytical techniques, such as risk assessment, brought in as management tools.
Outcomes	Limited agency action in hazardous waste management initiates innovative financing mechanisms in the area of waste cleanup, such as specialized funds and trusts. Increased liability as an outcome of laws and judicial review induces new initiatives in the insurance area. Laws that are vague with respect to the definition and measurement of risk spawn new analytical approaches, such as risk assessment.

Intensity and Direction of Government Activity

Government activity can be characterized in terms of the intensity and direction of its operations and where it concentrates its activity. Of particular importance is the rate at which risk-related laws are passed relative to lawmaking in general by the federal government, the rate of agency formation for risk regulation vs other forms of regulation, and the relative intensity of judicial review of chemical risk decisions. While intensity of activity, expressed often as simple trends, is only a gross indicator of what goes on beneath the surface, it provides a good introduction to the direction that government is taking in the risk area. Similarly, the direction of governmental activity and where emphasis is placed reflect changes in government's philosophy toward risk. Distinct directions in the philosophy and approach to risk management are discernible in government, rather than actions being merely random or unassociated.

Governmental Processes for the Institutionalization of Risk Management

Government's response to risk is reflected in how its organizations operate and make decisions. Along these lines, certain classical theories of how organizations behave are applicable to how organizations manage risk. These pertain to how organizations are structured and managed, how they make decisions, the nature of the process of regulation, and models of the policy process. These theories and approaches are presented in greater detail in Appendix 1.1. The relevance of these approaches to the risk problem and how they contribute to or pull together the processes at work are described briefly below.

Governmental actions have typically been described as lying somewhere between the two extremes of a highly bureaucratic process and an open or flexible process. *Highly bureaucratic* processes are usually characterized by standardized or well-defined work patterns, work that is specialized, a high degree of formalization or definition of work, and routinized work.[14] Synonyms for bureaucracy can be neutral, such as the term hierarchical organization, or pejorative, such as the term authoritarianism.[15] While the concept of bureaucracy can have a negative connotation, there are recognizably positive aspects to bureaucracy. Bureaucracies, for example, provide needed controls at certain points in the work cycle or for types of work that lend themselves to standardization. The common theme in all of these descriptors of bureaucracy is that bureaucracies are highly rulebound. Their functions and organization are governed by strict rules. Flexible or *open system* processes, on the other hand, are characterized by a low degree of definition of structure and process (Scott, 1981, 22–23). Few rules characterize their operation. Work is generally not standardized, routine, or formalized. Some synonyms for this type of process are organic systems and "adhocracies" (Mintzberg, 1979; Toffler, 1970).

When government manages chemical risks by setting up a registration or permit system for activities that pose risks and the standards for granting these permissions are very specific and prescribed, allowing few variances, government tends to be acting in a highly bureaucratic manner. In contrast, some programs are less formal and defined. When a government administrator makes decisions and sets requirements on a case-by-case basis, often negotiating with those regulated, decisionmaking is more discretionary, and hence, more open-ended.

These two extremes provide boundaries not only for how agencies are organized, structured, and managed but for how decisionmaking occurs in organizations as well. Theories of bureaucracy as applied to organizational operations and management appear in the work of Weber (1946), Taylor (1911), and others. As applied to organizational decisionmaking, the bureaucratic model is developed in terms of the rational decisionmaker or rational actor and how he acts (Allison, 1971). Theories that describe organization and management in terms of a more flexible system of governmental process are usually expressed in terms of environmental influences, both internal and external, on the behavior of organizations (Woodward, 1965; Lawrence and Lorsch, 1969; Pfeffer and Salancik, 1978; Thompson, 1967). Analogies to "open system" models for decisionmaking are the concept that organizations and decisionmakers engage in "muddling through" rather than rational choice (Dahl and Lindblom, 1953; Lindblom, 1968) and the "garbage can" model of organizational decisionmaking (Cohen, March, and Olsen, 1972; March and Olsen, 1976), in which various decisionmaking components randomly interact to produce outcomes. These theories see decisionmaking as an open system in the sense that decisions are largely unprogrammed. More recent open-ended theories of decisionmaking involve bargaining and negotiation (O'Hare, Bacow, and Sanderson, 1983; Talbot, 1983). The major theories that characterize the two extremes for organization, management, and decisionmaking are presented in more detail in Appendix 1.1 (Allison, 1971; Mintzberg, 1979; Morgan, 1986; Scott, 1981).

A couple of things have to be kept in mind in applying these models to a policy area such as risk. First, there are obviously many intermediate positions between the two extremes of a highly bureaucratic system and an open system, and the intermediaries rather than the extremes are probably more representative of how government operates. Allison (1971), for example, identifies an intermediate, called the organizational process model, where organizations may strive for total rationality but settle for limited rationality. Second, while the two extremes are presented somewhat simplistically here, each of the two concepts is actually defined in many different ways. For example, bureaucracies can be defined in terms of the way functions are defined, the way work is carried out, and so forth (Scott, 1981, 34–37). Open systems vary even more from one another in terms of the kinds of internal and external environmental factors they are influenced by, and how they are influenced by these factors (Mintzberg, 1979, 11 and 217).

One of the themes of this book is the way governmental management and decisionmaking for chemical risks, like government actions in general, move between the extremes of a high degree of bureaucracy and flexibility and the circumstances under which this movement occurs. The purpose of introducing this dichotomy is to see what the specific circumstances are under which a bureaucratic system is favored over a flexible one (or vice versa), and also to see whether the form affects the way risks are managed. Any generalization must be qualified, however, in light of the fact that both highly bureaucratic and flexible processes can operate simultaneously at different levels of an organization and for different processes. As will become clear in the discussions of the federal government's approach to risk management, the newness of an issue is often a critical factor in determining how bureaucratically an organization operates. Newer issues are managed more flexibly than older ones because of the greater uncertainty that newer issues pose.

There are several ways of expressing the highly bureaucratic/open system continuum. One way is in terms of the concept of discretion, and a second way is in terms of the openness and closure of decisionmaking and management.

Discretion/prescription. Discretion is the freedom that governmental organizations have or acquire and the factors that determine judgmental choice. The concept of discretionary authority has been recognized as being at the heart of governmental decisionmaking:

> Discretion must be considered a fundamental component of any attempt to explain administrative behavior and the formulation and implementation of public policy. (Shumavon and Hibbeln, 1986, 1–2)

Yet, few theories directly explain that process.

> Attempts to thoroughly define discretion, to explore the richness of the phenomenon, to observe how it is exercised differently in different settings and for different types of policies, and to develop analytic and theoretical frameworks for understanding variations in its exercise and ensuing consequences for public policy are essentially nonexistent. (Shumavon and Hibbeln, 1986, 1–2)

Applying the concept of discretion to degree of bureaucratization: highly bureaucratic processes tend to be characterized by prescription or little discretion, with few opportunities for choices allowed on a case-by-case basis. Open or flexible processes tend to have a high degree of discretion or choice in the way functions can be carried out.

Furthermore, the concept of discretion relates to outcomes of decision processes and management. There are four alternative ways that the degree of discretion in one action can influence the degree of discretion in a subsequent action or an outcome: prescription can lead to prescription, prescription can lead to discretion, discretion can lead to prescription, and discretion can lead to discretion. With respect to the normative question of how discretion affects outcomes, however, the concept of discretion is theoretically a neutral term. That is, discretion can have either positive or negative effects on outcomes.

Openness/closure. Another way of describing the bureaucratic/open system framework, somewhat related to discretion and paralleling bureaucratization and flexibility, is in terms of the degree of openness or closure of administrative and legal systems to new initiatives. ("Openness" as used here to describe a lack of closure of decisionmaking is different from use of the term to mean a low degree or absence of bureaucratic structure above.) This dichotomy has been explored by Zimmerman (1984) for decisionmaking in several chemical risk cases, in terms of toxic waste management decisionmaking by O'Brien, Clarke, and Kamieniecki (1984), and in controversies involving scientific information in general by Engelhardt and Caplan (1987). In the study by Zimmerman, closure refers to a circumscription or restrictiveness of information and is a decision process in which choices and options are limited or restricted.[16] Closure has been defined by Marrett (1987) in the context of the Three Mile Island incident as the circumscription of decisionmaking by "narrowing the terms of the debate" or limiting the way in which the problem is defined in a controversial situation. Openness, in contrast, pertains to a more fluid system in which there is continual or easy access at several points in a decision to new inputs (Zimmerman, 1984). In the analysis of what generated controversy in six toxic waste cases, O'Brien, Clarke, and Kamieniecki (1984) refer to open decisionmaking, similarly, as a process that enables the public to become actively involved so that the agenda becomes enlarged and expanded. They go further than simply using it as a concept that explains public behavior, and advocate that such a form of decisionmaking be encouraged.

Influence of Risk Management Outcomes on Future Action

Finally, the third analytical perspective, the outcomes of risk management processes, will be used as a way of characterizing how government reacts to chemical risks—in terms of either traditional mechanisms or new initiatives. This book emphasizes the process of risk management rather than a normative evaluation of its performance per se. In keeping with this theme, outcomes of prior actions are viewed in terms of how they influence subsequent actions of government (that is, how they influence the process of government in the course of managing chemical risks) rather than for their own sake or in terms of some standard of behavior.[17] A few performance measures are important to highlight in order to characterize governmental responses to managing chemical risks.

Mazmanian and Sabatier (1983) provide a useful set of general guidelines for defining performance in the context of policy implementation. Summarizing guidelines, both explicit and implicit, from their study, indicators of poor performance, or at least unresolvable conflict, are:

- Things take too long to get accomplished, but eventually get done (i.e., the process is inefficient).[18]
- Things never get done (the process is ineffective). The concept or design of either the initial goal or the plan is unrealistic and unachievable.
- The objective is not met, even though the plan is put in place, because it has negative/unintended impacts, does not accomplish what was needed, or misses the target group.

Hargrove (1983, 280) expresses the nature of performance in a way that uses the strength of the statutory base as a frame of reference for action. According to Hargrove, desirable levels of performance occur when "the actions required by law are carried out" and where "those actions encompass both formal compliance with the law and organizational routines consistent with compliance."[19]

Thus, adapting the examples above to outcome or performance as stimuli for governmental action, one arrives at the following composite set of outcome measures:

- the timing of action (actions get done, get done but not on time, or never get done)
- the extent to which actions actually meet the objectives
- the extent to which objectives and programs are well designed, in that programs are designed to meet objectives, resources match program needs, and potential negative impacts or side effects that can undermine programs are anticipated

MAJOR OBJECTIVES OF THIS BOOK

Within the broad framework discussed above of how government manages risk, a number of specific questions will be addressed. Among the more important of these are:

- How intensely has risk management activity been undertaken by government, as indicated by activities such as lawmaking, agency formation, and judicial review, and does this level of intensity differ from the way government traditionally undertakes these activities?
- How do questions of risk policy attract the government's attention and become part of the governmental process? (In terms of the policy process, this has typically been called agenda-setting or "issue creation" [Mitnick, 1980, 169].) Are crises largely driven by public interest groups getting risk issues on the governmental agenda, and if so, is this any different from the way public pressure influences government processes?
- What new initiatives have been introduced in the area of achieving control over risks through incentive systems as compared with regulation by directive or decree? Has the risk policy area generated and relied upon such initiatives?
- How does the traditional mode of governmental lawmaking through a decentralized committee system in Congress affect the way laws are passed governing environmental health risks? Does the committee system bureaucratize lawmaking for risk issues?
- How are agencies assigned authority over chemical risk problems by Congress, and how much discretion can they use in exercising this authority?
- Have judicial challenges to agency rulemaking in the risk area constrained agency choice and redirected agency actions?

ORGANIZATION OF THIS BOOK

This book is organized into several sections. The introduction sets the framework for how risk management in the context of government can be understood.

Part I looks at the substantive mechanisms for risk management: lawmaking and incentive systems. Part II concentrates on the processes by which risks are managed by the three branches of the federal government. The conclusion overviews the federal system of risk management in terms of tradition and initiatives.

Introduction

Chapter 1 has introduced the risk management concept as it applies to environmental health issues associated with chemicals. The chapter provided the framework for examining how risk management is undertaken by government. Two approaches were presented—one in which government uses its traditional mechanisms and processes to manage risks and the other requiring new initiatives. These two approaches were further defined in terms of the three perspectives of intensity and direction of governmental action, degree of bureaucratization of government institutions and processes, and government's response to poor performance. This matrix sets the stage for examining the components of risk management—laws, incentives, administration, and other governmental processes.

Part I

Part I is divided into two chapters, each dealing with a different approach to the control of risk. Chapter 2 explores risk management by directive or decree (that is, action that is mandated directly by government). Mandatory regulation is discussed in terms of laws and one of the key components of the law—standards. It focuses on the substance or content of the laws and standards as distinguished from the process by which they are created (the subject of Part II). Only one of the three analytical perspectives outlined in Chapter 1 is employed in characterizing the substantive aspects of laws and standards—intensity and direction of governmental action. This perspective, as applied in Chapter 2, covers the intensity of lawmaking; orientation of laws toward health risks, economic considerations, and long-term planning; and the direction of legal language toward greater prescription. Conditions underlying the content of laws are then identified with respect to attributes of the political and economic environment and the role of scientific uncertainty.

The second area covered in Chapter 2 is standards. Trends in standards are cast in terms of descriptors for risk, form of standard used, and levels or numerical values of risk embodied within standards. The type of standard used is related to the degree of complexity of knowledge characterized by different standards and the degree of discretion that can be exercised in applying them. While government generally uses standards in many of its activities, conditions that have led to the standards used for chemical risks are driven primarily by needs specific to risk policy.

Chapter 3 emphasizes management by incentive rather than direct regulation. Primarily, it emphasizes common law doctrines and financial strategies as ways of controlling chemical risks. As in Chapter 2, only the intensity and direction perspective is applied.

The common law is set forth as a framework for much of the legal basis for incentive systems. Trends in the use of common law include the relatively frequent use of common law, how common law relates to statutory law, and the role of federal common law in achieving risk management objectives. These trends are explained with respect to how the common law has evolved relative to statutory law.

Compensation is seen as central to incentives and disincentives for risk management. Trends in compensation are linked to trends in claims and court settlements within and outside the risk area.

Various financial mechanisms, such as insurance, taxation, marketable rights, and grants and loans are the heart of incentive-based systems for risk management. Trends in the fallout of the insurance mechanism are described in terms of insurance availability and scope of coverage in areas outside their application to chemical risks. Insurance industry conditions are linked to governmental processes such as judicial review of cases (not only in the environmental area but in medical malpractice). The risk management system has itself reacted with innovative attempts to counteract insurance problems. Unique forms of taxation and marketable rights are explained largely in terms of risk management initiatives but are drawn from a type of tax and marketable rights strategy that predates the risk issue. Finally, the use of grants and loans in risk management is shown to be largely driven by traditional governmental practices in this area, rather than by initiatives.

Part II

Part II concentrates on the roles, organization, and functions of the three branches of the federal government as they engage in risk management. Chapter 4 discusses the role of Congress in shaping the legislative basis of risk management. The trends in Congress's approach to risk management are explained in terms of traditional mechanisms and initiatives that this branch of government uses, such as voting, enactment of legislation and use of the legislative veto, the treatment of issues through a diverse and decentralized committee and subcommittee system, and oversight functions, such as approval of high-ranking agency personnel.

Chapter 5 looks at the organization of five major administrative agencies with risk management functions. Trends in the administration of risk management reflect how the number of agencies conducting these functions has grown and changes in their financial resource base. The degree of bureaucratization is examined in terms of the exercise of discretion in assigning positions of power to risk management functions, use of advisory bodies, setting of standards, and rulemaking in general. Trends are also examined with an eye to how the agencies react to the outcomes or performance of their risk management programs. Conditions that led to these trends relate to the general growth patterns of regulatory agencies and budgets, the context that the laws provided for these agencies, and the political climate affecting resource allocations and agency positions.

Chapter 6 evaluates the role of judicial review in shaping risk management. Caseload trends are discussed, along with which agencies and issues have dominated the judicial agenda, in which courts risk management cases are heard, the judicial philosophy expressed in decisions, the effect of this philosophy on case outcomes, and the overall consistency of judicial reviews. These are explained in terms of traditional characteristics of the judicial process: organization of the court system, citizen access to the courts, court procedures that have increased the caseload, and most importantly, the change in popularity in the use of bargaining and negotiation as alternatives to litigation. Conditions specific to risk policy include the volume and clarity of the laws underlying judicial review, intensity of agency rulemaking, organizational capacity of plaintiffs, and statutory determinants of where cases should be heard.

Conclusion

Chapter 7 concludes with an overall assessment of how government has managed risk, and whether in performing this function it has primarily drawn on traditional mechanisms or has undertaken major new initiatives in its operation. These relationships are portrayed as a series of stages in governmental decisionmaking, proceeding from lawmaking through implementation and back again to lawmaking.

APPENDIX 1.1

Theoretical Models

The roots of risk management as a governmental process lie in a much larger body of theory devoted to how decisions are made in organizations and in society at large. These approaches broadly encompass theories of organization, administration, and management and theories concerning regulation and other forms of decisionmaking in organizational contexts and broader social settings. They have their origins in a variety of disciplines, such as political science, sociology, policy analysis, decision analysis, public administration, business administration, finance, and law. It is the purpose of this appendix to identify some of the theories applicable to public organizations that provide a foundation for concepts used throughout this book's evaluation of the federal government's role in risk management. Examples of the more central concepts are degree of bureaucratization, discretion, closure of decisionmaking, and organizational performance or measurement of organizational outcomes.

While many of these areas are touched on in the discussion of risk management processes within government, only a few theories will be emphasized to explain what has been occurring over the past couple of decades to integrate chemical risk issues into the operation of government. These theories provide the basis for and cut across the areas of law, incentive-based systems, legislative or congressional initiatives, administration, and judicial review covered in each of the subsequent chapters. In particular, it is through these theories, as well as through more casual observation reflected in descriptive and inductive studies of decisionmaking, that one can begin to understand how risk management fits into governmental processes. In recent years this literature has been vastly expanded to encompass how organizations are designed for crisis management and how they behave under crisis conditions.

Theories of Organization and Management

Theories of organization and management have been used to describe and explain how both public and private organizations adapt to issues such as risk. Principles of organization and management can be used to characterize how Congress, the executive branch agencies, and the judiciary are organized to deal with risk problems and whether these patterns of organization are unique or depend more on how government organizes its functions in general.

Theories of organization and management range from highly structural theories of bureaucratic organization and behavior that emerged at the turn of century to models of organizations as much more flexible and dynamic systems expressed in terms of environmental influences and largely incremental and unstructured behavior.

The early classical theories of organization first appeared after the large-scale organizations of the late 19th century (Drucker, 1974, 23–24). These theories concentrated on how organizations are structured and how they function. One early theory that elaborated on hierarchical organizational structure and office routines was Max Weber's theory of bureaucracy from the field of sociology (Weber, 1946). Taylor's School of Management (Taylor, 1911) and Gulick and Urwick's theory of administrative management (1937) both emerged from the field of industrial psychology to explain organizational operations and the place of workers in the work setting. In addition to these classical theories were those that concentrated on human relations, such as Mayo's work on small group dynamics (1945), Barnard's theories of cooperation within organizations (1938), and March and Simon's introduction of decisionmaking concepts into organizations (1958). Finally, modern organization theory deals with themes such as the relationship of the organization to its environment, interorganizational networks and fields (Evan, 1966; Warren, 1967; Crozier, 1964; Pfeffer and Salancik, 1978), organizational power structures, such as Etzioni's "compliance theory" (1961), the organization as an open system (Katz and Kahn, 1978; Scott, 1981), and as an "adhocracy" (Mintzberg, 1979; Toffler, 1970).

As a whole, theories of organization and management focus on the parts of the organization and how they are organized and function together. Alternatively, they look at how systems of organizations function together. In organization theory, organizations are typically characterized in terms of such concepts as complexity, formalization, and hierarchy. More recent theories concentrate on organizational decisionmaking, communication, and organizational environments. Rarely do theories of organizations deal with specific situations or contexts such as risk. The literature on crisis management (Thompson and Hawkes, 1962; Smart and Vertinsky, 1977; Turner, 1976) is a rare example of the treatment of organizations adapting to threat and risk. In recent years this literature has been vastly expanded to encompass how organizations are designed for crisis management and how they behave under crisis conditions.

Theories of Decisionmaking

Related to theories of organization and management and developing in parallel with them is a large and varied literature on decision processes in both organization theory and broader political contexts. Many of these have application to the management of environmental risk. Allison (1971) organizes decisionmaking approaches into three basic categories. The first is rational decisionmaking, which stresses a systems theory view of the decision process. The second includes the Simon (1955, 1985) and Cyert and March (1963) models of limited or bounded rationality. The third set is comprised of various governmental and political decision models, including the Dahl and Lindblom (1953) and Lindblom (1968) theory of incremental decisionmaking ("muddling through" or the theory of "successive limited comparisons"). The March and Olsen "garbage can" model of organizational choice (1976) would fall within

Allison's third category. Decision theory is applied more frequently than management and organization theories to risk situations.

Some applications of decision theory to risk situations occur in the area of negotiation, bargaining, and conflict resolution (Mendeloff, 1986, 453; Harter, 1982, 17; O'Hare, Bacow, and Sanderson, 1983; Nelkin and Pollak, 1980). The negotiation literature has generated criteria or conditions for effective negotiation. Notably, some have observed (Mendeloff, 1986, 454) that the use of negotiation for risk problems at the federal level is rare, yet some recent legislation has built-in mandatory arbitration. In some of the decision literature, risk management has been portrayed in terms of the interaction of stages in decision-making and participants in the process. According to one scheme that is specifically tailored to risk institutions, the stages are seen as consisting of risk acknowledgment or recognition, risk engagement (where actions are taken), and risk resolution. The participants are similarly broken down into public groups, governmental actors, and private actors (Kawamura and Boroush, 1984; Kawamura, 1984, 7). Another scheme looked at the form of advisory groups and its influence on the degree of participation in science policy decisions in general, of which risk problems were a part (Nelkin and Pollak, 1980).

Theories of Regulation

Since risk management occurs largely in a regulatory context, theories of regulation are germane to understanding how risk management operates. Risk management draws largely from the tradition of "social regulation," which appeared as a movement distinct from economic regulation (Bardach and Kagan, 1982b, 3; White, 1981, 29–34). Theories of regulation run the gamut from complex models of the growth and decline of regulatory agencies and the forces behind these changes to models consisting of factors that describe regulatory behavior. These theories draw heavily from concepts developed in theories of organization, management, and decisionmaking. On the one hand, regulation can be thought of as a category within the theories of organization and management. Yet regulation theory covers many more elements not covered in those others, drawn from the implementation literature and elsewhere. It includes theories regarding the process by which standards and regulations are developed through regulatory decisionmaking as well as the organization and management of regulatory agencies (covered under theories of organization and management). One of the largest areas within the fields of both decisionmaking and regulation that has been applied to questions of risk is policy implementation, and a wealth of case studies exists in this area (Wilson, 1980; McCaffrey, 1982; Kagan, 1978; Mazmanian and Sabatier, 1983; Pressman and Wildavsky, 1984; and Thompson, 1982). These theories of regulation provide perhaps the most direct link to risk management. Bardach and Kagan's work (1982b) examines the broader context for these theories in social regulation.

NOTES

1. For brevity, the risks that chemicals in the environment pose to human health are referred to synonymously throughout the book as environmental health risks, chemical risks, or simply risks. It is important to point out that chemicals only pose health risks in the environment where they are present in sufficient concentrations to adversely affect human health and where there is sufficient exposure to them.

2. There are other functions of government that are equally important in managing chemical risks, such as compliance and enforcement. It is the lawmaking function and its administration within government, however, that provides a framework for these other processes and thus warrants special attention.

3. These actions are typically referred to as social regulation (as defined by Bardach and Kagan, 1982b, 3), and include protective programs in areas such as "worker safety, antidiscrimination, pollution control, fire prevention, product and housing quality."

4. A number of variations on this definition exist. Rowe defines risk as "the potential for realization of unwanted, negative consequences of an event" (Rowe, 1976, 24). According to Sage and White (1980, 426) risk is defined more quantitatively as "the probability per unit time of the occurrence of a unit cost burden"; it "represents the statistical likelihood of a randomly exposed individual being adversely affected by some hazardous event. Thus risk involves a measure of probability and severity of adverse impacts."

5. Aharoni (1981, 41) is an exception, however. He equates risk and uncertainty, defining both as undesirable outcomes that are uncertain.

6. The general idea of management can be defined from a variety of perspectives, i.e., from the viewpoint of individuals, social groups, organizations, or societies as a whole. Management also embodies a variety of concepts, which is why no one definition has been adopted (Stoner, 1981, 7).

7. In this definition of management, "the external environment or conditions" usually refers to those to whom the organization's products and services are sold or transferred and from whom resources are obtained. This external setting can also refer to the institutions of society. "Institutions" often means the characteristics of the social environment or setting, including its norms, culture, history, traditions, politics, and laws. See, for example, Selznick (1957, 5–12).

8. This meaning of risk management actually dates back to ancient times (Covello and Mumpower, June 1985).

9. Corporate risk management is a form that nongovernmental risk management has taken in the private sector, and has been addressed by Baram (1988) and Rowe (1976).

10. Baram (1982) uses three of these strategies as components of his "alternatives to regulation" and adds others as well.

11. These are reported by Vig and Kraft (1984, 4–5, 22), who recount the overt policies set by the Reagan administration to reduce governmental regulation of the environment by cutting its resources and the claims of environmental groups that Reagan had cut over 200 programs. Kenski and Kenski (1984, 99) cite the cuts that occurred at EPA and at other agencies with environmental responsibilities.

12. Mitchell (1984) makes a strong argument for this claim. A similar argument is presented by Kenski and Ingram (1986, 275): " . . . despite the efforts of the Reagan administration to roll back environmental regulations and return decision-making to the states, the political market compels strong federal regulation of pollution."

13. The regulatory budget approach and "legislated regulatory calendar" have been summarized by Mendeloff (1986, 446, 448).

14. These concepts are discussed and defined extensively for the organizational setting by Mintzberg (1979). While originating in the context of private, industrial organizations, these concepts are now applied widely to public organizations.

15. According to Mintzberg (1979, 86), another important concept describing management and decisionmaking that is not necessarily synonymous with bureaucracy is centralization. Organizations, according to Mintzberg, can be bureaucratic whether they are centralized or decentralized.

16. A study of over two dozen decisions observed that decisionmakers deal with complexity by factoring problems into smaller, more manageable but more structured pieces (Mintzberg, Raisinghani, and Theoret, June 1976).

17. A normative evaluation of performance is complicated by the difficulty of arriving at a consensus regarding evaluation criteria. MacCarthy, for example, summarizes the argument (in the context of OSHA's benzene and cotton dust decisions) that it is difficult enough evaluating environmental health decisions because of uncertainties in scientific and economic information. When one considers the possibility that performance evaluations also can depend on balancing judgmental factors and political values that change over time, the task of evaluation is made even more difficult (MacCarthy, 1987, 505–506).

18. Some have argued that delay can be an advantage. White, for example, argues that the delay in setting certain diesel emission standards had the advantage, in his opinion, of better public policy (White, 1981, 116–117).

19. However, Vig and Kraft (1984) provide a broad set of agency measures of performance that are specifically applicable to environmental problems. These are efficiency measures or what they call means-ends relationships, cost-effectiveness and cost-benefit tests, and process-oriented tests that are directed toward the functions of the agency. (These are drawn from a summary by Zimmerman, 1987b, 261.)

PART I

Legislative and Incentive-Based Systems for Managing Chemical Risk

The main parts of this book divide the management of chemical health risks into the *substance* or content of legislation and incentive systems (the subject of Part I) and the *process* of management (discussed in Part II). In addressing the substance of laws and incentives, Part I is the basis for Part II's analysis of processes that the federal government invokes to manage health risks.

Regulation is based on both laws and nonstatutory incentives (incentives can be suggested in statutes or have other origins). These two forms of regulation complement one another, and together they comprise a comprehensive system for risk management. Regulation by directive or decree (which incorporates statute-based requirements) restricts behavior directly by narrowing the choice of means to compliance (Mitnick, 1980), and once the means are chosen, choice is usually restricted further. Regulation based on incentive systems allows those targeted for control wider discretion in the choice of means. Strategies for direct regulation are generally tailored to the specifics of the risk issue and are more likely to be patterned after regulations from other governmental areas and thus traditional to government. In contrast, incentive systems, while often drawn from or modeled after incentives in other governmental programs, are more likely to represent initiatives in the kind of detail they involve.

The evaluation of regulation by directive and incentive pertains to one of the three analytical perspectives discussed in the introduction—the intensity and direction of governmental activity. The other two themes, governmental process and how outcomes influence future inputs, are applied in Part II, since they are more relevant to organizations and processes within the federal government responsible for overseeing lawmaking and incentives than they are to the laws and incentive systems themselves.

CHAPTER 2

Legal and Regulatory Directives and Decrees

INTRODUCTION

Regulation plays a central role in governmental management of chemical risks. Regulation, in general, restricts behavior in some way, and regulation by directive or decree (Mitnick, 1980, 342–343) restricts or directs behavior by means of rules. These terms and their origins are defined in Appendix 2.1.

This chapter covers two areas of federal regulation: statutory law and one of its major components, standards. The laws covered here pertain to federal statutes (as distinct from common law, covered in the next chapter). Standards are rules defining limits and conditions on activities that can result in some unwanted effect—in this case, a health risk. Standards referred to here are those defined in the laws directly or in agency rulemaking under the laws (as distinct from standards developed by professional or trade associations, which do not originate in legislation).

The analysis of the content of laws and standards is cast in terms of the intensity and direction of government activity (one of the three themes outlined in Chapter 1) and the extent to which this activity draws on traditional governmental processes or new initiatives. Under *lawmaking*, the intensity and direction of government activity specifically addresses (1) the intensity of lawmaking, in terms of changes in the number of laws over time; (2) the change in the legislative philosophy, notably the emphasis on health vs economic considerations; (3) the change in philosophy with respect to the emphasis on remediation vs planning; and (4) the degree to which requirements are specified (prescribed) in legislation.

These patterns and trends are then explained by examination of conditions already existing within government or, alternatively, new initiatives unique to the risk issue. The traditional governmental processes that influence the trends in the laws include (1) crises and interest group pressures on government as explanations for the growth of laws and their health orientation and (2) political forces that focus on economic conditions and emphasize economic tests as a basis for legislative design. The major conditions unique to the risk issue that represent new governmental initiatives include: (1) shifts in the burden of proof from government to industry, (2) advances in detection technology, and (3) underlying uncertainty in scientific evidence.

With respect to the *development of standards*, intensity and direction are discussed with emphasis on (1) the variety of terms used to describe risk in the standards, (2) the variety of risk levels assigned to the same substance from agency to agency, and (3) the shift in the types of standards used over time in the direction of referencing sources of chemicals rather than the quality of the ambient environment. These characteristics influence the discretion that agencies have in implementing the standards and the directness of causal proofs required for such implementation.

The conditions that lead to characteristics of standards are largely specific to the risk issue. These include (1) vague laws (with respect to the causal connections between chemicals and health risks) that influence the consistency of the definitions of standards and the basis for establishing risk levels and (2) scientific uncertainty.

THE LEGISLATIVE BASIS FOR RISK MANAGEMENT

Patterns and Trends in Intensity and Direction of Legislation

The laws dealing with environmental health and safety risks have, as a group, several striking characteristics that distinguish them from similar legislation in earlier time periods. First, there has been a dramatic rise in the number of risk-oriented laws within this century. Second, there has been a shift in emphasis of the laws beginning in the early 1970s away from exclusively environmental protection issues to environmental health risks, accompanied by a shift in emphasis of the laws of the middle to late 1970s to economic concerns. Third, laws have changed from emphasizing preventive strategies or long-term planning to emphasizing remediation and emergency response. Finally, there has been an increase in the degree of prescription in the legislation with respect to the number of risk agents specified and how compliance with the limits defined in standards is scheduled over time, though laws are generally less prescriptive about what analytical techniques to use in setting limits. Where prescription has increased, agency discretion has decreased in managing these risks (discussed in Chapter 5).

Increases in the Number of Risk-Oriented Laws

Dozens of laws currently provide a framework for public and private sector actions and management systems for risks to human health and safety. The major laws pertaining to such risks passed in this century and the dates they were enacted are listed in Table 2.1. A subset of these laws, pertaining primarily to environmental health rather than to safety or environmental protection (exclusive of human health), is listed in Table 2.2. Table 2.2 organizes the laws according to the environmental health areas they cover. As a group, the laws cover many different aspects of health-related risks. Some laws exercise control

Table 2.1.   Major Federal Legislation Dealing with Health and Safety Risk[a]

Statute	Dates of Enactments of the Statutes and Major Amendments
Asbestos Hazard Emergency Response Act (AHERA)	1986
Atomic Energy Act (AEA)	1954, 1957[b], 1959, 1962, 1966[b], 1974, 1975[b], 1977, 1988[b]
Clean Air Acts (CAA)[c]	1955, 1963, 1965, 1967, 1970, 1974, 1977
Clean Water Acts (CWA)[c]	1948, 1956, 1961, 1965, 1966, 1970, 1972, 1977, 1987
Coal Mine Safety Act (CMSA)[c]	1947
Coastal Zone Management Act (CZMA)	1972, 1976
Comprehensive Environmental Response, Compensation, and Liability Act (CERCLA)[c]	1980
Consumer Product Safety Act (CPSA)[c]	1972, 1976, 1977, 1978
Dangerous Cargo Act (DCA)	1877, 1975
Environmental Research, Development, and Demonstration Authorization Act (ERDDAA)[c]	1976, 1978, 1979, 1980, 1981
Federal-Aid Highway Act (FHA)—National Bridge Inspection Standards	1968
Federal Insecticide, Fungicide, and Rodenticide Act (FIFRA)[c]	1947, 1972, 1975, 1978, 1983, 1988
Federal Aviation Act (FAA)	1958, 1974, 1977, 1978
Food, Drug, and Cosmetic Act (FDCA)[c]	1906, 1938, 1958, 1960, 1968
Federal Meat Inspection Act (FMIA)	1906
Hazardous Liquid Pipeline Safety Act (HLPSA)[c]	1979
Hazardous Materials Transportation Act (HMTA)[c]	1975
Hazardous Substances Act (HSA)[c]	1960, 1966, 1969
Highway Safety Act (HSA)	1978
Lead Based Paint Poisoning Prevention Act (LPPPA)[c]	1971, 1978
Low Level Radioactive Waste Policy Act (LLRWPA)	1980, 1985
Marine Protection, Research, and Sanctuaries Act (MPRSA); Ocean Dumping Ban Act (ODBA)	1972, 1980, 1988
Mine Safety and Health Act (MSHA)[c]	1969, 1977
Motor Vehicle Carrier Safety Act (MVCSA)	1984
National Cancer Act (NCA)[c]	1971
National Dam Inspection Act (NDIA)	1972
National Environmental Policy Act (NEPA)	1969
National Traffic and Motor Vehicle Safety Act (NTMVSA)	1966, 1974
Noise Control Act (NCA)	1972
Nuclear Waste Policy Act (NWPA)	1982
Occupational Safety and Health Act (OSHA)[c]	1970
Pipeline Safety Act (PSA)	1979
Poison Prevention Packaging Act (PPPA)	1970, 1976
Ports and Waterways Safety Act (PWSA); Port and Tanker Safety Act (PTSA)	1972, 1978
Poultry Products Inspection Act (PPIA)[c]	1957, 1970
Radiation Control for Health and Safety Act (RCHSA)[c]	1969
Railroad Safety Act (RSA)	1970, 1976
Rail Safety Improvement Act (RSIA)	1975
Resource Conservation and Recovery Act (RCRA); Hazardous and Solid Waste Amendments (HSWA); see SWDA[c]	1976, 1984

Table 2.1, continued

Statute	Dates of Enactments of the Statutes and Major Amendments
Rivers and Harbors Act (RHA)	1899
Safe Drinking Water Act (SDWA); Lead Contamination Control Act (LCCA)[c]	1974, 1986, 1988
Safety Appliance Acts (SAA)	1893
Seat Belts Act (SBA)	1963
Solid Waste Disposal Act (SWDA); see RCRA[c]	1970, 1976, 1980, 1984

Superfund Amendments and Reauthorization Act
    (SARA) Title III—Emergency Planning
    and Community Right-to-Know Act (EPCRA)[c]
	1986

Surface Mining Control and Reclamation Act (SMCRA)	1977
Toxic Substances Control Act (TSCA)[c]	1976, 1986
Uranium Mill Tailings Radiation Control Act (UMTRCA)[c]	1978
Veterans' Dioxin and Radiation Exposure Compensation Standards Act (VDRESCA)[c]	1984

[a]This table does not include legislation pertaining to safety related to natural hazards, such as flood disaster protection legislation, nor does it include natural resource protection or conservation legislation. The list of statutes and their dates was drawn primarily from the *Code of Federal Regulations Index*, Table III ("Acts Requiring Publication") and supplemented by other sources of legislation as well. For a historical coverage of air and water legislation, see A. V. Kneese and C. L. Schultze, *Pollution, Prices and Public Policy* (Washington, DC: Brookings Institution, 1975) pp. 31–32. The dates given in the table are for major amendments only. The graphs in Figure 2.1, however, contain additional amendments as well as the major ones.
[b]Amendments to the Price-Anderson provisions of the Atomic Energy Act. The latest amendment was the Price-Anderson Amendment Act of 1988 (P.L. 100–408).
[c]Graphed in Figures 2.1 and 2.2.

over certain kinds of environments—air, water, food—and in that context emphasize routes of exposure, susceptibility of the population to exposure, and control of the effects once exposure occurs. Other laws exercise control over sources of risk—in terms of either types of substances or types of activities that could potentially create a health risk. Thus, different laws may apply to the same activity, but in different ways. As a result, these laws can interact with each other in complex and often redundant ways for a given activity.

The number of environmental and safety laws passed between 1970 and the late 1980s was larger than the total number of such laws that existed in this area prior to this period. Using the dates given for the laws and their major amendments in Table 2.1, just about twice as many laws and amendments were passed between 1970 and 1988 as in all of the years prior to 1970. The new laws that emerged covered risks in the occupational, consumer product, transportation, and general environmental areas.

A common way of measuring growth in the number of laws over time is to look at the cumulative rate of growth. The cumulative increase in the number of environmental health laws (pertaining to chemical risks and listed in Table 2.1) relative to overall lawmaking by the federal government since 1940 is

Table 2.2.   Categorization of Current Federal Environmental Health Risk Laws

Category	Legislation
By Type of Environment/Medium of Exposure	
Air	Clean Air Act
	Comprehensive Environmental Response, Compensation, and Liability Act
Water quality	Clean Water Act
	Comprehensive Environmental Response, Compensation, and Liability Act
	Safe Drinking Water Act
Food, drugs, and cosmetics	Food, Drug, and Cosmetic Act
	Meat Inspection Act
	Poultry Products Inspection Act
	Poison Prevention Packaging Act
Consumer products	Consumer Products Safety Act
	Lead Based Paint Poisoning Prevention Act
	Poison Prevention Packaging Act
By Type of Substance	
Toxic substances	Toxic Substances Control Act
Hazardous substances	Resource Conservation and Recovery Act
	Hazardous Materials Transportation Act
	Hazardous Substances Act
Pesticides	Federal Insecticide, Fungicide, and Rodenticide Act
Drugs, cosmetics	Food, Drug, and Cosmetic Act
Radiation	Atomic Energy Act
	Uranium Mill Tailings Radiation Control Act
	Radiation Control Act
By Type of Activity	
General working environments	Occupational Safety and Health Act
Mining	Mine Safety and Health Act
Manufacturing	Consumer Product Safety Act
	Federal Insecticide, Fungicide, and Rodenticide Act
	Food, Drug, and Cosmetic Act
	Lead Contamination Control Act
	Toxic Substances Control Act
Transportation	
General	Hazardous Materials Transportation Act
Pipeline	Hazardous Liquid Pipeline Safety Act
Energy	
Nuclear power plants	Atomic Energy Act
	Radiation Control Act

Table 2.2, continued

Category	Legislation
Waste disposal	Marine Protection, Research, and Sanctuaries Act
	Resource Conservation and Recovery Act
	Low Level Radioactive Waste Policy Act
	Nuclear Waste Policy Act
	Uranium Mill Tailings Radiation Control Act

shown in Figure 2.1.[1] Figure 2.1 gives two curves for environmental health laws—one with just major amendments and one that includes all the amendments, however small, for each law. Comparison of the growth trend in risk legislation with that for lawmaking in general must take into account how risk legislation is tabulated.[2] Taking this factor into consideration, it can be seen that on the one hand, major risk laws alone are growing at a rate about equal to or slightly greater than lawmaking in the country in general. But when one counts all amendments (including small technical amendments) as well as the major environmental risk laws, the cumulative growth rate is much faster than the growth rate for laws in general. The ratios graphed in Figure 2.2 underscore the trends in Figure 2.1 in a more concise way. It is important to point out, however, that annual growth levels do not show the precipitous rise for laws that cumulative figures do. Data from the Census and the *U.S. Code Annotated* reveal that the number of laws enacted per Congressional session has generally been declining since the turn of the century, with the exception of just a few active years (a trend that is also shown in the data presented by Johnson, 1985).[3]

To summarize, environmental risk legislation is growing somewhat faster than all legislation, but just how much faster is unclear and depends on the way the laws are counted. The cumulative total legislation or lawmaking in the country is growing at a comparable rate, and may in fact be a major factor driving the rate of increase in lawmaking in the environmental risk area. Similar growth trends have been observed in the issuance of regulations by agencies. This is often measured in terms of the dramatic rise in the number of pages in the *Federal Register*. The exponential growth trend in risk-related laws has been widely cited in the legal and policy literature.[4]

The growth in legislation dealing with environmental health risks seems to parallel a general increase in interest in risk. One measure of this trend is the increase in the use of the word "risk" in various academic and professional contexts, observed by Inhaber and Norman (1982, 119–120). Another measure of this interest is the large number of court cases involving risk, which is covered extensively in Chapter 6 (Oleinick, Disney, and East, 1986).

Shift in Emphasis Toward Environmental Health Risks

More laws deal with environmental health risks now than during the early part of the century. Morris and Duvernoy (1984, 457) point out that at the turn of the century only a few laws were related to risk, and these laws or court cases

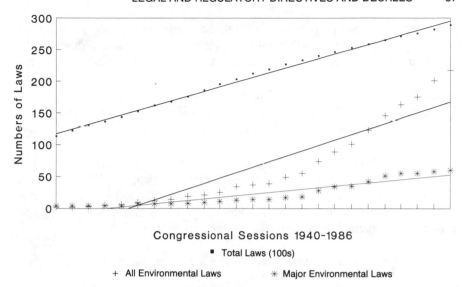

Figure 2.1. Cumulative growth in laws (risk laws vs total laws).

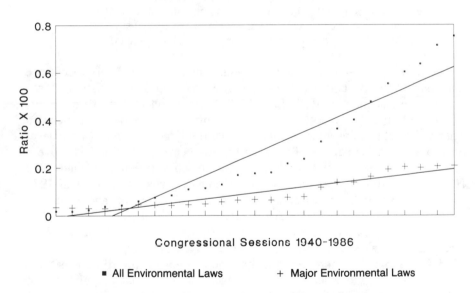

Figure 2.2. Cumulative ratios of major and all environmental laws to total laws.

pertained to air pollution, boiler explosions, and product liability. Furthermore, the early public health laws that did exist emphasized infectious disease and the living and social conditions associated with it, rather than health risks from chemicals.

There are now substantial differences in the way health risks are placed in legislation. First, as DeLong (1983, 5) has pointed out, the nature and level of risk to which agencies must respond is now explicit in legislation. Earlier laws did not have such direct references to risk.

Second, there are substantial underlying differences between the treatment of human health issues and environmental issues in legislation. When the health risk issue actively emerged in the legislation of the 1970s, Page (1978) observed the following differences in the way environmental health risk issues and the earlier environmental protection issues were portrayed:

- Current environmental risk issues involve long-term, life-threatening effects rather than diseases of short duration.
- The newer issues involve many chemicals, whereas earlier environmental protection statutes concentrated on a few chemicals.
- Much lower chemical concentrations are now of concern.
- Latent effects are now the focus of attention rather than near-term effects.
- Effects not easily recognized without the aid of sophisticated instruments and analysis are now of interest rather than very easily observable effects.

The focus of legislation on environmental health risks is a very recent phenomenon. O'Reilly et al. (1983, 2–5), discussing changes in legislation relevant to the chemical industry during the 1970s, zero in on the exact time period when a health risk perspective began to dominate environmental legislation. They point out that the environmental laws currently governing risk were developed over two time periods. The major environmental protection laws governing air, water, noise, and pesticides were passed between 1970 and 1972. However, O'Reilly et al. observed a subsequent period, beginning in 1976, that marks a dramatic increase in the complexity of risk agents and methods of control with the passage of the Toxic Substances Control Act (TSCA) and the Resource Conservation and Recovery Act (RCRA). The Safe Drinking Water Act (SDWA), passed in 1974, provides a bridge between the two periods. SDWA is somewhat more complex and more oriented toward human health risks than the 1972 legislation. Fewer chemicals were originally regulated under SDWA than under TSCA and RCRA when they were passed a few years later.

Legal Philosophy: Emergence of Economic Balancing

Existence and nature of economic balancing. While the previous section showed that laws of the 1970s and 1980s as a group were relatively more oriented to health risks than earlier laws, there are significant differences among the laws in terms of how health risks were considered relative to other factors, particularly economic impact. An important factor that shifted the focus of risk legislation to economic considerations was the passage of two federal executive orders: Executive Order 12044 during the Carter administration and Executive Order 12291 during the Reagan administration. These orders required regulatory impact analyses and cost/benefit considerations and provided increased review authority to the Office of Management and Budget to reinforce these new requirements.

Prior to 1970, two kinds of laws dealt with health risks. The first kind required that health be addressed without balancing it against other factors. Examples of pre-1970 laws of this type are the Mine Safety and Health Act of 1969 and the food additive provisions of the Food, Drug, and Cosmetic Act (FDCA) of 1958. (This characterization of the laws is based on the works of Field, 1980, and Doniger, 1978.) The second kind of law, while referencing health and environmental protection as major objectives, required that these objectives be balanced against the economic impact of achieving them. Examples of pre-1970 laws in this category (based on Field's 1980 categorization) are the National Environmental Policy Act (NEPA) of 1969; the drug, cosmetic, and medical equipment provisions of FDCA; and the Hazardous Substances Labeling Act. Brickman, Jasanoff, and Ilgen (1985, 34) summarized some of the laws appearing in the early 1970s that contained balancing provisions, including the requirement in the Federal Insecticide, Fungicide, and Rodenticide Act (FIFRA) amendments of 1972 (the Federal Environmental Pesticide Control Act) that EPA incorporate "economic, social and environmental costs and benefits" in banning or placing restrictions on pesticides; a judicial interpretation of the Occupational Safety and Health Act (OSHA) requiring that the health and safety goals pursued be feasible economically and technologically (*American Textile Manufacturers Institute, Inc.* v. *Donovan,* 452 U.S. 490, 1981); and a requirement that the regulation of toxic substances under TSCA take into account costs, benefits, and broader economic impacts on the economy. The SDWA uses economic tests in granting exemptions from compliance with national primary drinking water standards.

The two varieties of health risk legislation are consistent with a typology that Mitnick (1980) developed for public interest theories of regulation. The balancing concept in regulation is one of the forms of public interest theories that according to Mitnick "results from the simultaneous satisfaction of selected aspects of several different particularistic interests. The balancing result gives satisfaction to interests that may to some extent be contending or competing" (Mitnick, 1980, 93).[5] The statutes not requiring cost-benefit balancing would fit into Mitnick's category of national or social goals, since they are designed to override private interests. The emergence of these different kinds of laws over time also appears to be consistent with the association of regulatory trends and economic conditions put forth by Peltzman (1976) and verified by Shughart and Tollison (1985) for three federal agencies. Their theory associating economic condition and the direction of regulatory activity holds that during periods of relative economic decline or depression, regulation will tend to support producers, whereas during periods of relative economic prosperity, regulation will tend to emphasize consumer protection (Shughart and Tollison, 1985, 303).[6]

Trends in economic balancing provisions. A number of observations can be made with respect to the extent and importance of balancing provisions in environmental risk legislation:

- In the early 1970s more major laws were passed that exclusively focused on health risks, and did not require a consideration or balancing of economic factors.
- In contrast, later legislation—in the late 1970s—began to focus on the economic impact of controlling the risks and, with one exception, was accompanied by the virtual elimination of statutes exclusively oriented to health risks.
- This emphasis on economic balancing was not new, and had been characteristic of legislation passed all through the 1970s and earlier.
- Furthermore, the legislation of the 1980s retained economic balancing provisions but had a strong emphasis on the environmental health risk component.

In order to evaluate whether there has been a virtual elimination of health-risk-only statutes, the underlying philosophies of the laws have to be examined over the period between 1970 and the late 1980s. Existing typologies of these laws are a useful way of comparing laws according to their underlying philosophy and their coverage of and approach to a concept such as risk. A few typologies have been developed that attempt to explore how risk is defined and evaluated in environmental and safety laws passed since the late 1960s. The typologies help categorize legislation into at least two major categories: an orientation toward health risks only ("health-only" statutes) and an orientation toward balancing economic considerations against health risks ("economic balancing" statutes).[7] These typologies are described in Appendix 2.2.

Figure 2.3 organizes the major environmental health–related laws over time as they are categorized by Doniger (1978), Field (1980), and Morris and Duvernoy (1984). The authors of the typologies did not attempt to organize the laws over time, and their coverage of laws only goes through about 1980. Therefore, Figure 2.3 also characterizes laws passed through 1987 as they might have been categorized according to the criteria underlying the typologies. An examination of the figure shows that after 1972 there is a virtual disappearance of the health-only statutes, with the exception of RCRA and the amendments to the MSHA. RCRA may not even qualify as a true health-only statute placed in the time period from 1975 to 1980, since regulations implementing it did not appear until 1980 and the Hazardous and Solid Waste Amendments (HSWA) of 1984 changed many of its provisions.

Thus, the health-only statutes do appear to be supplanted by economic balancing requirements during the period from 1975 to 1980. It should be noted

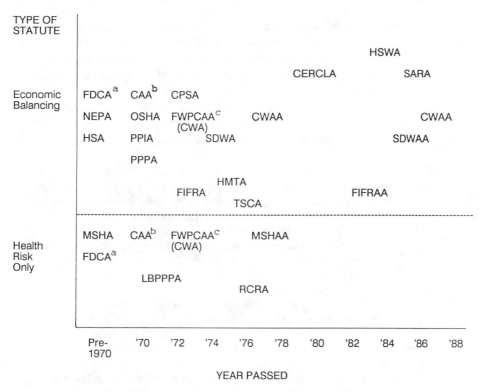

Figure 2.3.   Change in the orientation of major environmental health laws over time. Dates indicate the date that the act was first passed. Major amendments are indicated separately. Amendments did not alter categorization.

[a]The food additive provisions of FDCA were health risk–only. These provisions were incorporated in the famous Delaney clause. The drug, cosmetic, and medical equipment provisions were balancing statutes.

[b]The ambient standards (primary), according to Morris and Duvernoy, were health risk–only. The secondary standards (according to Morris and Duvernoy) and the New Source Performance Standards (according to Stever, 1988) were balancing standards.

[c]The ambient and effluent standards of FWPCAA were either health risk–only or technology-only, and other aspects of the statute required or suggested economic balancing. FWPCAA is one of the more difficult statutes to categorize, since there are provisions in the law relating to health and economics to override technology-based standards. Where all of the discharges into a waterway practicing "best practicable control technology" are still not meeting a surface water standard, the regulatory agency can require that more stringent standards be applied to the discharges until the standards are met.

that balancing statutes have existed all along, dating back as early as the appearance of the first health-only statute, FDCA. OSHA, one of the earliest health-related statutes, requires a test of economic feasibility under Section 6(b)(5), which Stever defines as threatening a whole industry rather than individual firms within it (Stever, 1988, 9–17). So the sudden introduction of economic criteria did not occur. Economic criteria were always there, and health criteria began to grow in prominence in the early 1970s along with economic criteria. Economic criteria regained prominence in the latter part of the decade.

Below, some examples of key provisions in post-1980 legislation highlight the continuing use of economic balancing, in spite of a parallel emphasis on health risks. As a group, economic balancing laws use economic tests in setting standards, developing remediation alternatives, and choosing among alternatives. The types of tests vary. Laws may specify cost-benefit analysis or cost-effectiveness tests. Sometimes, no particular test is required.

1. *The Hazardous and Solid Waste Amendments of 1984:* This act constituted major amendments to the Solid Waste Disposal Act and to RCRA. Several provisions of the amendments qualify requirements for the determination of risk with phrases such as "to a reasonable degree of certainty."[8] The use of the term "reasonable" has been interpreted to imply an economic test.

2. *Superfund Amendments and Reauthorization Act (SARA) of 1986—Cleanup Standards for Remedial Actions:* A new Section 121 added to the Comprehensive Environmental Response, Compensation, and Liability Act (CERCLA) provides for cost-effectiveness considerations in the choice of remedial actions. The section states that "the President shall select a remedial action that is protective of the human health and the environment, that is cost-effective, and that utilizes permanent solutions and alternative treatment technologies or resource recovery technologies to the maximum extent practicable."

3. *SARA—Short-Term Removal Actions:* Section 104(b)(2) describes a cost-effectiveness criterion for deciding how and whether to undertake short-term removal actions at Superfund sites. The section reads, "Any removal action undertaken by the President under this subsection (or by any other person referred to in Section 122) should, to the extent the President deems practicable, contribute to the efficient performance of any long-term remedial action with respect to the release or threatened release concerned." The Conference report clearly justifies this change by interpreting efficient performance as cost-effectiveness.[9]

4. *Safe Drinking Water Act Amendments (SDWAA) of 1986:* The standards mandated in the 1986 amendments to the SDWA are meant to take cost considerations into account. While a reading of the basis for the standards makes reference only to health-based evidence as an input to the standards, the Conference report on the amendments makes it clear that the change in the standards from Recommended Maximum Contaminant Levels (RMCLs) in the 1974 law to Maximum Contaminant Levels (MCLs) in the 1986 legislation involves an incorporation of the criterion of economic feasibility:

> The health-based standards of the Act—RMCLs—are changed to MCL goals. The Committee does not intend this to be a substantive change in the law. It merely clarifies the distinction between RMCLs and MCLs. MCLs must be as close to the RMCL as is feasible. "Feasible" means with the use of technology, treatment techniques, or other means which the Administrator determines to be the best available (taking cost into consideration). (U.S. Congress, House of Representatives, 1986a, 22)

In spite of the decline in legislation exclusively emphasizing health during the early 1980s, one can go through many of these laws (and later ones) and find examples of the persistence of health considerations:

- Certain key provisions of HSWA, such as the land disposal ban for certain chemicals, are based on a health risk (and environmental protection) criterion only.
- The 1983 amendments to FIFRA replaced the "unreasonable adverse effects on the environment" standard with an emphasis on adverse effects on humans, i.e., a "will endanger human health" standard (Dowd, 1983, 415A).
- SARA placed considerable emphasis on the conduct of health risk assessments on a site-by-site basis to justify the selection of remedial action alternatives.
- OSHA continued to require findings of "significant risk of material impairment" as the basis for setting standards (Oleinick, Disney, and East, 1986, 402).

During the middle to late 1980s, health-oriented laws reappeared but took a different form. Legislation that narrowly applied to a single chemical or activity was passed with a strong health orientation. Examples of these laws include the Asbestos Hazard Emergency Response Act, which required asbestos removal in a number of places, and the Lead Contamination Control Act of 1988, which required the removal, repair, and ultimate reduction of lead in contact with drinking water in coolers and other water distribution units to a specific level.

Shift from Long-Term Planning–Oriented to Remediation-Oriented Laws[10]

Another way of characterizing the legal philosophy and policies embodied in laws is the degree to which they emphasize long-term planning vs remedial or after-the-fact strategies that try to lessen the consequences of a risk once it has been realized. Whether laws emphasize plans or remediation is significant for the philosophy of risk management. Plans allow a more thorough evaluation of sources of risks and alternative control strategies. Statutes oriented toward remediation, which may be necessary to avoid further health risks from being realized, may not always be able to explore the full ramifications of the way risks are characterized or controlled.

Laws that emphasize planning components often organize regulatory requirements around plans consisting of an identification of risk agents or sources of risk, estimates of the likelihood of risks being realized, standards against which to evaluate risks, and strategies for risk reduction over time. They almost always involve long-term projections of the population likely to be affected by the plan and of the likelihood of the risk problem continuing. An evaluation of risks and benefits is usually included as part of the design of such plans.

One example of a planning-oriented law and its provision for planning is the State Implementation Plan (SIP) required under the Clean Air Act (CAA). Other examples are the Areawide Water Quality Management Plan and Facility Management Plan under Sections 208 and 201, respectively, of the Clean Water

Act (CWA). Still another example is the Solid Waste Management Plan under the Solid Waste Management Act of the early 1970s (the predecessor to RCRA). The use of these planning provisions peaked during the early to middle 1970s, and the last geographically and functionally comprehensive plans appeared in the late 1970s. The planning provisions appearing in subsequent statutes have a very different style and are much narrower in their focus. CERCLA's national contingency plan qualifies as long-term planning, but the provisions for Remedial Investigation Feasibility Studies are very facility-specific and are more remediation- than planning-oriented. SDWAA of 1986 provides for a state plan for wellhead protection and a funding program for localities to pursue wellhead protection planning. This is relatively more narrowly focused than previous planning provisions, focusing primarily on well systems rather than comprehensive water resources. Some of the laws with planning requirements contained provisions for continuing planning activities once the original plans were developed. The design of regulatory programs was keyed to provisions in each of the plans. While plans often did not mention the word "risk," they embodied the concept of environmental risk in the methods and approaches used to develop and implement the plans.

Another set of laws of a more recent vintage emphasizes remediation, often in a planning context. Remedial action usually concentrates on contingency and emergency provisions to stop an existing hazard from continuing. While they often contain planning provisions for emergency response, their time horizons are generally shorter than those for planning-oriented statutes. They are heavily oriented toward the application of existing technologies to abate a problem within a relatively short time frame. A comprehensive analysis of costs and benefits is usually bypassed at least in the short term, or is only used in a limited way to evaluate alternative technologies.

CERCLA was an example of one of the first of the risk-oriented laws to emphasize remedial action strategies. Section 104 of CERCLA provides for immediate removal actions for hazardous waste sites, a process that normally occurs within six months to one year. Prior to CERCLA, provisions to clean up spills of oil and hazardous substances existed under the Clean Water Act (for example, the spill prevention and countermeasure plans under Section 311). While RCRA ultimately contained extensive procedures for remediation of existing hazardous waste management sites, these requirements did not emerge in regulations until about the same time that CERCLA was passed. Contingency plans are a type of remedial action plan that is now a basic part of the siting and operation of hazardous waste management facilities under RCRA. Remediation under Section 311 of the Clean Water Act and emergency provisions under other environmental laws were implemented by a rather elaborate administrative system consisting of a network of response agencies organized nationally, regionally, and locally. Title III of SARA of 1986, separately titled the Emergency Planning and Community Right-to-Know Act (EPCRA), is an example of a remediation-oriented piece of legislation in which remediation is required in a planning context. EPCRA contains an extensive provision for regionwide emergency plans that identify facilities at which hazardous substances are present above a certain threshold quantity defined by EPA. In addition, EPCRA requires a comprehensive emergency response plan to be prepared at the local level, spelling out resources and responsibilities for managing emergencies.

Changes in Discretion and Prescription

When looking at the evolution of environmental law over the past two decades, two trends are apparent. First, it is clear that there have been increases in both the number of chemicals regulated and specified directly in legislation and the stringency of limitations placed in legislation and regulations for each chemical. By stating explicitly in legislation what the targets of regulation are, legislators can reduce the discretion of administrative agencies. Second, in contrast to the degree of prescription underlying the naming of pollutants and the specification of numerical limits, the choice of the methods for the estimation and description of risk is much more vague and allows considerable agency discretion.[11]

Kenski and Ingram (1986) identify a trend in the increasing stringency and prescription of legislation over time. They argue that prior to 1970 environmental legislation started out as being "non-coercive and nonregulatory, and that it was only with the failure of the milder, more noninterventionist pieces of legislation that policy escalation occurred" (Kenski and Ingram, 1986, 279). In other words, legislators began with highly discretionary legislation that eventually evolved into more prescriptive programs.

Definition of Discretion and Prescription in the Context of Regulation

Discretion in legislation (as distinct from administrative discretion) can be broadly defined as wide latitude in the interpretation and application of legal mandates, while prescription refers to the establishment of fairly well-defined, circumscribed procedures that allow little latitude. Legislative discretion can occur in a number of different forms. It occurs, for example, as (1) the absence of standards, (2) the existence of minimum standards above which an administering agency is free to make decisions or maximum standards below which agencies can exercise discretion, (3) the freedom given to an agency in legislation to choose the statutes applicable to a particular case, (4) flexibility in the decision to incorporate cost considerations in regulations, (5) the ability to exercise authority by interpreting laws through the promulgation of regulations, and (6) general flexibility in the exercise of other aspects of the law (Sabatier and Mazmanian, 1979; Kagan, 1978; Greenwood, 1984; MacIntyre, 1986; and others).

The degree of discretion or prescription within a given piece of legislation either is deliberate or may simply be a function of the degree of certainty and clarity in the legislative language. Sabatier and Mazmanian (1979) have developed a classification system for degree of discretion in legislation, as a function of the certainty of the legislation, that is useful for the characterization of risk-oriented laws.[12] Their scale, based on the degree of clarity of legislative objectives, is as follows (Mazmanian and Sabatier, 1983, 45):

1. *ambiguous objectives,* which give broad pronouncements about the purpose of regulation, such as the public interest, and do not specify or set priorities among alternative regulatory strategies
2. *definite "tilt,"* where a clear ranking of potentially conflicting objectives is given
3. *qualitative objectives,* a greater precision of objectives, as well as a ranking, but still restricted to a qualitative statement
4. *quantitative objectives,* which give precise levels of acceptable risk, methods of reduction, and/or timetables for compliance

Thus, less clarity implies more discretion, while more clarity implies more prescription. While this categorization is not explicitly intended to be normative, some scholars have taken an advocacy position with respect to whether discretion or prescription is more desirable in legislation and regulation. For example, Lowi (1969) has advocated prescription in legislation to the point of blaming regulatory ills on the lack thereof (deHaven-Smith, 1984, 414; Lowi, 1969, 297–303).

A different approach to a categorization of discretion was developed by MacIntyre (1986), who designed a comprehensive typology of reasons for discretion in legislation. In that typology, MacIntyre emphasized the sources of and forms in which discretion appears. These forms were divided into eighteen different categories. The broader categories in the typology included "information and time constraints faced by Congress, social and economic heterogeneity of the electorate, thresholds to political consensus-building (e.g., proliferation of veto powers and systemic need for compromise), absence of compelling reasons for legislative precision, concerns with implementation outcomes, and fallibility of language" (MacIntyre, 1986, 69). The reasons why Congress emphasizes discretion or prescription in legislation will be explored in more detail in Chapter 4, which deals with Congress's role in risk management.

Prescription in Specifying Chemicals and Target Dates for Compliance

Numerous examples can be cited of the increasing prescription in legislation aimed at regulating environmental health risks based on Sabatier and Mazmanian's criterion of legislative clarity. The language of much current legislation can be described in terms of Sabatier and Mazmanian's last category, "quantitative objectives," which allows the least discretion. This quantification takes the form of dates for compliance, specific chemicals targeted for regulation, and numerical limits for regulated chemicals. For example:

1. The Toxic Substances Control Act required EPA to ban the manufacture of polychlorinated biphenyls (PCBs) by 1978. Not only was a family of chemicals, PCBs, specified in the law, but the date by which the ban should take effect was also indicated. Section 4 of TSCA gave EPA one year to react to the Interagency Testing Committee's list of chemicals recommended for regulation by deciding what testing rules and methods were needed (Stever, 1988, 2–5).

2. The Safe Drinking Water Act of 1974 and its 1986 amendments reflect the growth in the number of chemicals that are subject to regulation.[13] In the early 1900s only about 12 general chemicals were specified for regulation in drinking water. In 1962, the Public Health Service published a list of chemicals recommended for regulation, and placed a total limit of 2 mg/L on organics in drinking water. Under the 1974 legislation, EPA had only issued a half dozen or so National Interim Primary Drinking Water Standards, which were never finalized. The list of recommended standards was longer. In 1979, a Maximum Contaminant Level of 0.10 was set for total trihalomethanes (TTHMs); this was actually the arithmetic sum of four organic chemicals— trichloromethane (chloroform), dibromomethane, bromochloromethane, and tribromomethane (bromoform). While various proposals or intents to propose regulations (Advanced Notices of Proposed Rulemaking) were issued by the EPA during the early 1980s (these appeared in U.S. EPA, March 4, 1982, 9350; U.S. EPA, 1983, 45502; and U.S. EPA, June 13, 1984, 24330), standards were not promulgated. In 1983, levels for 14 volatile organic chemicals were proposed, but in 1984, the new proposed rules were only for 9 of them (five were withdrawn because of insufficient information). Because the rate at which these standards were developed constituted a lack of progress, Congress required in the 1986 amendments that EPA enact standards according to a strict schedule, though no concentration limits for chemicals were placed directly in the legislation. Phase I required standards for eight organics and one inorganic by June 1987 and the development of a target list of 44 more by June 1988. Phase 2 required that standards be passed for at least 83 chemicals by June 1989. These were proposed in May 1989 (U.S. EPA, May 1989, 22064). At least every three years after that, 25 standards for chemicals on the 1989 list are to be prepared. The SDWAA specified granulated activated carbon systems as acceptable treatment technology for water systems. While alternatives are not entirely precluded, this option is strongly emphasized as at least a standard of performance that any alternatives must meet.

3. The Clean Air Act noted specifically which pollutants should be regulated as National Ambient Air Quality Standards. The CAA also contained a number of deadlines for automobile emission controls.

4. The Clean Water Act specified target dates for eliminating wastewater discharges from natural waterways and the level of quality certain discharges should attain, as well as the kind of general technology that should be used. The CWA also specified when wastewater treatment plants should meet a zero discharge goal or best control technology objective.

5. Section 201(d) of the Hazardous and Solid Waste Amendments of 1984 specified particular chemicals to be regulated and concentrations for these chemicals that would be prohibited from land disposal.

6. The Ocean Dumping Ban Act (though not identified directly as a health risk law)

specified 1992 as the date for the termination of ocean disposal of sewage sludge.

Chapter 5 contains more examples, discussed in the context of how these prescriptions limit agency discretion.

Discretion in Risk Measures and the Concept of Risk

Considerable discretion is allowed to agencies either in legislation or through judicial review in areas such as the use of methods of determining risk levels, the specification of additional risk agents (over and above those specified in the legislation), and the interpretation of what solutions and demands for knowledge and certainty are considered economically feasible. Many laws allow considerable discretionary action by rulemaking agencies and reviewing courts in the way they define and evaluate the existence of health risks. Thus, in contrast to laws that specify chemicals and timetables, many laws fit into the Sabatier and Mazmanian (1979) typology under "ambiguous objectives" because of their level of discretion with respect to risk measurement and definition of the concept of risk.

Morris and Duvernoy's 1984 typology of environmental risk–related federal statutes discusses the explicitness with which risks were characterized in those laws. They highlighted the following characteristics of statutory vagueness:

1. *Expression of risk:* Risk expression, according to Morris and Duvernoy, refers to the extent to which risk levels are specified. They argue that with the exception of the testing requirements under TSCA, no law specified how the risks were to be expressed, beyond making broad statements about human health and safety.

2. *Risk incidence:* Risk incidence is defined as the occurrence of risk with specific reference to some target population that will or can bear the risk in the present or the future. The specification of target populations, according to Morris and Duvernoy, varied in the laws they reviewed from animals and plants to humans. It also varied for human populations, referring to the public at large or to sectors of the population such as workers. Only about a third of the programs they reviewed even specified the target organism and the circumstances surrounding the risk exposure.

3. *Risk measure:* A risk measure, according to Morris and Duvernoy, is defined as "the types of consequences whose likelihoods are considered." Such measures are specified in almost every law. The categories of risk that Morris and Duvernoy found are worth summarizing here as an illustration of their diversity:

   - health, individual and public
   - safety/injury, personal and public
   - welfare
   - mortality
   - morbidity; irreversible, incapacitating illness; behavioral abnormality

- extinction
- environmental effects
- resource protection
- property damage

These findings imply that considerable discretion was built into the federal statutes implicitly by virtue of the lack of definition of risk and its outcomes.

Discretion and Prescription in Other Legislative Provisions

Economic balancing and agency discretion. In spite of the wide popularity of the balancing concept and the high degree of prescription within a given piece of legislation on this subject, different pieces of legislation have varied considerably with respect to how the requirement for balancing affects agency discretion. In the sense that they require agencies to do a certain kind of evaluation, i.e., balancing safety/health with economic considerations, they are prescriptive. O'Brien (1986a, 41) points out further that not only did the individual statutes substantially constrain agency discretion in the choice of a method of balancing, but the differences among the statutes opened the way for considerable judicial review of the statutes, which constrained agency discretion even further. Others argue that the statutes actually "provide only vague instructions on how to strike the final balance" [between economic and technical considerations and health risks] and further:

> The effect of these provisions is to delegate large policy decisions to the executive, arguably overstepping constitutional limits on the administration's lawmaking powers. The vagueness of these standards also undermines the capacity of executive agencies to develop authoritative interpretations . . .[and]. . . leaves them vulnerable to recurrent legal and political challenge. (Brickman, Jasanoff, and Ilgen, 1985, 59)

Discretion in planning vs remediation provisions. The changes from planning to remedial legislation discussed earlier also reflect a change in Congressional attitudes toward agency discretion. While the degree to which statutes containing planning or remediation provisions are discretionary depends more on the degree of detail to which the content of the plans or remedial efforts are specified, in general, provisions allowing for long-term planning in statutes allow more agency discretion than strategies for remediation-oriented planning.

Conditions Associated with a Risk Orientation in Legislation

A number of conditions external to government, producing both traditional responses by government and new initiatives, led to the rise in risk legislation as well as accounting for various characteristics of such legislation. Responses that tended to reflect more traditional governmental roles are as follows:

- The overall intensity of lawmaking in the country during the latter part of this century may have been driving lawmaking in the environmental health risk area.
- Environmental crises have been a major force in the passage of many risk-oriented statutes or parts of statutes and set the tone for a shift to more remedially oriented legislation.
- The number, size, popularity, and resources of public interest groups and industry lobbying efforts have risen.
- Political reactions to economic conditions in the country have changed; they are now viewed as reflecting not relative prosperity but hardship. First, relative prosperity contributed to the increased stringency in environmental requirements in the late 1960s and the early to middle 1970s; then, relative hardship contributed to attempts to loosen the very stringent environmental requirements of early 1970s legislation in legislation that was passed in the late 1970s and early 1980s (Bardach and Kagan, 1982b; Field, 1980). Slight improvements in economic conditions in the middle to late 1980s (which were by no means as good as earlier conditions) contributed once again to an expansion in legislation.

Other responses, though fewer in number, represented new initiatives on the part of the government.

- The burden of proof of causation has shifted from the injured party to government and polluters, and in recent years, away from government and toward polluters.
- The sophistication of analytical techniques and methods to assess environmental health risks has grown considerably. New and more refined analytical techniques allowed increased levels of detection, which led to more complex conceptualizations of environmental problems.
- Uncertainty in the scientific basis for risk decisions has had a persistent and pervasive influence on the degree to which risk has been defined in legislation, though developments in analytical techniques have reduced uncertainty in some areas.

General Intensity of Lawmaking

It was pointed out earlier that when the rates of growth in environmental health risk legislation are compared with rates of growth of laws in general in the United States (1) the rates of growth are roughly comparable, and (2) the growth in legislation in general preceded the growth of environmental risk legislation. One can infer from the second observation that environmental health risk lawmaking may have been initially driven by lawmaking activities in general. In addition, the momentum for the passage of health risk legislation in the 1970s originated from the momentum of general environmental protection legislation starting a decade or so earlier.

Crisis Events as Initiators of Laws

Researchers have recognized that crisis can initiate legislation (Stone, 1982, 179; Wilson, 1980, 370). Stone argues persuasively that one of the reasons crises initiate legislation is because legislators perceive that the costs of ignoring the problem will be high, and they want to reduce such costs and avoid losing public support.

According to Mitnick, crises can occur in the "issue creation" phase of public policy formation and implementation. He notes that "a 'triggering device,' which in the case of regulation is often a crisis in the subject activity, precipitates a controversy or conflict with some aspect of the activity as the issue, or subject matter, of the conflict." (Mitnick, 1980, 169)

The passage of several major pieces of risk-oriented legislation aimed at potential health effects from chemicals in the environment can be shown to have been prompted by crisis. To a large extent, the passage of such legislation in response to a crisis is the attempt of lawmakers to adjust to deficiencies in prior laws that presumably allowed the crises to occur. Legislation that emerged in part because of crises includes the National Environmental Policy Act; the Clean Air Act; certain sections of the Clean Water Act; the Safe Drinking Water Act; the Comprehensive Environmental Response, Compensation, and Liability Act and its amendments; the Occupational Safety and Health Act; the Food, Drug, and Cosmetic Act; and portions of other legislation (Zimmerman, 1986, 454–455). Some specific examples are listed below:

- The crisis that accompanied the passage of NEPA in 1969 was the Santa Barbara oil spill (Council on Environmental Quality, 1980).
- The passage of the Air Quality Act of 1967 has been associated with the New York Thanksgiving Day air pollution episode in that same year.
- Kelman (1980, 241) notes that the passage of the bill that produced OSHA in 1970 was prompted by a November 1968 mine disaster occurring in Farmington, West Virginia. The accident claimed 78 lives.
- The 1970 amendments to FIFRA were a reaction to Rachel Carson's *Silent Spring*. Changes made in 1983 were largely driven by a crisis created by the falsification of laboratory test data.
- Section 311 of CWA was a response to Torrey Canyon and Santa Barbara Oil spills (Stever, 1988, 6.1).
- The SDWA was passed in 1974 after an EPA survey uncovered widespread contamination of water supplies with organic chemicals. The survey was prompted by the discovery of these contaminants in the New Orleans water supply.
- TSCA's prohibition of the manufacture of PCBs was at least in part a reaction to the Hudson River PCB problem resulting from industrial wastewater discharges.
- The passage of CERCLA or the Superfund Act in 1980 has been attributed to the Love Canal incident.
- EPCRA, or Title III of SARA of 1986, was largely the result of chemical accidents at Bhopal, India, Institute, West Virginia, and other places.

Crises initiated environmental health legislation even prior to 1970. Air pollution episodes, primarily in New York City, triggered the passage of clean air legislation. Legislation was passed in 1955 after a New York City episode in

1953. Amendments were passed in 1963 and 1965 around the time of another episode in New York City in 1963. The legislation in 1967 and 1970 also followed the famous 1966 Thanksgiving Day air pollution episode in New York City. Other episodes that were estimated to have severe health effects occurred in Belgium in 1930, Donora, Pennsylvania in 1948, and London in 1952, 1953, and 1962. The Donora incident led to 17 excess deaths over those expected and the London 1952 incident resulted in 4000. These are documented in Fensterstock and Fankhauser (1968, 1).

Quirk (1980, 194–199) notes a steady pattern of motivation by crisis in food and drug legislation since the turn of the century. He notes, for instance, that the thalidomide disaster was the major impetus to the passage of the FDCA amendments of 1962 (Stone, 1982, 181–2).

In contrast to these laws originating in crises, other laws (particularly small amendments to major pieces of legislation) have emerged more gradually. Laws with less of a crisis orientation, however, are harder to find. One example is water pollution control legislation, which has evolved gradually since the passage of one of the earliest laws in 1948. This legislation is intended to protect the public from waterborne diseases contracted from drinking water or from waterborne contaminants concentrated in fish and shellfish that people consume. The gradual changes in the law in part reflected gradual evolutionary changes in water pollution control technology. The introduction of risk assessment methodologies in environmental health legislation may have similarly benefited from a gradual evolution in the theory for estimating cancer risks at low doses of chemicals. Hutt, for instance, traces this history in the context of food and drug legislation, noting "the concept of quantifying human risk on the basis of animal feeding studies had already been explored by scientists for 25 years" prior to the passage of the FDCA in 1972 (Hutt, June 1984, 173).

Thus, crises were used to create legislation and accounted for much of the increase in laws or specific provisions within the laws during the 1970s and 1980s. Furthermore, the urgency with which many of the crises were expressed may have contributed to laws becoming more heavily oriented toward remediation rather than planning.

The Role of Interest Groups

The blossoming of the environmental movement just prior to and following Earth Day is well known. This movement may have been influenced by even earlier developments in the 1960s, which have been characterized by Stone as the beginning of a period of intense "consumerism" (Stone, 1982, 32). Morgan and Rohr (1986, 214) point out that citizens' groups are a large proportion of the interest groups represented in Washington and cite the numerous mechanisms (of varying effectiveness) that have emerged to encourage such participation.

One only needs to look at the number of organizations, the size of membership, and the activity in terms of citizen-initiated court cases to realize this trend was a major one in potentially determining the direction of public policy toward environmental risks during the 1970s. Mitchell's history of national environmental lobbies reveals that at the turn of the century there were seven national and two regional conservation organizations, and by the end of the 1970s there were about 75. Furthermore, membership grew from a few thousand to almost 200,000 by 1977, and most of this growth occurred over about a decade (Mitchell, 1979, 93–96).

More recently, substantial growth in environmental organizations has continued, especially when the government enacts policies that appear to threaten the environment. For example, Vig and Kraft (1984, 5) observe that in reaction to Reagan Administration policies toward the environment, "membership in environmental organizations grew by leaps and bounds as public concern spread." Mohr (1989, 31) reported increased membership in environmental groups after the Exxon Valdez oil spill. This increase may have been in reaction to membership campaigns launched after the spill, but spokespersons for these groups point out that growth had been occurring for some time prior to the spill. Mohr cites a monthly increase of 8000 in new applications at the National Wildlife Federation since the latter part of 1988; during the 1980s, membership in the Sierra Club grew from 8000 to 500,000; the Environmental Defense Fund is cited as having an annual membership of 100,000, a doubling from previous periods. While several factors are used as indicators of size and growth in environmental organizations (membership, donations, magazine subscriptions), membership is the most commonly cited indicator.

This growth in environmental organizations has often been associated with a rise in the number of laws (Field, 1980). The relationship between the emergence of environmental groups and citizen access to the courts is discussed in Chapter 6.

Other interest group activity emerged as well, particularly in the labor sector. Labor has had a low representation rate among environmental interest groups, and unions have been declining since the 1970s. While the decline in labor union membership has occurred over three decades, the most dramatic declines, according to Farber (November 13, 1987, 915), have occurred between 1977 and 1984. In spite of these trends, however, the influence of labor unions in the environmental health risk area apparently is still considered to be quite strong (Brown, 1987, 257–258). Many labor unions maintain specialized units devoted primarily to occupational health issues. For example, the Amalgamated Clothing and Textile Workers Union maintains a director of safety and health, the Oil Chemical and Atomic Workers Union has occupational health specialists, the International Chemical Workers Union maintains a health and safety department, and the Industrial Union Department and the Department of Occupational Safety within the AFL-CIO are considerably active in environmental health activities.

Finally, industry as an important interest group influenced the direction of legislation. Field (1980) notes that the appearance of risk-only statutes reflects variations in the timing and ability of industry to lobby against health-related statutes; in the middle to late 1970s, industry was better prepared to lobby for the incorporation of economic benefits. Industry has initiated a major proportion of the environmental court cases that in many instances have changed environmental risk law.

Citizen action is intimately connected with the emergence of crises, which were identified earlier as a motivator of legislation. The initiating conditions in a crisis situation often exist for years before the crisis is identified as such (Zimmerman, 1986, 445). An impetus such as media attention or the motivations of an individual or interest group is often required to precipitate a crisis from the events.

While the influence of interest groups is pervasive, their mode of influence is not always formalized. In particular, influence is not often exercised within formal public participation programs. According to Kraft and Kraut (1988, 65), " . . . there has been little empirical study of public participation programs, their effect on regulatory agency decision making, and the way in which the behavior of agency officials affects the extent and type of public participation."[14] As discussed in later chapters, interest groups can exert informal influence through initiation of and participation in negotiated settlements.

Political Reactions to Changing Economic Conditions

During the period from the late 1960s through the early 1980s, the economy of the country moved from conditions of relative economic prosperity to periods of relative economic hardship. A number of indicators are often used to show the trends and characteristics of economic conditions, namely, wage and price inflation rate, the unemployment rate, and productivity. (See, for example, Bosworth, 1981.)

Unemployment rate. The percentage annual unemployment rate between 1958 and 1964 was consistently above 5.0% and ranged from 5.0 to 6.6%. In the late 1960s, however, the picture was quite different. Between 1966 and 1969, the rate was consistently below 4.0%, ranging from 3.4 to 3.7%. The unemployment rate began to rise again after 1970; after 1975 it was rarely below 6.0%, and during several years it exceeded 8.0%. 1982 and 1983 were the worst years, with unemployment rates of 9.5%. Since 1984 the rate has been dropping but still has exceeded 6.0 (Executive Office of the President, 1989, Table B-39, 352).

Gross national product. The annual percentage change in the gross national product (adjusted for inflation) is another indicator of economic condition. While the annual percentage changes are quite uneven between 1965 and 1985, a few distinct trends are apparent. The trends in the mid-1960s reflect a strong economy. The percentage change from one year to the next between 1962 and 1966 ranged from a growth of ≈4–6% per year. Between 1967 and 1969, there was a slowing to a range of 2.4–4.1% growth, with a downturn occurring in 1970. While 1971 through 1973 seemed to regain some of the earlier upward trends, the rate of change through 1987 was usually below 4% per year and was negative in a couple of years (Executive Office of the President, 1989, Table B-2, 69).

Thus, while periods of relative prosperity reigned during the 1960s, the 1970s saw gradual economic decline and the latter part of the 1970s was marked by periodic recessions. While the 1980s showed some recovery in terms of unemployment, the change was modest; gross national product did not show marked changes from the 1970s. Bosworth (1981, 8) summarizes the overall economic conditions that the country faced as it entered the 1980s, many of which became worse thereafter:

> Inflation is in excess of 10 percent on a sustained basis, unemployment is above the critical levels that triggered the expansionary policies of the early 1960s, productivity growth has fallen to about a third of its historical trend, industrial capacity is inadequate to employ the expanded work force of the 1970s, and the economy is extremely vulnerable to disruptions in world energy and food markets. All of this suggests that the 1980s will be dominated by severe constraints in every area and increased social and economic conflict. In sum, the coming decade will not be conducive to the expansion of programs limiting the external costs of industrial growth.

Many regulatory programs, and in fact the regulatory philosophy pursued by the federal government during various reform movements, were influenced by these economic conditions.

Bardach and Kagan's brief review casts the change in the concept of regulatory reform over time in terms of changing economic conditions or the political perceptions of those conditions (Bardach and Kagan, 1982b, 5). They argue that in the 1960s and early 1970s, reform occurred at time of relative affluence. Reform in this era meant passing some of the most stringent health and safety legislation of the century, with strict timetables for compliance. In the middle to late 1970s and early 1980s, the economic environment of the country was perceived to be different. According to Stone (1982, 33) this turning point was 1977. White (1981, 5) portrays this history and its relation to economic conditions in a similar way:

> As recently as the early 1970s, there seemed to be widespread sentiment that the goals toward which these regulations were aimed were important and that the fast-growing American economy could easily accommodate the costs. The dissenters were few. The middle ground was clearly in the area of more regulation. But, as the growth of the American economy slowed, as inflation became more serious and endemic, and as the costs became more apparent, opponents of increased regulation grew in numbers and vocal power. The midpoint of sentiment had clearly shifted.

Thus, during the late 1970s, economic conditions (more importantly, the political forces that made use of them) and growing experience with the cost of environmental controls changed the concept of regulatory reform to one of restraint, relative to the regulatory strategies pursued a decade earlier.[15] The political response to economic conditions was a major influence on the statutes and helps to explain the virtual elimination of health-only statutes of the early 1970s in favor of the currently prevailing balancing statutes.

Shift in the Burden of Proof

Administrative law usually requires that the burden of proof for any rule or regulation fall upon its proponent (Ricci and Molton, 1984, 380). As social or protective regulation for the public became more popular, the burden of proof in legislation began to shift in the direction of the originators of an activity. According to Kessler (1984, 1035), the shift in the burden of proof from government to industry occurred with respect to food additives as early as 1958. Under the original FDCA, the federal government had the responsibility to bring an enforcement action to show that an additive was deleterious or poisonous. But the 1958 statute, in the context of defining "food additive" and the way that an additive can be determined, placed the burden of proof on the user of the additive (Stever, 1988, 8-4 and 8-5). In particular, Section 409 of the act shifted the burden to industry in its premarket approval process for food additives. Thus, prior to approval the manufacturer would have to show an additive was acceptable under the statute.

Examples of this shift to government and industry exist in more recent risk legislation. Under Section 1414(c) of the SDWA Amendments of 1986, water suppliers have the burden of notifying customers of violations of standards and of their failure to monitor the water properly (Stever, 1988, 7–37). FIFRA requires whoever registers a pesticide to show that it is not harmful. Ricci and Molton point out that the Occupational Safety and Health Administration would not bear the burden of proof according to its regulations, but the courts required that it bear this burden in the benzene decision. In that case, the court required OSHA to demonstrate the health effects of benzene at the lower standard it was proposing (Ricci and Molton, 1984, 380–381). Under the Toxic Substances Control Act the burden to submit information is on the entity submitting a Premanufacturing Notice (PMN), though the extent of this information depends on a variety of factors, including what rules EPA has passed. Thus, EPA is required to list suspect chemicals among other responsibilities under TSCA (Stever, 1988, 2-20 and 2-21).

Analytical Methods and Detection

The 1970s saw a rise in the sophistication of analytical techniques used to evaluate environmental health problems. To summarize this transition briefly:[16]

the significant changes were the extension of methods of analysis from the pure sciences into applied environmental science, greater ability to calibrate environmental models as the number of applications grew, and the narrowing of the range of error estimates in environmental models, including risk assessment.

Simultaneously, there have been tremendous improvements in the ability to detect substances in the environment and in the human body (American Chemical Society, 1988; Zimmerman, 1987b, 235). This has enabled regulators to justify setting standards at lower levels than was ever possible, and to cover a larger number of chemicals. This ultimately increased the degree of prescription in legislation.

Uncertainty in Information

A pervasive undertone in almost any discussion of environmental health risk lawmaking and the administration of those laws is the fact that there is considerable uncertainty in the information that constitutes much of the basis for action.[17] It is important here to briefly describe the nature of this uncertainty, since it does underlie much of the subsequent discussion.

The heart of the scientific analysis of environmental health risks is the process of risk assessment. The steps in the procedure are useful for organizing the nature of uncertainty and where it occurs:[18]

1. characterization of a source of health risks, such as an industrial emission
2. estimation of the movement and changes in the chemicals from that source in the environment (called environmental fate and transport)
3. identification of an ultimate exposure point for human beings, such as a watercourse, a land area, or air
4. estimates of what happens, to whom, and to what extent. (Assessment of how much of an effect occurs is usually made using dose-response curves from animal data, since in most cases human data are lacking. This is combined with risk assessment models to extrapolate responses to regions of low doses where measurement is not possible even in animal experiments.)

Uncertainties occur throughout the process and are the source of much of the controversy surrounding how environmental health risks should be managed. First, sources of chemicals are often difficult to identify, especially where health effects have been discovered long after the original exposure occurred. In hazardous waste disposal sites in particular, the types and amounts of chemicals disposed of are difficult to identify after disposal.

The environmental fate and transport component embodies uncertainty in the prediction of how chemicals disperse in the environment. In the air environment, models employ dispersion coefficients that are combined with highly variable estimates of wind speed and direction. In the water and soil environment, parameters that define the extent to which chemicals attach themselves to soil vs moving down into water not only are a function of chemical character-istics, but vary considerably with characteristics of the environment such as soil moisture and porosity.

The use of dose-response curves constructed from animal data is associated with another set of uncertainties. In the range of doses and responses where empirical data can be collected, there are debates over whether the substitution of large doses administered to animals over short periods of time adequately portrays effects of small doses over long periods of time. Such an experimental design requires extrapolation of responses to low doses via risk assessment models. When these models were first used, the results of different algorithms often differed by four to six orders of magnitude.[19] Now with a growing consensus in that field and more refined models, the range is narrowing in a number of cases. When animal results are used for managing risks of chemicals to humans, different methods of extrapolation can produce different risk levels. The common methods of extrapolation are based on animal/human ratios of weight, volume, surface area, and length of lifetimes. Different risk levels also occur depending on which route of exposure is used to establish the dose-response curve, i.e., ingestion, inhalation, or dermal absorption. Often experiments via one route are used to characterize exposure via another route, and this can produce uncertainties. Finally, most of these techniques usually focus on one chemical at a time. The synergistic effects of several chemicals together at all of the stages outlined above can produce substantial variation in the results. When all of these uncertainties are put together, their combination magnifies the error even more. Thus, uncertainties in the way risks are characterized are a major consideration in the way risks are managed, and are perhaps the major factor in the choice of risk management strategies.

THE USE AND DEVELOPMENT OF RULES (STANDARDS)

As was the case with the development of laws, the development of standards reflects processes that are inherent in the way government typically operates, as well as new processes that are created specifically to address the risk issue. A standard is a rule that prescribes a particular level of exposure or degree of risk that a given activity may impose and is often accompanied by conditions that refer to when and where the limitations apply. It is defined by law or emerges from the regulations passed by agencies or their administrative procedures to comply with the laws. The concept of a standard is defined in more detail in Appendix 2.1.

The intensity and direction of developments in standards for the regulation of environmental health risks are affected by the following factors: (1) the variety of ways in which risk is qualitatively described in standards, (2) the implications of these qualitative descriptions for the ability to quantify levels for risk standards, (3) the shift in the types of standards used over time in the direction of characterizing sources of chemicals rather than the quality of the ambient environment, and (4) the relationship of this shift to information requirements and agency discretion in administering the standards.

The conditions that lead to these characteristics of standards include (1) the underlying vagueness of the laws regarding the causal connections between chemicals and health risks and its effect on the consistency of how standards are described, and (2) the general need for certainty on the part of regulatory agencies.

Trends in the Use of Standards

Variations in How Risk Is Described in Standards

In the last section, it was pointed out that the concept of risk is used and described in a number of different ways in legislation. In passing environmental health and safety laws, Congress used a number of terms to characterize health risk. These terms were often vague, and the linkage between the concept of an allowable level of control and the health risk that would result was also vague. In a number of instances definitions were never given in the statutes for the concepts used. This gave the responsibility of defining risk to the administrative agencies and the courts, which is the subject of Part II of this book.

Variation in risk qualifiers. The qualifiers or adjectives that precede the word "risk" in legislation and regulations are usually a clue to their meaning and use. The concept of risk is usually modified by adjectives that refer to different things: its economic justification (e.g., "reasonable," "feasible"), its acceptability (e.g., "generally recognized as safe"), or its level (e.g., "zero," "de minimis," "sufficient," "substantial," "significant," "ample margin of safety," and "adequate margin of safety"). Still another set of qualifiers refers to the extent to which risks are urgent or exist in an emergency or crisis setting. These terms include "imminent hazard," "grave danger," "endangerment," and "deleterious."[20]

Thus, risk qualifiers come in a wide variety of forms. The extent to which they are defined in the statutes and have consistent meanings where they are defined also varies. These qualifiers are usually defined in the context of setting standards or guidelines for the conduct of a given activity or for the level of a particular chemical. They are also used in the statement of policy or intent at the beginning of a law. Table 2.3 summarizes the use of many of these qualifiers in both standards and policy statements ("objectives") for some of the major pieces of risk-oriented legislation. These terms are central to many regulatory programs involving risk. Their meanings have evolved through the standards-setting process, other agency decisions, and judicial review of these decisions. Several categories of qualifiers are discussed briefly below as they have appeared in environmental health and safety legislation. These qualifiers are reasonableness, significance, and the set of emergency qualifiers listed above.

Examples of risk qualifiers. The following common qualifiers for risk in federal statutes, "reasonable" risk, "significant" risk, "imminent" risk, and "endangerment" exemplify the variability in the definitions and use of risk qualifiers.

Reasonable risk: Three laws that refer to risk as unreasonable or reasonable are TSCA, FIFRA (or its amendment, FEPCA), and the Hazardous Materials Transportation Act (HMTA).[21] A policy of TSCA is to protect the public against "unreasonable risk of injury to health or the environment" (15 U.S.C.A. Article 2601(b)(2)). Under TSCA, the concept of unreasonable risk is not defined explicitly but is defined instead by example. Some of the kinds of health and environmental effects that constitute unreasonable risk in TSCA are listed in the legislation, but only as potentially creating an unreasonable risk. The effects that may constitute unreasonable risk include "carcinogenesis, mutagenesis, teratogenesis, behavioral disorders, cumulative or synergistic effects" (Article 2603(b)(2)(A)). There are a number of reasons why the reasonableness of risk is specified for chemical substances and mixtures: it is used as the criterion for testing requirements under Article 4 (15 U.S.C.A. Article 2603); to establish a rule that will govern manufacture, use, and distribution of chemicals (Article 6); and as a basis for courts to seek injunctive relief against the activity (Article 5f(3)(B)). Under FIFRA/FEPCA (7 U.S.C.A., Article 136(bb)), "unreasonable adverse effects on the environment" is defined to mean "any unreasonable risk to man or the environment, taking into account the economic, social, and environmental costs and benefits of the use of any pesticide." This concept is used as the basis for deciding which pesticides to register under Article 136a of the statute.

Significant risk: The term "significant risk" appears in TSCA, RCRA, and FDCA. The term also evolved in the context of the judicial review of OSHA's decision to reduce the benzene exposure standard for workers. The court required OSHA to establish a significant level of risk or threshold for benzene exposure as a means of justifying the proposed lowering of the standard.[22] The variability in uses of the term originates from the fact that it implies some threshold of risk that is determined by the application of quantitative models. Different models, especially those involving extrapolations to regions where doses are not known, will have different results, and hence different significance levels. Whether the appearance of a significant effect is acceptable or not is a separate determination. Stever's reading of TSCA shows that significant and unreasonable risk both involve the same kind and degree of quantification. "Substantial" risk, according to Stever, is also a function of the same parameters that quantify significant and unreasonable risk, making the three terms somewhat indistinguishable (Stever, 1988, 2–35).

Imminent risk: The term "imminent hazard" in the Consumer Product Safety Act (CPSA), which is a basis for the commission to bring emergency actions,

Table 2.3. Examples of Terminology Used to Characterize Risk by Type of Legislation: Historical and Current[a]

Risk Terminology or Concept for Selected Laws	Type of Risk					
	Zero Risk	Threshold	Significant	Substantial	Unreasonable	Imminent
SDWA						
Objectives						
Imminent and substantial endangerment to the health of persons (Section 1421(a))				X		X
Standards						
No Observed Effect Level (NOEL)		X				
Lowest Observed Effect Level (LOEL)		X				
Suggested No Adverse Response Level (SNARL)		X				
Maximum Contaminant Level (MCL; RMCL)	X (some)	X				
Adequate margin of safety (National Primary Drinking Water Standards) (Section 1412 (b))		X				
FDCA						
Standards						
Delaney Clause	X					
Action Level/Tolerance Rule/Food Additive Regulation[b]	X (some)	X (some)			(X)[c]	
Generally Recognized as Safe (GRAS)			X		X	
CPSA						
Unreasonable risk of injury					X	
OSHA						
Standards						
Grave danger			X			
Permissible Exposure Limit		X	X		X	
AEA						
Standards						
As Low as Reasonably Achievable (ALARA)		X				

Table 2.3, continued

Risk Terminology or Concept for Selected Laws	Type of Risk					
	Zero Risk	Threshold	Significant	Substantial	Unreasonable	Imminent
CWA						
Objectives						
An unacceptable hazard to life, a significant loss of property, or an immediate, unforeseen, and significant economic hardship (33 CFR 325.2 (e)(4))			X		X[c]	
Standards						
Best Practical Control Technology (BPT)			NA[d]			
Best Available Control Technology (BAT)					(X)[c]	
Water Quality Standard			X			
CAA						
Standards						
Nondegradation (PSD)			X			
Adequate margin of safety requisite to protect public health, NAAQS-Section 109			(X)[c]			
Ample margin of safety to protect the public health, NESHAPs Section 112			X		(X)[c]	
Objectives						
Imminent and substantial endangerment to health (Section 313)				X		X
TSCA						
Objectives						
Section 4(f):						
• significant risk of serious harm			X		X	
• significant risk of widespread harm					X	
Section 6(a)—unreasonable risk for chemical or process					X	

Table 2.3, continued

Risk Terminology or Concept for Selected Laws	Type of Risk					
	Zero Risk	Threshold	Significant	Substantial	Unreasonable	Imminent
Imminent and unreasonable risk of serious or widespread injury to health or the environment (Section 7)					X	X
RCRA/HSWA						
Objectives						
Imminent and substantial endangerment to health or the environment (Section 7003)				X		X
Standards						
Alternative concentration limits			X			
CERCLA/SARA						
Standards						
Alternative concentration limits			X			
(Legally) Applicable or Relevant and Appropriate Requirement (ARAR)	wide range of risk types possible					

Source: A number of these provisions were written up in connection with emergency management in Zimmerman (1985a). The entries in this table are consistent with a recent synopsis in Ricci and Cox (1987), Table 1.

[a] Entries in this table include standards that have since been replaced by subsequent amendments. Entries are examples only and are not meant to be exhaustive. Where there are no entries following the risk terminology for a given law, the type of risk has not been made explicit in the law.

[b] Under FDCA, tolerance rules and their exemptions and revocations are issued under Section 408 for pesticide residues in food. Reasonableness is implied, because FDA can do cost-benefit balancing in establishing the rule. Food additive regulations, issued under Section 409, are governed by a criterion of safety. Safety contains cost considerations (Kessler, 1984, 1035). Action levels are set through discretionary action (U.S. EPA, Office of the General Counsel, 1984).

[c] Xs in parentheses indicate implied terms.

[d] NA = not applicable.

is not defined in the statute nor has it been defined in the course of judicial review (Stever, 1988, 4–8). In contrast, imminent risk or imminent hazard under TSCA is defined under Section 7(f) to mean a "chemical substance or mixture which presents an imminent and unreasonable risk of serious or widespread injury to health or the environment." The operational definition used for administrative purposes, however, is a risk that is likely to occur before a rule can be finalized under Section 5 of the statute.

Endangerment: The term "endangerment" has been used in the SDWA to characterize the risk of a contaminant for underground injection. In that context, it is defined first as an injection into the ground that may result in a contaminant entering a water body that supplies or that could reasonably be expected to supply public drinking water, and secondly, as a contaminant that is either not meeting the national primary drinking water standards or is otherwise affecting public health (Stever, 1988, 7–45, quoting SDWA, 42 U.S.C.A., Article 300h(d)(2)). The use of the phrase "otherwise affecting public health" is a curious one and implies almost any risk standard is possible.

Inconsistencies in the Quantification of Risk Levels

Since there are so many different terms or different qualifiers for the term "risk," and since the same term often has different meanings, considerable variation occurs in what numerical levels for risks are considered acceptable under different statutes. Because of this it is not surprising that numerical risk levels tend to vary from agency to agency and for the same chemical used in different situations. Acceptable risk levels used by different agencies as the basis for risk management often range from one in 10,000 to one in 1,000,000 (Cobler and Hoerger, 1985, 121). A review of 132 federal regulatory decisions made during the late 1970s and early 1980s revealed a considerable variety of risk levels for the same chemical in different settings, attributed in part to how agencies considered exposed populations in the estimates (Travis et al., 1987).

Even the risk levels used by a given agency can differ from what is considered a norm. For Superfund site cleanup thresholds, for example, the U.S. Congress's Office of Technology Assessment (OTA) has criticized EPA for using a one-in-100,000 risk level rather than the more conventional one-in-1,000,000 level that EPA uses in other circumstances (U.S. Congress, Office of Technology Assessment, 1988, 5, 7).[23] In contrast, the FDA regulates carcinogenic constituents of noncarcinogenic substances in food additives if they pose a risk greater than one in 10,000,000.[24] In a RCRA regulation proposed in 1988 as the basis for siting new solid waste disposal facilities, EPA proposed risk levels ranging from one in 10,000 to one in 10,000,000 (U.S. EPA, August 1988).

One reason the levels vary is because the health effect or endpoint of a risk measure can vary from statute to statute. Morris and Duvernoy (1984) have pointed out, for example, that the risk measure can refer to the risk of becoming ill (morbidity), or of dying (mortality). Each of these endpoints could imply a different risk level. A second reason is the different policies and politics dominating different agencies. For example, Fise (1987, 25) observed that the regulation of nitrosamines in certain similar baby products was split between two agencies, and the two limits differed from one another: baby pacifiers were regulated by CPSC and nipples by FDA. Nitrosamines in pacifiers were set at 60 ppb, and nitrosamines in nipples were set at 10 ppb. To some extent the difference represents a three-year difference in the setting of the levels, but otherwise the difference is difficult to explain.

Another reason for varying risk levels is that different agencies may be making different assumptions about the amount of money required to achieve the same level. A number of cost estimates attempt to quantify the cost of achieving certain risk levels. Using risk and cost figures for a variety of risks, Morrall (1986, 30–31) found that the cost per life saved ranged from $100,000 to $72 billion (1984 dollars) for a variety of regulated risks. Estimates for value of a life saved with respect to chemical health and safety risks over the past decade and a half generally range from less than $100,000 to $8 million per life.[25] The range of costs per life from all of these studies is roughly consistent with the range of estimates attributed to agencies by Morrall for about the first 20 out of the 44 regulatory programs he explored, rank ordered by cost. Morrall observed that "the most obvious implication of these figures is that the range of cost-effectiveness among rules is enormous." This is based on the observation that when regulations are rank ordered according to cost per life saved, annual risk estimates are not ordered in the same way as the costs.

Variations in Types of Standards Used

Standards as a group are by no means homogeneous. One kind of standard pertains to sources, activities, or products that pose a potential health risk. These standards typically appear in the form of controls on what materials are used in a product or process, how the process is designed, how it is operated, and how waste materials are managed. A second type of standard refers to desirable levels of quality in the environment, such as air or water quality. A third standard, less commonly used, refers to allowable levels of an activity or concentration of a substance that the human body can tolerate. Examples in this category would be allowable concentrations of radiation or heavy metals in the human body.

Table 2.4 summarizes many of the standards in risk-oriented laws organized according to these three categories. Extensive examples and discussions of these types of standards are found in Appendix 2.3. The application of the categories of environmental health standards to federal environmental health risk legislation reveals that there is quite a large variety of categories for standards—at least eight different ones.[26]

While conceptually it may be reasonable to vary the type of standards used to obtain different regulatory objectives or adapt to different regulatory climates, no such relationship between type of standard and regulatory objective is apparent in the area of environmental health risks. More typically, the type of standard used is a function of the amount of information available, the cost, and political expediency.

Order of Emergence of Standards: General to Specific

Within any given piece of legislation, a pattern emerges with respect to the order in which the different types of standards are used or appear. This order, it can be shown, reflects the reaction of lawmakers and administering agencies to the degree of uncertainty in the information on which the standards are based.

Ambient standards appear before source (or pollutant discharge) standards. Ambient standards that specify pollutant concentrations in the general air and water environment are usually developed before standards applicable to sources of health risk. For example, under the Clean Air Act, types of chemicals were identified and numerical limits were set for National Ambient Air Quality Standards (NAAQS) (ambient standards) earlier than for New Source Performance Standards or National Emission Standards for Hazardous Air Pollutants (both applicable to sources of health risks). NAAQS were first developed in 1971 shortly after the passage of the Clean Air Act in 1970 (with criteria documents appearing in the late 1960s),[27] whereas source standards for hazardous air pollutants began to slowly develop in the late 1970s. Similarly, under the Clean Water Act ambient water quality standards that set limits for pollutants in surface waters were initially set long before effluent guidelines and limitations were suggested for sources discharging into those waterways.

General ambient standards appear before specific ambient standards. Some standards are written in a general or narrative form,[28] while others are written in the form of precise numerical limits that specify a concentration or allowable weight of a chemical pollutant. Within the category of ambient environmental quality standards, very general standards (often in a form similar to objectives or goals) for broad categories of contaminants rather than specific contaminants usually appear earlier than specific numerical limits on a chemical-by-chemical basis. Under the Clean Water Act, for example, very general ambient standards appeared first as the requirements that no oil or grease exist in concentrations that would lend a visible sheen to waterways (no particular concentration was specified) or that no visible floating solids be allowed in natural waterways. These were followed later by numerical limits for more specific standards, e.g., for dissolved oxygen, suspended solids, and pH. In the 1970s and early 1980s the chemicals for which numerical limits were set for ambient water quality became far more numerous, and the numbers were often based on risk assessments.

Within the category of source (or pollutant discharge) standards, standards closer to the beginning of a process or activity appear later. For example, under

Table 2.4.   Examples of Types of Standards[a]

General Type of Standard	Legislation	Standard or Program Incorporating Standard
Standards Pertaining to Risk Agents or Sources		
Materials specifications	CAA	Sulfur limitations in coal
		Fuel additives; lead content in gasoline
	FDCA	Interim, emergency, temporary, and permanent standards for:
		• Food additives (general)—Delaney Clause
		• Food contaminants: aflatoxin, PCB, nitrosamines
		• Cosmetic additives
		• Drug additives
		• Medical devices
Design	CPSA	Packaging, construction, etc. (Section 7)
	RCRA	Standards for the design, construction, and installation of new underground storage tanks
Manufacture	TSCA	Significant New Use Rule (Section 5), the existence of which triggers the necessity for submission of a Premanufacture Notice
	FIFRA	Registration of pesticide manufacturing establishments
	FDCA	Registration and inspection of food processing establishments
	SDWA (LCCA[b])	Restrictions on the lead content of parts of drinking water coolers
Operation/waste discharge (output)	CAA	Motor Vehicle Inspection and Maintenance Program—vehicle emissions (Section 202)
		New Source Performance Standards (Section l09)
		National Emission Standards for Hazardous Air Pollutants (Section 112)
		Motor Vehicle Emission Standards
	CWA	Effluent Limitations Sections 301(b)(2), 304(b)(2), 307
		Water Quality Based Effluent Limits (Section 302)
		New Source Performance Standards (Section 306)
		Toxic and Pretreatment Standards (Section 307)
		Ocean Discharge Criteria (Section 403)
		Disposal of Dredge and Fill (Section 404)
	FIFRA	Rebuttable Presumption Against Registration (Section 3)
		Suspension of Registration (Section 6)
	RCRA	Hazardous waste generator standards (Section 3002)
	HSWA	Hazardous waste transporter standards (Section 3003)

Table 2.4, continued

General Type of Standard	Legislation	Standard or Program Incorporating Standard
		Hazardous waste facility standards (Section 3004)
		•Interim status standards
		•General status standards
		•Interim standards for new facilities
	CERCLA, SARA	Remedial Action Cleanup Standards (Section 121)
	SDWA	National Interim Primary Drinking Water Standards (Section 1412)
		Maximum Contaminant Level Goals
Product use	CPSA	Consumer Product Safety Standards (Section 2056a)
		Voluntary Standard
		Consumer Product Safety Rule (Section 2058(b)(1))
		Labeling requirements
	FIFRA	Generic registration standards for active ingredients
	HMTA	Restrictions on chemical transport
	HSA[c]	Labeling requirements; voluntary standards; rules
	OSHA	General Industry Standards, Part H
Ambient Environmental Standards	CAA	National Ambient Air Quality Standards—primary and secondary (Section 106)
	CWA	Water Quality Criteria (Section 304 (a))
		Water Quality Standards (Section 303)
	OSHA	Emergency Temporary Standard (Section 6 (c))
		Permanent Standard
Human Exposure Standards	AEA[d]	Radiation standards

Source: This classification of standards into source, ambient, and receptor categories initially appeared in a more general form in Zimmerman (1987), Table 8-3.
[a]The incorporation of OSHA standards as ambient environmental conditions really should be interpreted as the environment of the workplace rather than the general environment.
[b]Lead Contamination Control Act.
[c]Hazardous Substances Act.
[d]Atomic Energy Act.

the Clean Air Act (CAA), a standard for the content of a pollutant in a raw material applies to the beginning of a process or activity. Standards for such raw materials, e.g., for lead in gasoline or sulfur in coal, appeared in the middle to late 1970s. (The CAA is one of the few statutes that contains raw material or input standards.) In contrast, emission standards under the CAA were developed much earlier, though the process has been a continuing one. Another example appears in the water quality area. The order in which water quality standards (pertaining to sources) tend to appear is as follows: discharge or effluent standards emerge first, followed by performance standards for wastewater treatment systems; finally, pretreatment standards (the analogy to a raw material standard) follow.

Figures 2.4 and 2.5 both portray the order in which standards generally emerge in the regulatory process as discussed above. In interpreting these figures, it must be kept in mind that not all laws contain standards of all types. Since examples of human exposure standards are too few in number to allow generalizations, they have been omitted from these two figures.

Types of standards and causal proof. Figure 2.4 portrays how standards set at different control points vary according to the extent of information required to establish them. Regulations for controlling environmental contaminants can be expressed either directly, in terms of the amount of a chemical in the environment, or indirectly, in terms of the impacts that different activities may have. Indirect controls inherently involve more information than direct ones. An instructive example is the control of the amount of sulfur to which people are exposed in the air from coal combustion. A standard that expresses the amount of sulfur in the ambient environment requires (relatively) the least information. It mainly involves a knowledge of the level of sulfur in the air below which health effects can be avoided. A second standard sets a prescribed level for discharges (emissions) of sulfur through a stack into the ambient air. This standard also requires a knowledge of how the discharge is translated into an acceptable environmental concentration, a task usually performed with the aid of dispersion models. A third standard sets an allowable level of sulfur in coal. In addition to the information needed for the other two standards, it requires knowledge of how much of the sulfur in the coal will be converted into air emissions and end up in the stack.

Types of standards and discretion. Figure 2.5 shows that once a source standard is developed, a regulated entity has relatively less discretion in designing its process for standards that are closer to the beginning of the process. For example, a regulated entity operating under a raw material standard will generally have less discretion in designing its process than one operating under an operation and maintenance standard or a discharge standard. Similarly, a regulated entity operating under a design standard will have less discretion (i.e., more constraints imposed on it) than one subject only to a discharge standard.

Consolidated and Generic Standards

The large number of chemicals and sources of chemicals under the scrutiny of regulation, along with the inability to make fine distinctions within many of the categories of chemicals and sources of chemicals, has led to some consolida-

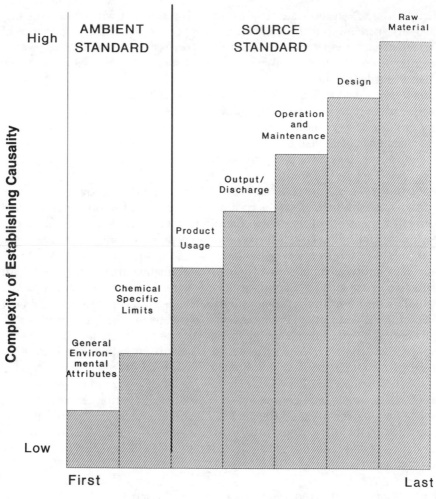

**Order in which Standards Emerge
in the Regulatory Process**

Figure 2.4.   Order in which standards emerge in the regulatory process and complexity of establishing causality between chemicals and health effects.

tions in the design and application of standards. Recent legislation tends to consolidate the use of standards more than older legislation. For example, CERCLA/SARA and RCRA for the most part refer to standards under the Clean Air and Clean Water Acts and the drinking water legislation rather than developing new standards. In fact, under SARA, parties responsible for hazardous waste site cleanups must first try to apply legally available and applicable requirements (ARARs) before developing new cleanup standards. Revisions to

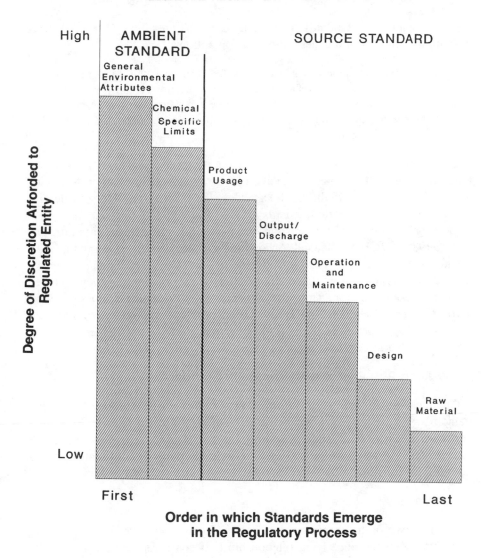

Figure 2.5. Order in which standards emerge in the regulatory process and degree of discretion afforded to regulated entity and regulatory agency.

SDWA have taken into account the need to use drinking water standards under CERCLA and SARA (Stever, 1988, 7–47). A more unusual form of consolidation, which was an outcome of political considerations, occurred under the pesticide laws. Pesticides are exempt from the standards of TSCA, FDCA, and SDWA, and the pesticide legislation is considered as exclusively dictating these substances.

The most common form of consolidation is the use of a generic approach to the development of standards. Brickman, Jasanoff, and Ilgen (1985, 42) have defined the concept of a generic procedure as follows:

> Generic procedures represent a framework for decision making that falls conceptually between broad legislative statements of objectives and particularized agency decisions on specific chemicals. These procedures seek to standardize decision making by establishing general principles for evaluating the adequacy of evidence and assessing risk.

Examples of the use of generic procedures occurred in the area of pesticide regulation and in the general regulation of carcinogens. In reaction to having to reregister over 35,000 different pesticides, the FEPCA of 1978 provided for the development of generic standards for active ingredients in pesticides in lieu of developing standards for each and every pesticide (Brickman, Jasanoff, and Ilgen, 1985, 37; U.S. General Accounting Office, 1980). Perhaps the most important examples of a generic approach to health risk standards are the generic cancer policies promulgated by practically every federal agency involved in health risks regulation. These are summarized in Appendix 2.3 under "Standards Governing Human Exposure." Generic procedures are a common approach to the preparation of environmental impact statements under state versions of the National Environmental Policy Act.[29]

Conditions Associated with Trends in Standards

Chemical risks place large information demands on the process of setting standards. Vague laws combined with a complex and often uncertain information base underlie the standards-setting process. These factors are the major ones driving the way standards are set, because alternative forms that standards can take are sensitive to legislative and scientific uncertainty.

Vague Laws

The first part of the chapter pointed out that many laws have been vague with respect to what the underlying relationship between a risk level and a health effect is, whether such a relationship has to be determined, and how it should be determined. The vagueness in the laws in part contributes to the considerable variability in terms used to qualify risk, which in turn contributes to variability in the use of quantitative levels for the same chemical under different programs. O'Brien underscores the confusion created by vague laws, noting that "Congressional mandates for regulation of toxic and carcinogenic substances actually provided agencies with disparate, crosscutting, and overlapping criteria and responsibilities for setting regulatory standards" (O'Brien, 1986a, 41). The vagueness originated primarily in the scientific uncertainty underlying all aspects of estimating the health effects of chemicals on humans, discussed earlier.

Need for Certainty in Order to Regulate

The order in which standards emerge can be explained in terms of the need of both lawmakers and regulatory agencies (developing standards through rule-making) for certainty in light of the vagueness of causal relationships underlying the standards. The progression of standards over time from the general to the specific is consistent with Mitnick's model of regulatory behavior. According to Mitnick, when agencies first extend their reach, standards tend to become more general. Then agencies tighten and constrain rules as they become more bureaucratic (Mitnick, 1980, 33). Regulatory agencies have to create manageable boundaries around the risk problem and they do so by defining standards that are specific to sources they can control.

CONCLUSIONS

Laws

Legislation dealing with environmental health risks has experienced steady growth; if major laws and their amendments are counted, the growth has been precipitous. Yet, relative to the overall growth in legislation in the country it does not seem as pronounced. Furthermore, major shifts in the substance or philosophy of risk legislation occurred relative to what was typical of environmental legislation in general. The 1970s statutes governing environmental health risks from chemicals departed from the direction that 1960s legislation took. The 1970s statutes heavily emphasized health risks over environmental risks. The legislation of the 1960s and early 1970s emerged during periods of relative economic prosperity and allowed health considerations to take on a new prominence relative to economic considerations. In contrast, the legislation of the late 1970s was more sensitive to balancing costs against risks and reflected the response of political interests to a relative reversal of economic conditions in the country. During this period, the health-only statutes were virtually eliminated. Yet, health requirements continued to be pervasive in the legislation of the 1980s, along with a continued emphasis on economic impact. Another philosophical shift occurred in the relative emphasis on long-term planning vs remediation. Most of the long-term planning applicable to health risks occurred during the 1970s, while remedial and emergency planning tended to dominate in the legislation after that.

The degree of prescription in legislation took the form of greater specification of chemicals to be regulated and when they should be regulated. This contributed to a highly bureaucratized system of regulation in the sense that it fostered less agency discretion. However, discretion was promoted by virtue of the vagueness in the underlying information about the causal connections between the chemicals being regulated and the health risks that were the target of the regulations. This uncertainty led to vague terminology used to describe risk and highly variable risk levels.

Conditions external to government itself, but not unusual in terms of what influences governmental processes, contributed to the rise in this legislation and to the changes in the direction of the laws. A growing populist movement, which could not be dampened by the more conservative environmental policies of the early 1980s, and the strategy of using crises (not unique to environmental regulation) to raise issues high on the regulatory agenda (Stone, 1982, 183) both contributed to a rise in the amount of legislation. The changing political climate promoted agency balancing of costs against risks. While balancing initially may have been discretionary, court cases could circumscribe discretion in this area. Technological advances in detection were initiated that were in part driven by the needs of the environmental health risk policy area and contributed to the increase in the degree of prescription of statutes. Finally, the relative uncertainty and complexity that characterized the scientific basis for determining chemical risks led to initiatives in the form of risk assessment methods that influenced the nature and form of risk regulation.

Standards

An intensive analysis of the substantive nature of standards under environmental health risk statutes shows a large number of different kinds of standards in use, variability in the risk levels, shifts in emphasis from ambient environmental standards to standards pertaining to chemical sources, a high variability in the amount of information they require for their development (assuming that estimates of effects on human health have to be the basis for the standard), and variations in discretion allowed to the regulated entity in complying with the standards (depending on the type of standard).

A number of conditions have contributed to these attributes. One condition already discussed extensively in the first section is the vagueness of original statutes. A second condition, recognized over and over again in the regulatory literature, is lack of information or uncertainty of available information. It is in this context that the standards regulating environmental health risks have emerged.

A major reaction to legislative ambiguity and complexity in the risk area has been a move toward or a renewed interest in regulatory alternatives in the form of incentives for regulation. This is the subject of the next chapter.

APPENDIX 2.1

Definition of Terms

Regulation

Regulation as a general process has many different definitions. Bardach and Kagan define it as an activity to control socially harmful behavior (Bardach and Kagan, 1982b, 3). Mitnick defines regulation a little more specifically, as "the intentional restriction of a subject's choice of activity, by an entity not directly party to or involved in that activity" (Mitnick, 1980, 20). He construes regulation as encompassing the formation of a new agency, rules, or standards set by an existing agency, or any other form of "external choice restriction" (Mitnick, 1980, 80). Stone (1982, 10) narrows the definition somewhat further by describing the nature of the restriction, as "a state-imposed limitation on the discretion that may be exercised by individuals or organizations, which is supported by the threat of sanction."

Regulation by Directive or Decree

A particular set of legal and regulatory controls were covered in this chapter—those that are initiated by directive[30] or decree.[31] Regulation by directive is defined by Mitnick (1980, 342) as "circumscribing or directing choice in some area—i.e., making rules for behavior that may be transmitted as instructions (e.g., in directives or regulations)." This is distinct from regulation by incentive[32] which is the subject of Chapter 3.

Standards and Criteria

In the context of regulation, a standard is a type of rule. In fact, the three terms—standard, rule, and regulation—are often used interchangeably. Davis (1972) discusses the dilemmas of defining a "rule," because it is usually used interchangeably with the word "regulation." He gives the following dimensions of the concept of a "rule" as an approach to defining it: A rule is the product of rulemaking. It is designed in such a way as to not apply to a particular named party or individual. It can be obeyed without the need for enforcement proceedings. The Administrative Procedure Act, which sets forth federal rulemaking and adjudication procedures, defines rules under Section 551 as:

> the whole or a part of an agency statement of general or particular applicability and future effect designed to implement, interpret, or prescribe law or policy or describing the organization, procedure or practice requirements of an agency . . ." (Davis, 1972, 125).

O'Reilly et al. define a rule as "a written policy determination by a federal agency" (O'Reilly et al., 1983, 4-3).

The definition of a standard has been both broadly and narrowly construed, and its meaning is often given by example rather than explicit definition. EPA's Criteria Document for Lead refers to a standard as being "prescriptive" or prescribing "what a political jurisdiction has determined to be the maximum permissible exposure for a given time in a specified geographic area" (U.S. EPA, September 1984, I-1). Baram defines "standard" in the context of industrial voluntary standards (as used by the Federal Trade Commission) as "a descriptive technical document developed in the private sector for providing guidance to manufacturers, sellers, users, and consumers of a product or a system" (Baram, 1982, 53).

The technical foundations of many standards are an information base known collectively as criteria. The development of criteria is often a statutory requirement, as was the case with the Clean Air Act and the Clean Water Act. One definition of criteria used in implementing part of the Clean Air Act is that in contrast to standards, "criteria" are "descriptive": "they describe the effects that have been observed to occur as a result of external exposure at specific levels of a pollutant" (U.S. EPA, September 1984, I-1). Similarly, in the development of water quality criteria, the concept of criteria has been defined as follows:

> . . . criteria are derived from scientific facts obtained from experimental or in situ observations that depict organism responses to a defined stimulus or material under identifiable or regulated environmental conditions for a specified time period. (U.S. EPA, July 1976, 1)

Criteria represent a base of knowledge that attempts to describe cause-and-effect relationships to the extent that scientific knowledge can identify such relationships. Standards, on the other hand, represent a decision about the level or range of acceptable levels of effects that are desirable. Such a decision then implies the acceptability of a certain level of an activity that has resulted in the accepted level of effect.

The use of the term "standard" is not consistent from law to law. The word "standard" has been used to describe a specific numerical limit such as the concentration of a chemical substance. Conditions are often specified along with these limits, such as applicability of the limit over a particular time period, in a particular geographic location, or for a specific set of activities. These conditions can get rather elaborate and can incorporate a time schedule for compliance with the numerical limits. They are actually a part of the standard along with the numerical limits. "Standard" has also been used interchangeably with the words "guideline" or "limitation," which may be structurally similar to a standard, but typically are not as rigorously applied. "Standard" is also often used interchangeably with the word "regulation," as in the case of the National Interim Primary Drinking Water Regulations. In this case numerical limits are specified as they would be in standards, but are referred to under the broader term "regulation."

The word "standard" was used in this chapter to signify a restriction expressed first as a numerical limit. It also refers to any special conditions placed upon on a risk agent or any conditions placed on the environment that result in risk agents being transported or attenuated. It thus covers concepts that have been called guidelines, limitations, limits, and regulations.

APPENDIX 2.2

Typologies of Legislation Based on Health Risk–Only vs Economic Balancing Concepts

Doniger (1978)

In 1978, Doniger proposed and discussed a typology of environmental health and safety laws in the context of the regulation of vinyl chloride. In that discussion, Doniger's two categories were (1) laws that contain "balancing rules" and (2) laws that are "health-only rules" (Doniger, 1978, 156). Balancing rules pertain to requirements for the incorporation of economic considerations such as cost when levels of safety are being devised. In this category, he grouped TSCA, FIFRA (and its amendments, FEPCA), and HMTA. Doniger argues that a major criterion for classifying a law as a balancing statute is the use of the concept of "unreasonable" risk. The health-only laws do not contain a provision for the balancing of costs, and consider public health only when safety levels are established. Under this category, Doniger lists the 1958 Food Additives Amendment to FDCA and sections of CAA and FWPCA.

Subsequent uses and extensions of this initial typology have been put forth by the Federal Regulatory Council (1979), Field (1980), the U.S. Congress Office of Technology Assessment (1981, Chapter 6), and Morris and Duvernoy (1984). The typologies of Field and of Morris and Duvernoy are relatively more extensive and are discussed below as to how they categorize the legislation pertaining to environmental health risks.

Field (1980)

Field developed a fourfold typology for the regulation of chemical risks. This typology categorizes laws according to the way risks are balanced, along with other considerations such as economic effect. The categories Field used were (1) risk only (not allowing any benefits to be considered that might outweigh the risks), (2) technology only (using only a technological feasibility criterion and not including the balancing of costs of such technologies), (3) implicit balancing (involving a balancing of risks, costs, and benefits, but not giving specific factors and methods of balancing), and (4) explicit balancing (where factors and methods of balancing are given or suggested).

A summary of the application of this typology to 31 laws is as follows:

Risk only	7
Technology only	2
Balancing (implicit, explicit, or unspecified)	22

This distribution shows the preponderance of balancing provisions in these laws relative to risk-only or technology-only provisions. Field primarily explains the classification of laws in terms of hypotheses about changes in the nation's economy and the pattern of influence of different interest groups.

Morris and Duvernoy (1984)

Morris and Duvernoy developed a categorization of 33 components of 23 laws along a number of dimensions pertaining to risk. The risk assessment elements used in their typology to describe various characteristics of risk are (1) "hazard" (the type of threat), (2) "causative event" (type of incident leading to the threat), (3) "action" (triggering the need to evaluate the risks), (4) "party responsible" (for the risk), (5) "risk measure" (the consequences of risk being realized), (6) "risk incidence" (the occurrence of the risk in a specific population), (7) "data and methods" (used to characterize and measure risk), (8) "expression of risk" (severity of risk), (9) "evaluation standard" (acceptable level), and (10) "value of information" (specificity of the information). One purpose of the typology and its application was to evaluate how the use of risk assessment procedures in agency decisionmaking has been spelled out in the laws. Each of the above elements was assigned any one of four values for the degree of specificity in the law: none, low, moderate, or high. The definitions of each of the elements are as follows:

- *Hazard:* Hazards are threats that may be expressed in terms of substances, activities, or systems.
- *Causative events:* Causative events are actions, recurring accidents, or catastrophic accidents that "can give rise to a risk."
- *Action:* The action triggering the risk assessment refers to the potential activity that triggers the need to assess a risk, such as the application for a permit, the development of regulations, and so forth.
- *Party responsible:* This can be either the originator of the hazard, the regulatory agency, or a third party.
- *Risk measure:* Risk measures are "the types of consequences whose likelihoods are considered."
- *Incidence of risk:* Risk incidence is the occurrence of risk defined with reference to the person, group, or population (present or future) that is likely to bear the risk.
- *Data and methods:* This encompasses the primary data and methods of collection and analytical procedures such as parameter construction techniques and models.

- *Expression of risk:* This refers to levels of risk and their degree of specificity.
- *Evaluation standard:* The evaluation standard is the level at which the risk is considered acceptable.
- *Value of information:* This refers to the extent to which information of a certain type, such as a risk assessment, is specified and valued.

APPENDIX 2.3

A Typology of Standards

Standards for Risk Agents or Sources of Risk

Standards exist that specify the materials used and the design, construction, operating, and waste discharge characteristics of activities or *sources* that can potentially lead to the release of a risk agent. Such a source can be a factory discharging noxious chemicals or an activity producing and distributing a dangerous consumer product. The *risk agents* from such sources, in turn, can be chemicals being discharged from a factory or chemicals that can be released by the use of the consumer product.

The fate and transport of the risk agent in the environment and the potential for human exposure may have been used as a basis for establishing standards for such sources of release. Many complex models exist for the purpose of deriving source (or risk agent) standards from environmental and exposure characteristics that are considered protective of human health.

Standards aimed at the risk agent are often based upon the level to which state-of-the-art technologies can reduce the amount of the risk agent, regardless of the cost of such a technology. In other words, standards are often set at the level at which it is technologically possible to achieve a certain level of risk reduction.

Source- or risk agent–based standards can be further classified in the following way.

Raw Material Processing Specifications

An example of a raw material standard is the limitation on the percentage of sulfur allowed in coal used as a fuel in certain power plants. The purpose of this limitation is to reduce the rate of air emissions of sulfur dioxide at the source, and hence the risks to those exposed to the emissions. A second example is a standard for the amount of lead allowed as an additive in gasoline. In the building trades an analogous example is the specification of the composition of steel or concrete to maintain a given level of strength to maintain the structural stability or integrity of bridges and buildings.

Raw material standards can also specify what kinds of raw materials are permissible rather than what substances are permissible in them.

Design Standards

One example of a design standard is performance standards for the processing efficiency or waste removal efficiency of waste treatment systems. This is often expressed as the percentage removal of a given set of pollutants that the treatment system must attain. A second example is the set of design standards for nuclear power plants published by the American Nuclear Society, which includes "Design Objectives for and Monitoring of Systems Controlling Research Reactor Effluents" and "Guidelines on the Nuclear Analysis and Design of Concrete Radiation Shielding for Nuclear Power Plants." Many design specifications in the nuclear power industry appear in the form of design criteria rather than standards, and are published by the American Nuclear Society. Other examples include the design standards for landfill liners for landfill-like hazardous waste disposal facilities; the design standards for incinerators so that they achieve certain discharge levels; and the extensive design standards required for underground storage tanks.

Construction or Manufacturing Standards

An example of this type of standard appears in the area of consumer product safety. The Consumer Product Safety Act (CPSA) (Section 7(a)(1,2)) specifies, among other requirements, that construction standards can be used to regulate the safety of consumer products. An amendment to the Safe Drinking Water Act, called the Lead Contamination Control Act of 1988, regulates drinking water coolers and has deemed lead-lined water coolers, defined by the EPA as "imminently hazardous consumer products" under CPSA, subject to CPSA's replacement, repair, or recall provisions. The 1988 act prohibits the manufacture and sale of water coolers that are not lead-free, and requires the reduction of lead in school water coolers or their removal. The act defines lead-free drinking water coolers as those in which no part of the cooler coming in contact with drinking water can contain more than 8% lead or where materials such as solder contain no more than 0.2% percent lead (Section 1461 (2)).

Operation, Discharge, or Output Standards[33]

In the broadest sense, operation standards aim at the safety of a system, its operators, other workers, consumers, and the general public. Operation standards can take the form of describing how a facility should be operated or describing what the output of the system should be (most commonly in the form of waste discharges). Under OSHA, standards that describe how a worker should conduct activities in the workplace in order to avoid exposure to chemicals are a kind of operation standard. Examples of waste discharge standards are emission standards for discharges into the air or effluent standards for wastewater discharges.

Product Usage Standards

Prior to the advent of many of the pollution control laws, some laws required manufacturers and distributors to disclose product information to make the users of the ultimate product aware of the potential risks of its use or misuse. The most common approach to the provision of such information or warnings was labeling.[34] The labeling approach was and still is used for pesticides, consumer products, and food products. Posting requirements in the workplace indicating the dangerous conditions, facilities, and substances that should be avoided are a kind of product usage standard.[35] This standard is still popular in spite of increased complexity of the information to be covered by such standards and the liability issues associated with leaving information out or providing the wrong information. The logic behind product usage standards such as labeling is that by requiring manufacturers and distributors to admit to the dangers of a product and include a warning, the user is given the opportunity to avoid the risks. A major debate in the use of this type of standard is exactly what should be disclosed (Breyer, 1982, 162). The labeling requirements in early pesticide laws were weak in that they required disclosure of how the product should be used to avoid harm but did not inform the user of what the harm actually was. Later revisions remedied this weakness to some extent (National Research Council, 1980). Breyer points out that the advantage of a product disclosure standard is that these standards do not have to be fine-tuned. But "despite this major advantage, disclosure will not work unless the information is transmitted to the buyer in a simple and meaningful way" (Breyer, 1982, 163). The effectiveness of this type of standard also depends on the extent to which the user of the product has other alternatives and chooses to heed the warnings (Breyer, 1982, 164). This type of standard also presumes that the user of the product will understand the warning and heed its message. Another major problem with disclosure provisions is the risk to third parties (e.g., those that come in contact with the risks but are unaware that they exist).

The communication of risks from products as well as activities has expanded well beyond labeling. During the 1980s there was a virtual explosion of techniques and approaches for communicating risks. The use of warnings took the form of an expansion in the workplace hazard standards that began under OSHA and state "right-to-know" laws and the duty-to-warn provisions of Title III of SARA.

Requirements for the Administration of Standards for Sources or Risk Agents

Distinct administrative processes are conducted for standards that apply to sources or risk agents. These administrative processes take the form of approvals either before or after the risk agent is on the market. The standards can be invoked prior to the manufacture or marketing of the activity or product (premarket approval). Alternatively, the standard can be invoked once the product is already on the market and about to be put into use (postmarket approval). Each type of administrative procedure has its own advantages and limitations.

Premarket Approvals

The premarket approval tests and registers a substance prior to its being allowed on the market. The most common examples, as shown in Table 2.2, are the approval of the manufacture of a toxic substance under TSCA (Premanufacture Notice), the approval of new products by the FDA, and the registration of pesticides prior to marketing under FIFRA by EPA. Quirk highlights some major problems with this form of regulation (Quirk, 1980, 200–204). He argues that the agency can never anticipate all of the problems associated with a given chemical or process (the thalidomide disaster is used as a case in point); and the agency responsible for the premarket testing inevitably will miss certain effects that will arise postmarket, since the population used for premarket testing is so much smaller than that for postmarket testing. Because of these problems, agencies will tend to be extremely conservative in granting premarket approvals. Therefore, premarket approvals may not be sufficient regulatory devices for managing risk unless they are accompanied by stringent postmarketing surveillance and provisions for withdrawing approvals if problems arise. Critics of the premarket approval strategy maintain that because a new product or substance still needs postmarketing regulation once approved, the expense of the premarket testing and approval to the extent that it is currently carried out is not warranted.

Postmarket Approvals

The postmarket approval usually takes the form of a permit, license, registration, certification, or other form of approval for a product or activity, containing strategies for meeting the standards as conditions for continuation of the approval. The existence of source-based standards speeds the issuance of such approvals. Yet, a major problem regulatory agencies face in postmarket approvals is meeting time limits for application review. Agencies can circumvent these time constraints in a number of ways. They can request voluntary extensions from regulated parties. Alternatively, they can find missing information in the application. This determines that the application is incomplete, and thereby stops the clock for application processing until the information is provided.

Standards for the Condition of the Ambient Environment

Examples of ambient environments are the atmosphere and natural waterways. The most common examples of ambient standards are those for water quality in the form of allowable concentrations of chemicals or agents of disease in natural waterways and similar concentration limits for chemicals in the atmosphere that define a given level of air quality (e.g., the National Ambient Air Quality Standards). A less common set of examples refers to concentrations of substances in other media to which humans might be exposed, such as dust, soil, and fish, shellfish, and other wildlife consumed by humans. Ideally, the development of ambient/environmental standards is based on the tolerance of human beings to the risk agents for which the standards apply.

Ambient/environmental standards have appeared in several different forms in agency rulemaking. The first one is usually a general statement about the condition of the environment, and can be mistakenly interpreted as a goal or objective statement rather than a standard. Examples of such general statements include a requirement that oil and grease shall not be present in waterways in sufficient amounts to produce a visible sheen, or prohibitions against substances that can produce odors. These general statements are not often accompanied by numerical limits or even a specification of the chemicals of concern.

A second type of ambient standard is one in which chemicals or chemical, physical, or biological environmental conditions (such as pH, turbidity, and suspended solids) are specified numerically. The numerical limits are usually set separately for different geographic areas. For example, under the Clean Water Act, numerical limits for a given pollutant will depend on the use for which a waterway is classified—water supply, shellfishing, fishing, swimming and other contact sports, or recreation. Under the Clean Air Act, air quality control regions are defined geographically, and each region has separate numerical limits for each of the National Ambient Air Quality Standards. In fact, the geographic classifications are typically subsumed under the concept of an ambient standard.

Standards Governing Human Exposure

In addition to source-based and environmental condition standards, some standards are expressed in terms of the tolerance limits of the person exposed to the risk agent. For chemicals, examples of these standards are allowable exposure levels for ingestion, inhalation, or absorption through the skin. Limitations on radiation exposure are another example of this type of standard. Another form in which these standards can appear is as allowable concentrations of substances in particular organs or systems of the human body, such as the blood, bones, or fat tissue. One way that these standards are developed is by finding the threshold at which no effect on a human being or surrogate (such as a laboratory animal) is observed. Examples of these kinds of thresholds can be found in drinking water regulations: "suggested no adverse response levels" (SNARLs), "No Observed Effect Levels" (NOEL), and "Maximum Contaminant Levels" (MCLs).

Cancer Policies as Surrogates for Human Exposure Standards

Tolerances of human beings to concentrations of chemicals are not known for every chemical. In the absence of such information, federal agencies have issued policies governing guidelines for the assessment of such human health impacts.[36] Brickman, Jasanoff, and Ilgen (1985, 42) refer to these policies as "generic procedures" which are "a framework for decision making that falls conceptually between broad legislative statements of objectives and particularized agency decisions on specific chemicals." When environmental factors were first suspected of causing cancer, agencies responsible for legislating environmental contaminants suddenly had to grapple with the effects of chemicals at low concentrations rather than the traditional high-concentration effects they had been used to. The role of environmental agents as carcinogens became apparent in the 1950s with the concern over the effects of radiation (National Research Council, 1983, 54), and gradually spread to other chemicals in the early 1970s. The risk element for carcinogens is distinct from that of other kinds of chemical effects. Cancer effects develop over long periods of time at low exposures. Effects are expressed in terms of a probability rather than as an immediately observable response.

Federal policies for the regulation of carcinogens have appeared in the form of formal agency rulemaking. Each agency has issued its own guidelines independent of any other.

EPA. The U.S. EPA issued a carcinogen assessment policy in draft form in 1976, which guided the agency's decisions for almost a decade (U.S. EPA, May 1976). New guidelines for carcinogens were issued in draft form in 1984 (U.S. EPA, November 1984) and the final version was issued in 1986 (U.S. EPA, September 1986). These proposed guidelines were combined with the agency's rulemaking on risk assessment procedures. The U.S. EPA, under Administrator William D. Ruckelshaus, issued risk assessment policies in the form of endorsements or encouragements of the risk assessment approach to environmental problems (U.S. EPA, December 1984c). The application of risk assessment procedures in EPA goes beyond carcinogenesis to birth defects and mutations in general. In addition to being integrated into the agency's policies on carcinogens, risk assessment has been applied to the evaluation of the effects of hazardous wastes on landfills (U.S. EPA, February 1984, 5857–5858) and to the effects of alternative methods of sewage sludge disposal.[37] Prior to and also concurrent with EPA's efforts were several unsuccessful attempts to develop risk assessment and management procedures in legislation for governmental operations in general by Congressman Don Ritter.

OSHA. OSHA's final rule dates from 1980 (U.S. Department of Labor, OSHA, 1980), after having first appeared in draft form in 1977. In OSHA's policy, the considerable uncertainty in establishing the threshold for various carcinogens in humans led to the definition of an exposure level that was the lowest that was technologically feasible. The purpose of OSHA's policy was to "streamline the process of risk assessment, to speed up regulation, and to reduce the workload of agency staff" (National Research Council, 1983, 59). Several changes have been attempted since 1977 but are not in final form.

FDA. The Food and Drug Administration's cancer guidelines tend to be more specific to a given level of acceptable risk and specify in considerable detail how to measure risk levels and establish the sensitivity of tests. The guidelines were never formally adopted by the agency. The guidelines were primarily applicable to drugs, since for foods, the Delaney Clause dictated the approach to carcinogenicity—not allowing any level at all (U.S. Department of Health, Education, and Welfare, 1979; National Research Council, 1983).

CPSC. The Consumer Product Safety Commission's carcinogen assessment guidelines appeared in 1978 primarily as a vehicle for making their procedures and principles for hazard identification known (U.S. Consumer Product Safety Commission, 1978). The validity of the guidelines was challenged, and they are no longer in use (National Research Council, 1983, 60).

Interagency Regulatory Liaison Group (IRLG). After the first set of agency guidelines on carcinogens was issued in the middle to late 1970s, the agencies got together under the aegis of the IRLG to develop a consistent set of guidelines. The IRLG guidelines appeared in the *Federal Register* in 1979. Since the guidelines had no official status, and the IRLG was abolished soon afterwards, the new wave of cancer guidelines of the early 1980s has been issued once again on an agency-by-agency basis (National Research Council, 1983).

Office of Science and Technology Policy (OSTP). In 1984, the OSTP issued its draft risk assessment policy for regulatory agencies. The policy's purpose is to "articulate a view of carcinogenesis" that is commonly held by scientists and to develop a series of general principles to guide carcinogen risk assessments (Office of Science and Technology Policy, 1984; National Research Council, 1983)

NOTES

1. This figure gives a trend for (1) total legislation and amendments only and (2) all risk legislation and amendments. The listings of the environmental health laws and all of their amendments were drawn from the *U.S. Code Annotated.* The listing of total federal legislation over time was obtained from the U.S. Bureau of the Census (1975, 1081) through 1970, and supplemented by Congressional Quarterly, Inc. (1982, 350) and the *U.S. Code Congressional and Administrative News* (annual).

2. That is, tabulations tend to differ on what is counted as a law—major initiating legislation and major amendments, or all amendments. The tabulation of total laws was obtained from historical tabulations, and it was uncertain how consistently amendments were counted from law to law. In order to take that into account, the two extreme measures of total environmental risk laws were used—just major statutes on the one hand and major statutes plus all amendments on the other hand.

3. Johnson (1985, 457) drew a comparison among hazard laws (including safety as well as environmental protection) and total laws, but compared annual figures for the laws rather than the cumulative trend. Using this approach, he found that as a proportion of all laws, hazard legislation remained practically stable or slightly increased between 1957 and 1978. He points out that the spurt in growth of hazard laws is not apparent from the perspective of number of laws passed annually. A cumulative curve is used to develop the data in Figure 2.1 because laws are often not replaced entirely by amendments—the contribution of lawmaking is, in substance, cumulative.

4. Similar trends appear in O'Brien (1986a, 40; 1988, 128). The trend is also identified by Rushefsky (1984, 134), who quotes Dodge and Civiak (1981). Still another reference to the trend can be found in U.S. Congress, Congressional Research Service (1983, 6). Bills developed in Congress show a similar emphasis on risk issues. Barke points out that "the word 'risk' appeared in the abstracts of 177 bills submitted to the 96th Congress (1979–80)" (Barke, 1986, 23).

5. Balancing is one of the five public interest concepts that Mitnick identifies. The others are orientations that (1) compromise, (2) trade off among various factors, (3) emphasize national or social goals, or (4) are particularistic, paternalistic, or personally dictated.

6. Shughart and Tollison (1985) find this consistent with Peltzman's theory of "political wealth" (1976) as influencing regulatory policy.

7. The typologies have been developed by Doniger (1978), Field (1980), and Morris and Duvernoy (1984). One of the typologies by Field adds two subcategories to these and another category pertaining to technology. To accurately characterize the legislation according to these two categories, the legislation ideally has to be divided into two areas: (1) the setting of standards and (2) technological or other requirements for attaining the standards. Very often, the same legislation will apply the health-only or economic balancing approaches differently to standards vs pollution controls. The typologies are often not refined enough to identify such differences.

8. Section 201(d) and (e), for example, refer to determining, to a reasonable degree of certainty, the migration of chemicals for land disposal and the migration of solvents and dioxins from disposal units or injection zones, respectively.

9. The conference report reads, "To the maximum extent practicable, the Agency should avoid wasteful, repetitive short-term actions that do not contribute to the efficient, cost-effective performance of long-term remedial actions" (U.S. Congress, House of Representatives, 1986b, 190).

10. For a discussion of the planning orientation of environmental legislation, see Zimmerman (1985a; 1987b, 257).

11. An important exception to this trend is the Hazardous Substances Act. It was first passed in 1960 as the Hazardous Substances Labeling Act. While it was extremely specific in the kinds of risks it was attempting to avoid, it was not as specific in

terms of chemicals identified. On the other hand, it was very specific about methods of determining risk levels. As Hadden points out: "The HSA was one of the last federal statutes to define specific standards of risk, including numbers of test animals killed by a substance, in the text of the law. Later statutes tended to delegate these determinations to agencies to promulgate as rules, instead providing more general goals and guidelines within which agencies were to act" (Hadden, 1986, 80).

12. Other more generalized classification or categorization schemes exist. Kagan, for example, argues that legislation ranges from minimum prescription, i.e., the so-called expert model, to considerable prescription, which he terms the legal model (Kagan, 1978, 13–14).

13. This history is drawn from the Conference Report to the 1986 Amendments to the Safe Drinking Water Act, p. 21.

14. They cite Mazmanian and Nienaber (1979, 66): Models of citizen access identify public support as acting outside of a statutory framework and formal access provided by legislation as acting within the statutory framework.

15. Many scholars of regulation have identified and evaluated this condition in addition to Bardach and Kagan (1982b), White (1981), and Field (1980). See, for example, Baram (1982), Breyer (1982), Stone (1982, 32–33), Graymer and Thompson (1982), and Magat (1982).

16. This summary is largely drawn from Zimmerman (1987b, 234–5 and 262–3) and the references provided therein.

17. Uncertainty of information has often taken a central place in discussion and analysis of environmental health risk policy. See, for example Greenwood (1984). Several professional conferences have also focused primarily on uncertainty.

18. These steps have been formalized by the U.S. EPA in their 1986 regulations for carcinogens, mutagens, teratogens, and other selected health threats. See U.S. EPA (1986b).

19. Schneiderman (1980, 33–34) has illustrated the range of these early applications of models for saccharin and vinyl chloride. For saccharin, the estimated number of cases per million exposed ranged from 0.001 for the two-hit model to 1200 using the one-hit model. An even greater range of estimates resulted from the application of models to vinyl chloride. Ricci and Molton (1985, 475) cite a difference of six orders of magnitude for models in the low-dose range for ethylene thiourea; however, when the one-hit model is eliminated (since it does not have a good fit with the data) the difference is only slightly more than one order of magnitude.

20. For a review of risk qualifiers pertaining to emergencies in legislation oriented toward chemical risk management, see Zimmerman (1985a); Skaff (1979); and Ricci and Cox (May 1987, 77–96, Table 1).

21. See, for instance, the Toxic Substances Control Act (Article 6), the Federal Pesticides Control Act (Article 2(bb), and HMTA (Article 104).

22. *Industrial Union Department, AFL-CIO* v. *American Petroleum Institute*, 448 U.S. 607 (1980).

23. This criticism was made regarding the cleanup of the Davis Liquid Waste Site in Smithfield, Rhode Island and the Re-Solve, Inc. site in North Dartmouth, Massachusetts.

24. Stever (1988, 8–9, footnote 54.2) quoting 48 *Federal Register* (April 2, 1984, 13018).

25. Cantor and Bishop (1987, Table 1) have summarized a number of these studies that derive these estimates and are based on either the "contingent-market" approach or the "implicit-valuation" approach.
26. These are raw material, design, construction, operation (output), and product use standards for sources; general and specific standards for the ambient environment; and human exposure standards.
27. In spite of the early appearance of the first set of NAAQS, a U.S. General Accounting Office study (December 1986a) has criticized the standard-setting process for updating NAAQS as proceeding much too slowly. They attribute the slowness to the large number of steps now involved in setting such standards. Berry's analysis showed that while the first set of NAAQS appeared within four months of the CAA, revisions and the addition of lead took anywhere from over two years to over seven years (Berry, 1984, Table 6–1).
28. The U.S. Congress, Office of Technology Assessment (1984, Volume I, 101) has defined narrative standards as standards that "describe limits but do not specify concentrations (e.g., a non-degradation standard requiring that concentrations be at or below natural background levels) or even necessarily individual contaminants (e.g., a standard prohibiting the discharge of toxic, carcinogenic, teratogenic, or mutagenic substances into groundwater)."
29. An example of the generic approach within the area of environmental protection but outside of the environmental health issue is the use of "generic" environmental impact statements. These generic statements are normally performed for similar programs or facilities that differ in location.
30. This is a term that has been used by Mitnick (1980, 396) to denote regulation that is based on rules, standards, and other requirements.
31. Aharoni (1981, 50) uses the term "decree" to denote a control mechanism similar to the concept of a directive.
32. Mitnick (1980, 342–343) defines regulation by incentive as involving a strategy of changing the attractiveness of alternatives, though he recognizes the illusiveness of the term.
33. Output standards are commonly referred to as performance standards. See, for example, Hadden (1986, 35).
34. For an extensive history of the use of labeling see Hadden (1986, Chapter 1).
35. These are given extensive treatment in the OSHA General Industry Standards (29 CFR 1910), March 11, 1983.
36. For an extensive discussion of cancer policies by agency, see National Research Council (1983, 54–62).
37. Proposed regulations under Section 503 of the Water Quality Act of 1987 (U.S. EPA, February 1989).

CHAPTER 3

Incentive-Based Approaches

INTRODUCTION

A number of shortcomings have been attributed to the use of regulation by statutory directive as a means to control the sources and adverse outcomes of risk. Some of the shortcomings often cited pertain to the high direct costs of regulation and the inability to balance these costs against benefits (Breyer, 1982, 2–4, and others):

- Regulation is costly, regardless of how its costs are measured. The costs of regulation have been measured directly, in terms of governmental expenditures to achieve regulatory goals, and indirectly, in terms of the amount of money those being regulated have to spend to comply with regulations. The federal government now requires economic justification for regulatory programs in the form of explicit identification and quantification of costs of the programs. The Regulatory Reform Act and Executive Orders 12204 and 12291 require Regulatory Impact Analyses on major rules that have a major effect on the economy, result in major increases in consumer or industry costs or prices, or have significant effects on various economic factors related to production, such as competitiveness and employment.
- Regulation creates complex and often unwieldy procedures that are difficult and time-consuming for those being regulated. Regulatory complexity can contribute to both high costs and an inability of affected parties to participate effectively in accomplishing regulatory goals.

Another set of shortcomings of regulation identified by Havender (1982, 36) pertains to the way regulatory agencies process information in the course of making decisions about risk.

- Havender first claimed that agencies tend to become overly conservative; one example of this is the use of worst-case scenarios in analyzing the impacts of risky activities.
- Agencies are said to oversimplify information, creating ambiguity and uncertainty.

- Agencies tend to ignore economic considerations or at least fail to subject economic impacts to the same rigorous analysis as environmental impacts.

Many counterarguments exist in support of regulation (e.g., the argument that any cost-benefit calculus will be biased because the costs of regulation are usually more quantifiable than the benefits). Regardless of how the debate is resolved, strategies based on various incentives and disincentives have been designed to circumvent the problems (however defined) attributed to direct regulation, or to be management tools in their own right. These alternatives are now popular as a means of averting, avoiding, or reducing risks from chemicals as well as from other sources of risk.

The alternatives that are covered in this chapter are primarily financial incentives and disincentives. They include (1) the use of common law rather than statute-based laws to induce behavior that avoids costs associated with damages sought by plaintiffs; (2) the existence of monetary sanctions against risk-producing activities in the form of compensation to victims under statutory law, common law, and insurance from a variety of sources (fines, jury verdicts, out-of-court settlements); and (3) various financial incentives and disincentives such as insurance, taxation, marketable rights, and requirements for financial responsibility. Two of these alternatives, the use of common law and the use of financial mechanisms, have often been cited as alternatives to regulation (Baram, 1982; Breyer, 1982; Mitnick, 1980). According to Baram (1982) and others, alternatives to direct regulation also typically include a number of categories not covered in this chapter: the use of professional standards, licensing, and certifications that are not mandatory but are strongly backed by those involved in the market for risk-producing activities; various government and industry education campaigns; and reliance on bargaining, negotiation, mediation, and arbitration as a means of reducing or avoiding risk or reducing its consequences.

In this chapter, the way in which these incentive systems have evolved as management strategies for environmental health risks are examined, along with the conditions that contributed to their use in risk management. It will be shown that many of these mechanisms, along with their advantages and disadvantages, had roots in other traditional areas of governmental control. Furthermore, it can also be shown that the application to environmental health risks has refined and extended many of these more traditional mechanisms. In doing so, it has created new demands on governmental agencies to integrate these into regulatory programs.

RISK MANAGEMENT VIA COMMON LAW[1]

Common law refers to the settlement of disputes regarding injury or damages ("torts") by judges. In fact, common law has been referred to as the "law of judges." These injuries or damages are attributed by one party to another. The settlements under common law are achieved through the court system, though not all settlements are trial verdicts. Verdicts and settlements under common law rely heavily on monetary incentives and disincentives. Common law is a vehicle for risk management in that the threat of having to provide monetary compensation or its equivalent to injured parties theoretically acts as a deterrent for the conduct of risk-producing activities.[2] While decisions under common law often influence developments in statutory law, the common law is considered to be an alternative to regulation by directive in managing risks because it is based on sanctions that potentially induce behavior in the direction of more socially desirable ends. It does not prescribe or mandate a particular type of behavior the way statute-based regulation does.

There have been a number of developments in the application of the common law doctrine to environmental health risks. Some of them are traditional uses of common law, while others are ground-breaking developments that may influence other governmental processes. The trends pertain to the popularity of using common law, the shifts in the kinds of common law doctrines used for chemical risk cases, and developments in a federal common law doctrine.

Developments in Common Law Doctrines

Popularity of Common Law for Health Risk Cases

The common law became a popular mechanism for risk management in spite of whether cases were resolved through jury verdicts or out-of-court settlements. This has occurred because of the relative advantages that common law affords to parties injured from chemicals. Individual doctrines, however, differ in the degree to which they afford such advantages. This variation occurs primarily because common law, as Katzman (1985, 21) has observed, emerged primarily from state laws and through the judicial processes within each state.

Common law doctrines were used prior to and during the period when the environmental statutes of the 1960s and 1970s were in force (Baram, 1982, 207). Nevertheless, environmental health risk problems have posed unusual challenges to the application of common law to obtaining compensation for health injuries from chemicals. One scholar in the area of environmental law has argued that "there is no formal, coherent common law of environmental risk" and that the general trend in the use of common law in the area of environmental risk has to be derived from case law and legal developments (Silver, 1986, 71). While a number of statutes have provided alternatives to judicial review under common law, bringing a case under the common law is not precluded by statutory law. As Katzman (1988, 41) points out, RCRA and CERCLA by omission do not preclude common law actions.

To understand how and to what extent common law is used for chemical health risk cases and what the different doctrines are, one must understand how they work and their comparative advantages and disadvantages.

Baram summarizes the five major theories encompassed by common law that are alternatives to the control of risk via regulation. These doctrines are (1) negligence, (2) strict liability, (3) nuisance, (4) trespass, and (5) product liability.[3] The five doctrines often differ from one another in their relative advantages and disadvantages with respect to:[4]

- the kind of standard, test, thresholds, or conditions required to establish the existence of an injury
- the extent to which a given party can or needs to be identified as the cause of the injury
- the extent to which causation has to be established

Table 3.1 compares each of the doctrines in terms of their relative requirements. While these doctrines are discussed separately below, in any given case the principles of several of the doctrines may be operating simultaneously.

Negligence. In negligence law, a given party needs to be identified as the cause of injury and the costs associated with the injury have to be quantified. Because the use of the negligence doctrine is based on assignment of blame to a party, it is termed a fault-based liability system. In order for a negligence suit to be brought, a standard of reasonable or non-negligent behavior should be established. This standard has also been referred to as a standard of due care (Katzman, 1985, 23).

The application of negligence law requires the existence of several conditions. First, some accepted standard of behavior should exist and be agreed on by or acceptable to the plaintiff, the defendant, and the courts. Second, a violation of that standard of behavior has to be established.[5] Third, a relationship between the violation and the injury it caused should be established. Fourth, substantial losses or damages have to be established.

The advantage of negligence law for a polluter is that in order to stem the course of a suit, all the negligent party has to do is try to meet the standard of behavior. An example would be a polluter introducing mitigation measures to avoid negligence litigation. Use of negligence law means the judicial process can proceed more smoothly where there is a short time period between an event and its effects (Katzman, 1988, 23–24). Its relative advantage as a form of litigation is reflected in a review of the type of litigation in 41 asbestos-related cases in the United States. According to the review, negligence claims are considered the most common type of claim in tort litigation. The review reported that negligence and gross negligence accounted for most (34) of the cases (Cantor and Bishop, 1987, Table 3), reflecting the common use of negligence law in that area of litigation during the 1980s.

Some of the disadvantages of negligence law are:

Table 3.1.   Comparison of Characteristics and Conditions of Common Law Doctrines

Type of Doctrine	Characteristics or Conditions for Application[a]		
	Party/source of harm must be identified[b]	Degree of proof and knowledge of causation required	Kind of test
Negligence	Yes	Yes	Fault-based; must define reasonable behavior;[c] establish deviation from reasonable behavior; deviation must be linked to injury.
Strict liability	Yes (for both source and activity)	Only potentially and generally	No-fault; risky activity and source must be linked, risk established.
Nuisance	Yes	Yes	Fault-based; requiring unreasonable use of property; usually involves balancing-of-interests test.
Trespass	Yes	Only for negligent trespass, otherwise the question is what constitutes an invasion of property.	Fault assignment unnessessary if trespass intentional; fault assignment necessary only if trespass was unintentional and resulted from negligence or a dangerous activity.
Product liability	Yes	Yes	In most states, strict liability; in some states, negligence.

[a]Since common law is primarily state law, its characteristics and conditions will vary accordingly. This table only presents a summary. In reality, there are many variations and diverse tests.
[b]The party or source may be an individual or a class of people. There is considerable variation.
[c]For hazardous waste activities, reasonable behavior includes duty to warn (Cantor and Bishop, 1987, 10).

- According to Baram (1982, 11), it assumes that the injured party has the resources, the know-how, and the will to recover damages. In deciding whether it is worth bringing a case under negligence law, the litigant should calculate the relative cost of litigation vs the expected return to the litigant. In making such a calculation, the litigant can balance the claim against the assets or insurance of the negligent party. This system is often used for immediate, short-term, easily recognized injuries and causes.
- It can be difficult to define who the negligent party is for the purpose of calculating the assets on which a settlement will be based. For example, there is often a question as to whether one includes the assets of parent companies when the subsidiaries are proven to be negligent. These problems arose in the case of Kepone contamination in the James River in Virginia. In that instance, Allied Chemical claimed that only its subsidiary, Life Sciences, was liable for the damages produced by the contamination. A similar parent company/subsidiary problem arose in the mercury contamination case in Berry's Creek, New Jersey. Ultimately, the assets of the parent company were included in both the Kepone and the mercury cases (Zimmerman, 1985b).
- As summarized by Katzman (1985), it is often difficult in practice to define a standard of reasonable behavior that is acceptable to the courts, the plaintiff, the defendant, and society at large.

Bazelon has summarized general disadvantages of negligence law in the following way:

> Negligence law is simply inadequate to deal with . . . emergent health problems. To cite only one example, environmental pollutants are often impossible to trace to their source. Their health effects are uncertain, and their harms may not show up for years or even generations. Such uncertainty makes a finding of negligence liability unlikely, inappropriate, and therefore ineffective in protecting society from these dangers. (Bazelon, 1981, 792)

Strict liability. In strict liability law, the originator or source of an allegedly injurious activity is liable regardless of what the motive was or who was at fault. The standard or test on which a strict liability case rests is the establishment of a linkage between the activity that produced the risk and the party that is considered liable. (An example of such a linkage is the connection between a manufacturer and the production of a harmful chemical.) Strict liability is a no-fault system in that the originator of the activity is still liable even though he/she may have been extremely careful and cautious in practicing the activity. This type of liability is usually restricted to unusually risky activities or an activity that occurs in an unlikely place. As summarized by Katzman (1985, 24–25), there are several criteria used to determine the applicability of strict liability: (1) where injuries have resulted from an improper or "nonnatural" use of land, (2) the conduct of an "ultrahazardous activity" (defined differently by different courts), (3) the conduct of an activity that meets several tests of risk incorporating risk measures, (4) propensity to cause harm, and (5) risk relative to benefits.

Two important conditions for bringing a strict liability suit are (1) knowledge of the source of the alleged injury and (2) the association or linkage of an abnormally dangerous activity with the source of alleged injury.

The advantage of a strict liability suit is that only the source of danger and the fact that a particular activity conducted by that source was a dangerous one have to be established. Therefore, the plaintiff does not have to actually show damage. For example, in some states the mere storage of chemicals can constitute a dangerous activity, sufficient to bring a suit under strict liability statutes. There is no requirement to establish that the storage of the chemicals has produced some damages. Existence of an imminent or potential harm or the use of strict liability was not always a sufficient or viable basis for bringing a nuisance suit.[6] The advantage of not having to show negligence is making strict liability a popular mechanism for controlling risk in the courts (Havender, 1982, 52).

Disadvantages associated with the conditions for a strict liability suit include the fact that locating the source of the harmful activity and establishing that the activity can be harmful may not always be an easy task. In addition, strict liability suits require that the plaintiff litigate. Other limitations are that awareness on the part of the defendant that the activity in question was hazardous may be required. Baram (1982, 27) sums up some limitations peculiar to applying strict liability to risk management:

> . . . the common-law strict-liability doctrines are useful in terms of deterring risks undertaken by systems' operators or designers in the absence of regulation. However, the application of these doctrines to different systems is somewhat limited by the necessity for the activity or system to be abnormal (out of place) and ultrahazardous (not just hazardous). In addition, section 519 of the Restatement [(Second) of Torts] limits the liability without fault exclusively to harm that is within the foreseeable scope of the abnormal risk created by the activity. [Furthermore,] no liability without fault will be attributed to a defendant in the absence of a statute unless the defendant has been aware of the abnormally dangerous activity or condition and has voluntarily engaged in or permitted it.

In other words, the application of the common law–based strict liability doctrine may have to be associated with proof of negligence-like actions in some instances. Baram (1982, 28, quoting the *Restatement (Second) of Torts*) points out that a statute-based strict liability provision can, if properly designed, circumvent many of the shortcomings of a strict liability doctrine based only in the common law.

In some cases federal legislation has attempted to address what some have considered politically as weaknesses in state law. It is argued that state law relies too heavily on strict liability, making it too easy for liability cases to be brought.

Nuisance.[7] A nuisance is defined in common law as an unreasonable use of one party's property that places in jeopardy the life, health, or property of another party. The injury must be assigned to a particular party (fault-based) in order for an action to proceed. Nuisance law can be applied whether the nuisance is intentionally or unintentionally created. The tests under the common law of nuisance overlap with the other doctrines. That is, the tests of strict liability and negligence are often tests under nuisance law. In addition, the application of nuisance law requires that a standard of reasonable use of one's property be established. Katzman (1988, 22) has pointed out that courts can evaluate nuisance in light of another test involving the balancing of interests of the injured party against the activity's utility to society.

A major advantage of nuisance law is that it is simpler to apply than regulatory statutes to nuisances that occur in a specific locality, affect a limited and identifiable number of people, and affect a particular source. Prior to the passage of certain statutory laws covering hazardous waste disposal, nuisance law was considered the major vehicle of compensation for personal injuries from hazardous waste disposal activities.

A disadvantage of nuisance law is that it is not useful in cases where the source cannot easily be identified or causation (in terms of a direct link between injury and the alleged source of injury) cannot easily be established. This is particularly true of diseases with long latency periods. It is also not readily adaptable to a situation where multiple sources of harm are discovered. Another disadvantage is that damages may have to be high to justify the suit, since costs of litigation could be high. Actions have to be brought quickly, since nuisance statutes often have short statutes of limitation, often less than the latency period of a disease. Katzman (1985, 22) adds that plaintiffs must show ownership of the affected properties, and in the case of private nuisance cases, an interest that the general public does not generally have. Furthermore, Katzman points out that the plaintiff has to show that he/she unknowingly came into proximity with the risk.

Trespass. Trespass is defined as injury resulting from the invasion of one party's property by another. Unlike negligence or nuisance law, the assignment of fault depends on whether the invasion was intentional or unintentional. If the invasion was intentional, then no fault need be assigned; if it was unintentional, then fault is assigned only if the party was negligent or the activity was abnormally dangerous. The tests for trespass as summarized by Katzman (1988, 21) are proof of negligence by the defendant, knowledge by the defendant that harm was likely, or intent by the defendant to trespass (but not to commit harm). The law of trespass depends on a standard of what constitutes an invasion. Its application to environmental health risk or safety occurs where the invasion of property is created by the migration of either an unhealthy substance or an unsafe activity. There are three forms of trespass: negligent (or unintentional) trespass, where "proof is required of actual damage"; intentional trespass, where "there is no need to prove actual damages"; and continuing trespass (Baram, 1982, 39).

An advantage of using trespass doctrines is that the burden of proof is somewhat less complicated than in a nuisance case. The plaintiff may only have to show invasion of the property and rights to the property. The plaintiff can collect a certain amount for damages without showing injury.

The disadvantages to the use of trespass for environmental risks are that not all courts interpret the invasion of invisible risk agents as a physical invasion of a property. Katzman (1985, 21) adds that the trespass doctrine is not easily adapted to latent injury, since it is necessary to identify a single polluter as the cause.

Product liability. Product liability involves elements of both strict liability and negligence, as well as the law of contracts. In order to bring a suit under product liability one has to establish (1) that the manufacturer of the product could have in fact avoided the damage and (2) that the damage was attributable to the product and not to some other cause.

Limitations of product liability suits are that only the direct users of the product are covered, not bystanders. This is in contrast to strict liability, where parties indirectly injured can bring suit. Damages must be able to sustain the costs of litigating a product liability claim by exceeding them, and a link between some particular cause and some damage has to be established. The court can include the financial condition of the source of the nuisance and the state of the art in abatement techniques in determining the settlement, all of which count against the plaintiff.

Shifts to Strict Liability

There has been a shift toward the use of strict liability over and above other common law doctrines. Strict liability has been combined with a showing of negligence where the source of harm and the relationship between the activity and the harm are known, as in the case of asbestos. The no-fault provision and the absence of a need to show intent in the strict liability doctrine are well adapted to the absence or uncertainty of information about cause and effect created in part by the latency period between the cause and the effects.

The popularity of strict liability arose out of the government's need to assign liability in abandoned hazardous waste site cases prior to the passage of CERCLA. The strict liability doctrine allowed the government to bring cases where the source could be identified, but harm could not easily be proven in specific cases and cause-and-effect relationships were hard to establish. The government only had to show that the hazardous waste site was capable of causing harm. In spite of the fact that CERCLA and its amendments, SARA, provided the statutory base for government action, these statutes do not preclude the use of strict liability as a common law doctrine.

One outcome of the shift to strict liability has been an apparent shift, at least implicitly, in the burden of proof from those injured and seeking damages to those allegedly causing the injury.

Re-emergence of Federal Common Law

Environmental issues have produced groundbreaking applications of federal common law. These applications were brought about under FWPCAA and CER-CLA. The re-entry of federal common law since 1938, according to Grunbaum's 1988 summary, has been justified on the basis of the interstate effects of the migration of environmental contaminants and, in the case of CERCLA, the unique federal interests created by the federal financial investment in Superfund.

Common law has traditionally been considered the domain of the states. According to Grunbaum (1988, 169–171), federal courts were not allowed to develop a federal common law in the case of *Erie R. Co.* v. *Tompkins*, 304 U.S. 64 (1938). However, cases pertaining to environmental health and safety risks have been brought under federal common law. The justification for federal common law has been the existence of either interstate concerns or a strong federal interest not covered by state common law (Taylor, 1988).[8] There are a number of instances in which federal common law has appeared at the forefront of judicial decisionmaking on environmental health risks. The federal common law doctrine of nuisance was considered to be a foundation for decisions under CERCLA and the 1980 amendments to RCRA (Hinds, 1982, 12). The way that the federal common law doctrine operates under CERCLA is to allow individual parties to share the harm that is caused by each of them individually, where the harm can be reasonably divided (Grunbaum, 1988, 170, quoting *Restatement (Second) of Torts*). In reviewing the emergence of federal common law under CERCLA, Grunbaum cites *United States* v. *Chem-Dyne Corp.*, 572 F.Supp. 802 (S.D. Ohio 1983), as having initiated an exception to the 1938 *Erie R.R.* case on the basis that specialized federal common law is warranted where unique interests are at stake, such as the monetary investment in Super-fund. Grunbaum (1988) summarizes further developments in the use of federal common law under CERCLA in a later case in terms of the need for the United States to have a comprehensive approach to hazardous waste management, to prevent generators from disposing of wastes in the states with weaker laws, and to protect the federal financial interest in the cleanups.

Conditions Under Which Common Law Doctrines Have Emerged for Risk Management

Use of common law doctrines for risk management emerged as a result of factors that are both common to government's use of common law and unique to the risk issue. The common law is used to supplement coverage by statutory law in the area of environmental health risks in a manner not unlike how it is used elsewhere in the legal system. Also, the existing framework of common law has provided a way that case law on risks could eventually develop into or be integrated with statutory law. Again, this is not unlike the role it plays in other areas.

The application and extension of common law in the risk area is unique (1) in its emphasis on strict liability, and (2) in the resurgence of the federal common law doctrine. The emphasis on strict liability has risen largely from the absence of conclusive cause-and-effect information and the difficulties faced by the government under RCRA and CERCLA in identifying, after the fact, the sources of risk. Shifts in the philosophy of who bears the burden of proof have also led to reliance on strict liability. The resurgence of the federal common law doctrine is largely a function of the unique characteristics of the risk issue that pertain to the enormous financial interest of the federal government (in the form of a multi-billion-dollar Superfund) and the interstate nature of many of the contamination problems.

LIABILITY CLAIMS AND PAYMENTS

The size of payments for in-court and out-of-court settlements and the publicity they receive can be in and of themselves an important risk management tool in acting as a deterrent for activities involving risks to human health and safety. This, of course, depends on the extent to which behavior of the risk producer is influenced by the incentives created by court awards and other settlements. Some studies indicate that this linkage in fact exists (Cantor and Bishop, 1987; Atiyah, 1982), while others imply that it does not (Kelman, 1981). Compensation for actions that produce injuries or the threat of injury is based in both statutory and common law, as well as settlements associated with insurance (discussed in the next section) and voluntary negotiation. The extent to which monetary compensation for damages can be and has been obtained for those claiming injury under environmental statutes varies considerably from statute to statute, for different levels of government, and over time. The right of the general public to obtain compensation under statutory law depends first on the rights of citizens to sue polluters ("standing to sue"). Second, even given the right to sue, statutes vary with regard to whether citizens can actually sue for damage payments.[9]

Developments or trends in liability claims and payments are discussed in terms of how many claims are being brought and how large the payments are. Explanations for these trends can be shown to lie within the domain of the system of judicial review of cases and the nature of the insurance industry—both factors being discussed in later sections of the book.

Number and Size of Claims and Awards

General Increases in Size and Number of Awards

The number of tort cases filed in both federal and state courts is thought to have increased (Harrington and Litan, 1988, 741). Information on both the number and size of claims in environmental health cases is not available directly on any consistent basis for several reasons.

First, many settlements are reached out of court, and reporting the settlement amount is not necessarily mandatory. In the mid-1970s, it was estimated that about 95% of all tort litigation cases were settled before going to trial (Cantor and Bishop, 1987, 12, quoting U.S. Department of Commerce, Interagency Task Force on Product Liability, 1977).

Second, both mandatory and nonmandatory compensation derived via nego-tiation, arbitration, and other forms of bargaining or administrative procedure have gained in popularity as a means to deal with a facility's negative conse-quences that influence community acceptance of a project (O'Hare, Bacow, and Sanderson, 1983). In fact, some federal environmental laws now require that such negotiation occur. Compensation under these arrangements can be in the form of direct payments or in-kind services or facilities.

Third, there is currently no systematic reporting of claims or awards made for pollution (U.S. General Accounting Office, October 1987, 4). The amount of compensation for out-of-court settlements is not consistently reported; in fact, a condition for settlement is often that the exact amount be kept confidential.

In spite of the limitations on estimating settlement trends, it is possible to discern some tendencies.

First, the results of a number of surveys and studies as a group point to growing numbers of claims and increasing size of individual awards. These studies are those of the Jury Verdict Research, Inc.; Rand Corporation's Insti-tute of Criminal Justice asbestos cases, trends in claims in Cook County, Illinois (as illustrative of the nation), and general tort litigation trends; and the National Center for State Courts. The trends have been summarized by the Executive Office of the President, Tort Policy Working Group (1987).

Second, characteristics of personal injury cases pertaining only to chemical-related injuries tend to corroborate growth in number of cases and size of claims. Fifty personal injury cases have been analyzed here, covering only chemical accidents and a few slow releases that resulted in sudden and cata-strophic damages. These cases comprised the entire record of this type of injury in the files of Jury Verdict Research, Inc., between 1978 and 1988. The results of this analysis, shown in Figures 3.1, 3.2, and 3.3, can be summarized as follows:

- Only 50 personal injury settlements involving chemicals (but excluding con-sumer product cases) were reported to Jury Verdict Research, Inc. over a 10-

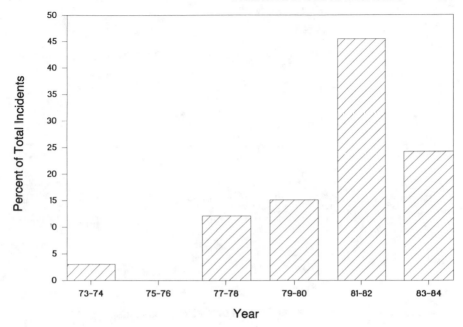

Figure 3.1.   Chemical-related injuries: year of incident.

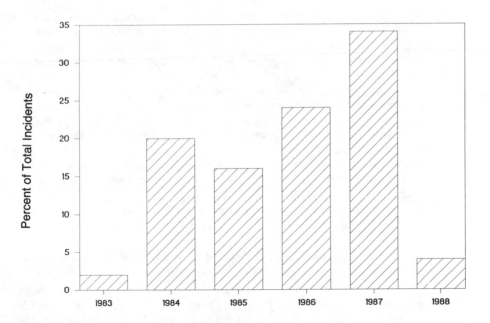

Figure 3.2.   Chemical-related injuries: trial date.

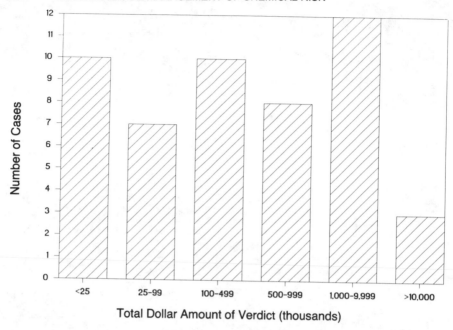

Figure 3.3.   Chemical-related injuries: personal injury verdicts.

year period. While lack of reporting can be a significant factor, this small showing is also indicative of out-of-court settlements and claimants taking routes other than personal injury claims to obtain settlements. Personal injury cases do not include cases brought by the government or other parties for legal infractions.

• For cases in which the date of the incident was recorded, there was a larger concentration of cases in which the incident had occurred recently. In addition, while some incidents were recorded as occurring back in the 1970s, trial dates were concentrated in more recent years, from between 1984 to 1988. This trend is portrayed in Figure 3.1.

• Within the period from 1984 to 1987 when most trials occurred, the number of trials increases as time passes. (Data were incomplete for 1988.) The data for this observation are portrayed in Figure 3.2.

• Many of the awards have been well over $1,000,000; however, the size of awards tends to vary considerably. (See Figure 3.3.)

Third, historical data on accidents reported from a variety of sources reveal a large number of cases with large claims and/or settlements in the early 1980s. In 1983 alone, Smets estimated the costs of various pollution incidents at relatively high levels, often well into the millions per claim. These are shown in Table 3.2 (Smets, 1987, Table 4.13). The damage claims and settlements listed in Table 3.3 are illustrative of claims and decisions in several risk cases in Japan during the 1970s that influenced the nature of regulatory controls in the United States.

Table 3.2. Magnitude of Claims for Selected Pollution Incidents in 1983

Oil—	
Tankers	5 incidents exceeding $10 million
Offshore	4 incidents exceeding $10 million
Waste disposal	13 incidents exceeding $20 million
Air pollution: plant	3 incidents exceeding $10 million
Water pollution	7 incidents exceeding $10 million
Radioactive pollution	1 incident exceeding $20 million

Source: Henri Smets. "Compensation for Exceptional Environmental Damage Caused by Industrial Activities," in *Insuring and Managing Hazardous Risks: From Seveso to Bhopal and Beyond*, P. R. Kleindorfer and H. Kunreuther, Eds. (Heidelberg: Springer-Verlag, 1987), p. 99, Table 4.13.

Table 3.3. Damages Sought in Environmental Risk Cases, Japan

Nature of Damage	Location	Defendant	Time Period	Amount Involved (millions of Yen)	
				Claimed	Decided
Mercury poisoning	Agano River	Showa Denko Ltd.	1967–71	530	270
Mercury poisoning	Minimata Bay	Chisso Ltd.	1969–73	1500	937
Itai-Itai (cadmium)	Jintsu River	Mitsui Metal Ltd.	1968–71	62	57
			1971–72	151	148
Asthma related to air pollution	Yokka-Ichi	Petroleum Combinat	1967–72	200	88

Source: M. Tatemoto, et al., Eds. (1980) *Encyclopedia Economics* (Toyo-Keizai-Shimposha).

There are several other examples of the large magnitude of both claims and court-awarded settlements in the United States during the 1980s for chemical risks in a wide variety of settings. Some of the more visible ones are shown in Table 3.4 (chemicals) and Table 3.5 (drugs and consumer products). Details of some of the key cases are worth highlighting.

Chemical manufacturing. In October 1987, Monsanto was required to pay $16 million in damages to 65 claimants for failure to warn of the dangers of a chemical spill that had occurred in 1983. According to a newspaper account, "the plaintiffs originally sought $100 million in punitive damages and $35.4 million in compensation" and smaller additional amounts for actual damages and property damages (*New York Times,* October 25, 1987).

Monsanto was required by a federal court to pay $108 million to the family of a worker at its Texas chemical plant who died of leukemia. The worker had been working with benzene for five years prior to his death. Of the $108

Table 3.4.   Claims, Costs, Settlements, and Fines in Major Liability Cases Involving Chemical Releases

Company and Date[a]	Number of Claimants	Risk Factor	Cost[b]	Reference
Johns Manville (1982)	52,700	Asbestos	$2 billion	*Business Week* (1984)
7 chemical companies[c] (1984)	40,000–50,000	Agent Orange—defoliant	$180 million	*Business Week* (1984)
Ayer (1983)	350	Various chemicals	$16 million[d]	Baram (1987, 419) Cantor and Bishop (1987)
Velsicol (1986)	5	Various chemicals	$7.5 million[e]	Cantor and Bishop (1987)
Monsanto (1986)	1	Benzene	$108 million	Reinhold (1986)
Monsanto (1987)	65	Dioxin	$16 million	*New York Times* (1986)
Monsanto (1988)	170	Rubber additive	$1.5 million	*New York Times* (1988)
W.R. Grace (1986)	Residents of Woburn, MA	Various chemicals	$8 million	Maeroff (1986)
Unifirst (1986)	Residents of Woburn, MA	Dry cleaning fluids	$1 million	Maeroff (1986)

[a]Dates refer to the approximate date of announcement of a settlement or actions reflecting the costs or cost estimates in claims where settlements had not been reached.
[b]Cost figures are given for the year indicated.
[c]Dow Chemical, Monsanto, Hercules, T.H. Agriculture and Nutrition, Diamond Shamrock, Uniroyal, Thompson Chemical.
[d]Compensatory damages; this award was recently slightly reduced to eliminate claims for psychological damages.
[e]Punitive damages.

Table 3.5. Claims, Costs, and Settlements in Major Liability Cases Involving Drugs and Consumer Products

Company and Date	Risk Factor	Cost[a]	Reference
A. H. Robins[b] (1984, 1988)	Dalkon Shield	$259 million $2.47 billion	*Business Week* (1984) Freudenheim (1988b)
Merrell Dow[c] (1984)	Benedectin	$120 million[d]	*Business Week* (1984)
American Cyanamid (1983)	AM-9 (Grout)	$290,000[e] $50,000[f]	Cantor and Bishop (1987)
Lederle (1982)	DPT vaccine	$1.13 million[e]	Cantor and Bishop (1987)
Chevron Chemical (1982)	Paraquat	$60,000[e] $80,000[f]	Cantor and Bishop (1987)
Kerr-McGee[g] (1979)	Plutonium	$500,000[e] $10 million[f]	Cantor and Bishop (1987)
Lederle (1979)	Varidase	$50,000[e] $100,000[f]	Cantor and Bishop (1987)
Abbott Lab (1976)	DES	$50,000[e]	Cantor and Bishop (1987)
E. R. Squibb (1974)	Delatin	$1.5 million[e]	Cantor and Bishop (1987)
Eli Lily (1974)	DES	$500,000[e]	Cantor and Bishop (1987)

[a]Cost figures are given for the year indicated.
[b]7700 cases were resolved as of 1984. On July 27, 1988, a revised settlement for 195,000 claims was reached that represented an intermediate between A. H. Robins' figure of $1.2 billion and the claimants' estimate of $7 billion (Freudenheim, 1988b).
[c]800 claimants.
[d]Tentative settlement as of 1984.
[e]Compensatory damages.
[f]Punitive damages.
[g]1 claimant.

million, $100 million was for punitive damages (collectable under workers' compensation) and $8 million was for actual damages (not collectable under workers' compensation) (Reinhold, 1986).

As part of an agreement, the Monsanto company awarded $1.5 million to 170 workers in its Nitro, West Virginia plant. The workers claimed that exposure to a rubber additive resulted in bladder cancer (*New York Times*, June 9, 1988).

W. R. Grace & Co. paid $8 million for various costs related to pollution of well water in Woburn, Massachusetts, and the alleged leukemia and other health effects associated with that pollution episode. The settlement from W. R. Grace & Co. follows a settlement of $1 million that the plaintiffs won from Unifirst, a dry cleaning establishment, for its actions related to the pollution episode (Maeroff, 1986).

In the Bhopal case, claims against Union Carbide ranged from $350 million in early 1986 (Lewin, 1986) to $3 billion by the end of 1986. The $3 billion estimate was what the Indian government felt was needed to cover the 520,000 claims for the 2347 deaths and 30,000 to 40,000 injuries related to the accident (Hazarika, 1986). The final settlement was estimated at $500 million.

Asbestos claims. The tables do not list asbestos claims, which have accounted for a large number of chemical risk claims over the years. Kakalik et al. (1984) have estimated that through 1983 about 24,000 claims had been filed (Cantor and Bishop, 1987, 40). Two major reviews of asbestos cases indicate individual settlements totaling several hundred thousand dollars each. One study of the claims from 513 asbestos-related cases between 1980 and 1982 found that the average jury award was $205,000 for a "medium" asbestosis injury, and the settlement for the same injury was $47,000 (Kakalik et al., 1984). A later study of 41 cases found the mean to be $325,400 in compensatory damages and $349,580 in punitive damages, with a standard deviation of the same order of magnitude as the mean. (The estimate for the "medium" injury is about the same as that derived by Kakalik et al. [Cantor and Bishop, 1987, Table 3, 42].)

In 1987, U.S. Mineral Products Company had to pay $520,000 for asbestos-related injuries to a worker in a federal building in downtown Cleveland that had been sprayed with an asbestos-containing fireproofing material. The basis of the award was a failure on the part of the manufacturer of the material to warn of its hazards. Some federal workers had been compensated under workers' compensation claims (*New York Times,* October 8, 1987).

Nuclear power plant claims. The Nuclear Regulatory Commission estimates average awards of $100,000 for early injuries and latent cancers; for early fatalities the estimate is $1 million. These estimates were used by the General Accounting Office in its recommendations for revising the Price-Anderson Act (U.S. General Accounting Office, June 1987, 15).

The Three Mile Island accident was the first time claims had been filed for damages by the public. According to the U.S. GAO (June 1987, 26), the largest damage claim filed against a nuclear facility under government auspices was $271,000. The private insurance claims that were paid totaled $41 million as of January 1987, but 2000 personal injury claims were still pending (U.S. General Accounting Office, June 1987, 27).

Hazardous waste sites. An initial award of $16 million in damages was awarded to 350 residents in Jackson Township, New Jersey for the polluting of their water supply by a municipal waste disposal facility (*Ayer* v. *Jackson Township*, 189 N.J. Super. 561, 461A.2d. 184 (1983)). The amount was reduced by a state appeals court to $5.6 million, and was taken to the State Supreme Court, which restored the original amount minus the amount for psychological injury. In this case, no injuries or deaths were reported, and some of the original award monies were for medical surveillance (Baram, 1987, 419).

A survey by the U.S. General Accounting Office of pollution claims in 200 cases that had been closed by the end of 1985 revealed that payments totaled $6.6 million, the average was $33,040, and the median was $5000. Insurers, however, maintain that these claims are only a small number of the 11,900 that were active and represented the more easily settled and lower claims (U.S. General Accounting Office, October 1987, 4).

Other contexts. Product liability verdicts have in general shown a rapidly rising trend since 1979, according to Jury Verdict Research, Inc. surveys. The number of verdicts of over $1,000,000 increased from 9 to about 90 per year between 1975 and the early 1980s (Executive Office of the President, Tort Policy Working Group, 1986, 40). The Dalkon Shield case was settled on July 27, 1988 in federal court for $2.47 billion. The A. H. Robins company, manufacturers of the Dalkon Shield, claimed that $1.2 billion would be adequate to settle the claims; the claimants wanted $7 billion (Freudenheim, 1988a). The company's insurer, Aetna Life and Casualty Company, paid $425 million of the settlement. There were 195,000 claimants (Freudenheim, 1988b).

The way in which claims are presented and negotiated can have an impact on how the settlement is perceived, and hence the degree of incentive or disincentive for managing risks it provides. First, the plaintiff produces a damage claim (called the demand). This is followed by (or is produced simultaneously with) the defendant's estimate of cost of cleanup and/or payment of the damages (called the offer). Where a case goes to trial, there is a jury award or verdict. Court-awarded payments vary according to the type of coverage of the award. That is, an award can be for compensatory, punitive, actual, or other costs or damages. Sometimes there are wide disparities in which type of award is made by a court. In a derailment in Sturgeon, Missouri in the mid-1980s that involved a spill of pentachlorophenol owned by Monsanto, the jury concluded that there was no injury (therefore awarding only $1 collectively for compensatory damages) but awarded $19 million punitive damages because of negligence (D. W. Stever, Jr., personal communication, 1988). Finally, there is the actual payment made for both in-court and out-of-court settlements. Jury awards are typically larger than the final settlement (Shanley and Peterson, 1987).

In the course of bringing a suit, each one of these figures is likely to be publicized. Of all of these kinds of estimated or demanded costs and payments, it is difficult to sort out which has the greatest impact on regulators, insurers, and the general public. The size of the jury award may be what influences insurers the most in setting premiums (Harrington and Litan, 1988, 740). However, the behavior of the general public and risk takers may be most influenced by the first figure that is claimed, i.e., the plaintiff's demand. It is important to remember that the final settlement is usually not equivalent to what is awarded; furthermore, the claims are not necessarily reflective of the extent of injury. O'Brien has underscored this latter point, reporting that "fewer than 1% of those injured sue to recover damages" (O'Brien, 1988, 62, quoting Lieberman, 1981, 49 and O'Connell, 1979).

The design of claims, jury verdicts, and out-of-court settlements potentially influence incentives for managing risks. This can only happen, however, if there is some correspondence between awards, payments, and injuries, and some indication that pay-outs influence future behavior.

Conditions Leading to Growth

Court Cases and Insurance

The *number of claims* has been largely influenced by the growth in court cases in general, which is discussed in Chapter 6, and the influence of developments in the area of insurance claims (discussed in the next section).

The *size of claims and liability payments* has in part been attributed to the increasing number, size, and cost of insurance claims, fines, and various out-of-court settlements, especially in the area of environmental risks (Executive Office of the President, Tort Policy Working Group, 1986, 40).

The Role of Fines

The size and number of court claims may bear some relationship to the nature of the system of fines imposed under statutory law. Plaintiffs other than the government may be using the system of sanctions in the form of fines as the basis for establishing the magnitude of their claims. This has received little attention in the past, probably because fines have generally been considered insignificant relative to court-awarded verdicts (Havender, 1982, 52). In addition, fines in the past have been relatively small. Nichols and Zeckhauser (1977, 54) observe that in the early years of OSHA, in spite of the fact that only 21 to 23% of the firms inspected were considered to be in compliance, "the actual fines levied [were] very small, averaging $37.49 per violation and $188.22 per citation in 1976. Fines levied under OSHA-approved state programs have averaged still less." A 1985 GAO report noted that state fines for the illegal disposal of hazardous wastes were set well below what could have been assessed. Twenty-eight cases reviewed by the GAO that were resolved between 1981 and 1984 revealed that while the range of fines imposed was from none to a maximum of $100,000, the range of fines that could have been imposed was from $25,000 to $15.3 million (U.S. General Accounting Office, February 1985, 39–40).

As is the case with claims for damages and other forms of relief, the system of fines based in federal law varies considerably from statute to statute. Under the CPSA, civil penalties range from $2000 per violation to a maximum of

$500,000 (Stever, 1988, 4–9, quoting CPSA, Section 2069(a)). RCRA's 1980 amendments provide for a fine of $5000 per day of noncompliance for willful violations of Section 7003. The fine must be levied by a court, and requires the issuance of an administrative remedial order (Hinds, 1982, 23). RCRA's Section 3008 allows EPA to invoke substantially greater sanctions (U.S. General Accounting Office, June 1988, 9):

- civil actions: $25,000 per day
- administrative actions: $25,000 per day
- criminal actions: general, $50,000 per day; committed with knowledge of the violation, $250,000 for individuals and $1,000,000 for organizations

Some examples are noteworthy. Union Carbide settled with the federal government for a fine of $408,500 for violation of OSHA rules in connection with its West Virginia plant. This settlement came a year after the government sought $1.37 million in fines (Shabecoff, 1987). Earlier, a fine of $4400 had been assessed for a leak at the West Virginia plant under OSHA, which was a substantial reduction over the $32,100 OSHA originally sought (*New York Times*, 1986).

In 1985, the Nuclear Regulatory Commission levied a $900,000 fine (a record at the time) against the Toledo Edison Company for an accident at its Davis-Besse nuclear power plant on June 9, 1985 (*New York Times*, 1985). As of 1988, the largest fine that the Nuclear Regulatory Commission had issued was $1.25 million against the owners of the Peach Bottom nuclear plant for laxity in management (U.S. Nuclear Regulatory Commission, Public Affairs, personal communication, May 1989).

FINANCIAL MECHANISMS FOR RISK MANAGEMENT: SOME GENERAL COMMENTS

Financial mechanisms are commonly used to discourage conduct of risky activities and to provide incentives to reduce or avoid the consequences of a risk once it is realized. These mechanisms were popularized in the late 1960s and 1970s in connection with pollution control programs, and interest in this area persisted with renewed vigor in the 1980s.

Some of the more popular financial mechanisms used as incentives to reduce risks are (1) insurance and other forms of compensation, such as workers' compensation and victims' compensation, (2) taxation, which provides the govaernment with resources to reduce risks or consequences, (3) marketable rights that allow a redistribution of risky activities over space or time, while attempting to hold the total amount of risk in a given area constant, and (4) grants and loans to provide those who conduct risky activities with the resources to reduce risks.[10]

Many of these mechanisms are commonplace in society, and have traditionally been used or supported by government. Adaptation of these mechanisms to manage environmental health risks, however, is relatively new. The degree to which environmental risk management has initiated or taken advantage of financial incentives is the subject of this chapter. These actions will be evaluated in each of the financial incentive areas that have been explored for environmental risk management—insurance, taxation, marketable rights, and grants and loans. The extent to which the uses of the mechanisms are simply commonplace applications or new initiatives will be emphasized.

INSURANCE AND COMPENSATION

Advantages of Insurance for Governmental Chemical Risk Management

Insurance is a traditional way of reducing the financial burden imposed on those who conduct risky activities (and indirectly, those who are the victims of such activities) through the redistribution or sharing of the expense among a number of individuals or entities facing similar risk.[11] Insurance can theoretically provide the financial resources to restore property damages or lost pay, reverse adverse health impacts (to the extent possible), or provide compensation for loss of life or health when a risk has been realized. More importantly, the traditional role of insurance for the redistribution of risk could be extended into the area of risk reduction. That is, depending on its design, insurance might also prevent the risks from existing in the first place. According to Havender, the very design of the insurance mechanism theoretically has advantages as an alternative to the direct regulation of risk: since insurance depends on balancing premiums and pay-outs, premiums can be held to a minimum if risks are reduced, thereby building into insurance an automatic incentive to reduce risks (Havender, 1982, 54). As Doherty, Kleindorfer, and Kunreuther (1988, 6–9) have summarized, government has targeted the insurance mechanism for a number of different roles, namely, as policeman, regulator, and source of financial resources ("deep pockets") for relieving the financial burden on individual firms as well as providing financial guarantees for insured parties ("insureds").

One of these roles—government expectations that insurers will directly or indirectly act as regulators—takes several different forms. The first is a more conventional role, in the absence of mandatory requirements, of having insurers use statute-based standards and professional standards as a basis for designing and granting coverage (Doherty, Kleindorfer, and Kunreuther, 1988, 6–7) and also having them use generally accepted methods of analyzing risks, e.g., risk assessment, to estimate the probability and magnitude of losses.

A second role has been advanced in the form of financial responsibility requirements under RCRA and Superfund. Mandatory financial responsibility requirements (insurance being one type) imposed on regulated entities (treatment, storage, and disposal facility [TSDF] owners and hazardous waste transporters) would theoretically induce both the regulated entities and the insurance companies to ensure environmentally safe siting and operation of hazardous waste treatment facilities. Insurance companies could set premiums in such a way as to induce safer operations. RCRA requires the owner/operators of hazardous waste TSDFs to have liability coverage for third party claims in the amount of $3 million per occurrence and $6 million in total annual coverage for nonsudden accidental occurrences (where the facility is a landfill or some other land treatment facility) and $1 million per occurrence and $2 million in total annual coverage for sudden accidental occurrences (U.S. General Accounting Office, October 1987, 11). Superfund requires similar coverage. While it was anticipated that commercial insurance would provide most of this coverage, owners of TSDFs were allowed to provide it through self-insurance where they could demonstrate sufficient resources to do so.

A third role for insurance as a regulatory device is in the form of a subrogation mechanism (Kunreuther, 1987, 194; Abraham, 1986, 52, 54). Under such a mechanism the insurance companies pay for damages (administered through a damage fund that rapidly compensates victims) but can presumably control firm behavior in the way they set premiums. Fourth, as suggested and summarized by Katzman (1988, 82), the insurance industry could act as regulator by suggesting various risk reduction methods that could be employed and by providing insureds with information on how different technologies produce differences in the frequency and severity of accidents.

The tradition of using insurance as a regulatory mechanism by imposing financial responsibility requirements dates back at least a couple of decades. Federal laws requiring financial responsibility include the Clean Water Act, the Trans-Alaska Pipeline Act, the Deep Water Ports Act, the Outer Continental Shelf Lands Act Amendments of 1978, RCRA, and CERCLA (Parker, 1986, 14; Katzman, 1988, 82). Regulations under the Motor Carrier Act of 1980[12] require certain transporters of particular kinds of hazardous wastes to carry minimum insurance (Parker, 1986, 14).

Several factors theoretically reduce insurance's effectiveness in its government-imposed role as a regulatory mechanism and risk management tool, though a number of these deficiencies could be remedied. First, if victims are readily compensated for injury through insurance, without additional sanctions against the insureds, it can encourage risk-taking. Berliner (1988, 15) refers to this as the so-called "moral hazard" issue and Doherty, Kleindorfer, and Kunreuther (1988, 15) have defined "moral hazard" as "intentionally careless behavior on the part of an insured, knowing that if an accident occurs it is covered by insurance." A growing literature is appearing, however, that explores the opposite trend in the behavior of insureds—moral imperative: "the drive for an individual to produce more safety when insured than when uninsured" (Wu and Colwell, 1988, 101) assuming that insurance conditions encourage such behavior. For example, if the pricing of insurance were to take into account previous claims and conditions observed at the time of writing the policy, then insureds might be more risk-averse. Furthermore, insurers could refuse to grant coverage or pay claims if certain safety and health standards were not being followed at the time of an accident. Abraham (1986, 46) has suggested two formal mechanisms that reflect these approaches and states that "only if liability insurance premiums are feature rated (to reflect the safety level of an activity's operations) or experience rated (to reflect the insured's actual liability record) will something approaching optimal deterrence be achieved." Not enough is known yet about the influence of insurance on the behavior of insurance holders to know whether such behavior actually occurs in response to insurance.

Second, long time periods for settling claims can decrease the effectiveness of the insurance mechanism in providing resources when they are needed, especially those resources needed for victims. New administrative mechanisms discussed in this section are expediting the payment process.

Third, the existence of insurance may preclude careful analysis (through such techniques as risk assessment) of factors contributing to risk in particular cases as a basis for preventive measures, though this point has been debated.[13] In fact, risk assessment could be the basis of more careful design of standards into insurance policies.

Thus, legislation originating in the early 1970s covering environmental health risks from chemicals reflected government's expectations that conventional insurance could be used for their regulation. For a variety of reasons discussed below, these expectations have not been entirely fulfilled. As a result of these limitations, modifications of conventional insurance were undertaken and their success is yet to be realized.

Limitations of Traditional Insurance for Chemical Risks

In spite of the ostensible attractiveness of conventional insurance for risk management, it has fallen short of chemical risk management needs because of problems associated with the cost, availability, and coverage. Many theories as to why these problems have occurred have been reviewed (Harrington and Litan, 1988; U.S. General Accounting Office, October 1987, 12; Huber, 1987). One set of theories pertains to practices within the insurance industry that allow some areas an increasing competitive advantage for insurance coverage over areas involving technological risk; the accounting system the industry uses to determine its profits and losses; the fact that the insurance industry went into a general state of decline between 1979 and 1985 (Morais, 1986, 170–1); and other factors. Another set of theories is based on various legal and judicial changes, such as changes in the statute of limitation for bringing a suit and the way the time period was computed; reliance on strict liability, which changed the nature of proof required by plaintiffs; the allowance that retroactive risks be covered and pollution not be excluded under certain policies;[14] reliance on evidence of causation based on statistical association rather than causal proof; and use of the burden of the defendant to warn of risks as a basis for a case (Baram, 1986, 568).[15] Factors associated with risk events believed to affect at least the availability and cost of insurance include the rise in the number of accidents in which insurable risks have been realized, the large number of claims and the large size of individual claims (discussed in the first part of the chapter), and the increased stringency of governmental regulations pertaining to these risks (Executive Office of the President, Tort Policy Working Group, 1986, 40; 1987, 32, 44).

While practices within the insurance industry and judicial reviews have had a considerable impact on the insurability of chemical risks, it is important to understand one critical aspect of the problem. Traditional insurance mechanisms have been difficult to apply to environmental health risks because these risks do not meet several conventional tests or criteria for insurability. Berliner (1982) has set forth criteria generally used to evaluate insurability that have been widely applied (Kleindorfer, 1986; Katzman, 1988, 83–87). Briefly, criteria that have been cited for judging the insurability of environmental risks are:

1. The risk must be identifiable in terms of its magnitude and severity (in order to determine whether the insured is liable and if so, how liable).
2. The frequency and severity of losses and the total expected losses have to be estimable (in order to determine the level of the premium).
3. There should not be a strong positive correlation among the risks and losses of the insured population in order to avoid too much exposure at one time, but there should be enough of a correlation to justify pooling.
4. The condition of moral hazard (that is, incentives for insureds to act in a way that does not reduce risks ["to undertake less than due care"]) should be absent (Berliner, 1988, 14) as discussed above.
5. Katzman (1988, 83) has added to this list the need for a loss to occur within a definable time period, which is an aspect of the extent to which losses can be estimated.

The first two criteria as applied to environmental health risks from chemicals are discussed below.

Difficulties in Identifying and Assigning Risks

A number of judicial and legislative developments have complicated the ability of insurers to identify the risks they face. First, there is considerable uncertainty as to who is liable, for how much, and when. For example, the "joint and several liability" provisions of CERCLA allow one party to be liable for the costs of other responsible parties simply because the government only has to find one party, though SARA (Section 113(f)(1)) has introduced a greater degree of equity (LaGrega et al., 1987, 11). The provision has existing parties pay for damages that have occurred over many years, and the need for historical coverage introduces many uncertainties into estimating the distribution of losses. This uncertainty arises with respect to the identity of the parties that are part of the distribution, what the risks are, and the magnitude of the known risks. The Jackson Township case was one where the amount of liability was highly uncertain. The issue of whether an insured can be liable for additive damages arose in that case, in which 97 well contamination incidents were allowed to be treated as separate claims or occurrences (Kleindorfer, 1986, 18–19; *Jackson Township Municipal Utilities Authority* v. *Hartford Accident & Indemnity Company*, 186 N.J. Super. 156, 451 A.2d 990 (1982); and *Township of Jackson* v. *American Home* et al. [Docket No. L-29236, Law Div., August 31, 1984]). Second, scientific uncertainties in establishing the existence, extent, and source of disease make it difficult to identify risks for insurance purposes. The science of risk estimation, however, seems to be getting more accurate, or at least consensus appears to be growing on what techniques are acceptable in estimating risks within a certain range. Third, the "trigger" for liability appears to be highly variable across cases and depends on which legal doctrines (e.g., strict liability vs nuisance or negligence) prevail. In some cases, such as Jackson Township, many triggers can prevail (LaGrega et al., 1987, 15).

Difficulties in Estimation of Loss

The basis for loss estimates should ideally be a distribution of losses for known incidents, and that distribution should have a small standard deviation. Such an estimate has been complicated by characteristics of the industries that potentially pose risks from chemicals as well as by judicial responses to these risks. The number of potential sources of hazardous chemicals is substantial enough, which is important for the estimation of premiums. However, some claim that the frequency of chemical accidents is too small for the estimation of losses from these sources (Katzman, 1988, 83–85), while recent studies have estimated the number of chemical accidents occurring in the United States since the early to middle 1980s as being over 11,000 (Cummings-Saxton et al., 1988). The more significant factor in loss estimation is the variability of the risks potentially posed by chemicals. Health risks posed by chemicals are highly variable from chemical to chemical, for different chemical concentrations, and for different human exposure scenarios. Looking to trial and out-of-court settlements as an indicator of loss provides no solutions either. Recent judicial history has been characterized by widely varying and generally escalating claims, settlements, and verdicts, especially for third party claims. This shift to third party claims (Huber, 1988, 8) has been difficult for the insurance industry to accommodate. In addition, the judiciary has been gradually embracing a strict liability policy in settling claims. This is believed to have raised the magnitude of losses. These political and judicial factors are major inputs into the insurance industry's estimates of loss, and if they are uncertain, then estimates of loss will be uncertain (Kleindorfer, 1986, 19–20).

Implications of the Limited Utility of Conventional Insurance

There are several outcomes or implications of the fact that environmental health risks associated with chemicals do not meet many of the tests for conventional insurance. First, the availability of insurance in terms of the number of carriers willing to offer coverage has decreased. Second, where insurance is available, the amount or scope of coverage, often for the same premium, has declined. Third, the cost of the available insurance has increased considerably. Fourth, conditions of coverage for available insurance have become more complex, and establishing the validity of a claim can be time-consuming and expensive.

Limited Coverage Provided by Conventional Insurance

The most common type of coverage for environmental risks in the past was the standard commercial (or comprehensive) general liability insurance (CGLI) (U.S. General Accounting Office, November 1986). CGLI is a type of property and casualty insurance. One form of property and casualty insurance, third party liability insurance, has been applied to environmental risks by covering claims by a third party who has been injured by some action on the part of the insured. Such insurance covered injuries to a third party connected with a product or service provided by the insured.

Several questions arose in connection with the degree of coverage provided by CGLI for chemical risks: whether injuries are covered after the expiration of a policy, whether several policies could cover the claims from a single case ("stacking"), and whether an insurance company is liable for claims for each occurrence—in other words, whether there is any limit on the aggregate amount of coverage.[16]

In response to these problems, a number of provisions in conventional insurance existed or were introduced during the 1980s to deal with the uncertainty in exposure of insurers to liability. Three kinds of changes are noteworthy—restricting coverage to the period during which the policy was in force (claims-made policy), restricting the kinds of events covered through pollution exclusion clauses, and shortening the statutes of limitation on the claims to basically exclude coverage of latent diseases. While on the one hand, these changes (discussed below) limited the exposure of the insurers so they would theoretically be encouraged to provide insurance, they also restricted the extent of coverage for those insured.

Emphasis on claims-made policies. In reaction to a fear of unlimited claims over time, insurers developed claims-made policies. This policy only allowed claims to be paid that (1) were filed during the period that the policy was in effect and (2) were for incidents that occurred while the policy was in effect. In contrast, the traditional occurrence-based policies allowed claims to be filed whenever an injury occurred or was discovered even if the policy had lapsed, although the incident that caused the occurrence must have occurred when the policy was in force (U.S. General Accounting Office, November 1986, 10–11; Kunreuther, 1987, 181; Doherty, Kleindorfer, and Kunreuther, 1988, 21; Katzman, 1988, 87). Thus, while both kinds of policies require that the incident to be covered had to occur during the period when the policy was in effect, the claims-made policy requires that the claim be filed during that period as well, while the occurrence-based policy allows the claim to be filed either during the policy period or any time afterwards, i.e., when an effect of the incident appears. However, the claims-made policy can be modified to extend the time period of coverage after the policy period by providing "tail coverage." Such coverage continues the policy for a specified time period after the policy has lapsed (but the incident creating the injury still has to occur within the policy period) (U.S. General Accounting Office, November 1986, 13). This is often used to provide limited coverage where a policy has been interrupted. The longer the time period or "tail," the more reserves are required to cover potential pay-outs, and the more sensitive the system is to unpredictable changes in the rate of pay-outs (Huber, 1987, 31, 32).

The claims-made policy was developed by the Insurance Services Organization, a nonprofit organization representing insurers, as a revision to CGLI in 1986, and the changes were approved by many states (U.S. General Accounting Office, November 1986, 9).[17] A number of studies have observed that the claims-made policy provides more protection for the insurers than occurrence-based policies (though indirectly benefiting insureds by encouraging insurers to issue insurance), while claimants may have less coverage unless they have uninterrupted policies (Katzman, 1988, 88; U.S. General Accounting Office, November 1986, 2–3). However, an advantage of the claims-made policy for insureds is that it clarifies which policies are applicable and when, whereas under occurrence-based policies this is often not clear. Clarity can be an advantage for both the insurer and the insured, since it makes the responsibility for coverage more certain.

Exclusions.[18] Provisions for conventional insurance policies were developed that excluded certain kinds of activities from coverage. One mechanism used in connection with claims-made policies was the "laser endorsement," which excluded certain activities from coverage under a general insurance policy. This allowed an insurer to eliminate from future policy renewals incidents occurring under the current policy that were considered risky. Another mechanism was the pollution exclusion. The pollution exclusion arose, according to the U.S. General Accounting Office (November 1986), because of the insurers' perception (from court settlements) that they could no longer restrict coverage to sudden and accidental releases of pollutants, but had to cover gradual chemical releases as well. As a result of this, insurers have introduced pollution exclusions into conventional coverage policies. While all of these changes do not necessarily preclude coverage for pollution, they make it more difficult.

Statutes of limitation. The time period over which people are allowed to receive compensation for injury-related claims relative to when these injuries first occurred, were caused, or were discovered strongly influences the viability and timing of insurance claims and other forms of compensation. Several new developments in the area of statutes of limitation, however, altered the constraints and conditions imposed thereby.

Dore identifies three rules that have governed the criteria state courts use in determining the timing of a claim relative to when the injury occurred, known as the accrual period (Dore, 1988, 12-2–12-8). These are the exposure rule, the manifestation rule, and the discovery rule. Prior to the development of these rules, statutes of limitation often existed that bore no relation to the characteristics of the injury.

As described by Dore, the exposure rule states that the cause of action is measured from the date the claimant was exposed to the cause of harm. States vary as to whether this date was the time of first or last exposure. This criterion is not popularly used, since exposure may not always correspond with time of injury. Under the manifestation rule, the statute of limitation begins to accrue when the effects manifest themselves (either when the effect occurs or when it is observed). The discovery rule defines the accrual period in terms of time of discovery of an injury rather than exposure or manifestation of the effect.[19] By 1987, the discovery rule was adopted by some 31 states, but the form of the rule varies from state to state, depending on how much has to be known or discovered before the statute of limitation period begins. The discovery can pertain to the injury alone, the injury and the cause, or the injury, cause, and liable parties. Discovery rules have often replaced preexisting statutes of limitation. For example, Baram has described how New Jersey's statute of limitation of two years for health and six years for property has been overridden by the discovery rule, i.e., that the statute of limitation does not start until the injured party has discovered the injury (Baram, 1987, 422).

Thus, comparing the three rules, the discovery rule appears to be more favorable to the claimant, since it effectively lengthens the statute of limitation the most. In rare instances, state law has provided for additional time periods for substances with long latency periods.[20]

The Superfund Amendments and Reauthorization Act of 1986 contains a statute of limitation based on when a plaintiff knew or should have known that an injury or damage was triggered by a particular action or incident. This provision can override state statutes of limitation for hazardous waste–related injuries and illnesses if the state statute contains a date that is earlier than the federal commencement date (Dore, 1988, 12–5).

Other coverage issues. A number of court cases have clarified the circumstances under which insurers have to pay a claim. For example, in the area of cleanup costs, the insurers were said to have won a considerable victory over Shell Oil in a case where Shell knew of the source of contamination during the coverage period. The suit pertained to whether Shell was insured for a $500 million contribution toward the cleanup of pesticides from an Army site that Shell had leased. The basis for the insurers not being required to pay the cleanup cost was that Shell had known of the pollution at its plant.[21] Other court cases also pertain to the circumstances under which insurance companies have to provide coverage. In 1988, the California Attorney General, in conjunction with several other states, sued major insurers throughout the country, in response to the refusal of major insurance companies to cover the liability associated with the activities of municipal governments nationwide.

High Cost of Insurance

The costs of coverage have been rising astronomically for many of the reasons discussed above. This trend is reflected in the several-hundred-percent rise in liability insurance premiums in the mid-1980s (Executive Office of the President, Tort Policy Working Group, 1986, 1) or more specifically, an increase from an estimated $6.5 billion to $19.6 billion between 1984 and 1986 (Harrington and Litan, 1988, 737, quoting A. M. Best Company of Oldwick, New Jersey). In the chemical industry, it has been estimated that insurance rates rose 200–400% between 1985 and 1986 alone (Huber, 1987, 31).

Reduced Availability of Insurance

The availability of insurance has been seriously impaired by the reduction in the number of companies offering special kinds of insurance aimed at chemical risks and changes in the conditions and amount of coverage.

Kleindorfer characterized the change in insurance for facilities posing environmental risks: "By 1980, coverage limits in the $25–30 million dollar range were widely available. In the period 1980–84, however, claims made against these policies grew exponentially, leading to severe net losses in this line of insurance and reinsurance" (Kleindorfer, 1986, 2).

Environmental impairment liability (EIL) insurance was one of the major forms of insuring risks from gradual pollution episodes that was developed to cover the limitations in conventional insurance imposed by the pollution exclusion clause. However, during the 1980s the number of insurers willing to write EIL insurance has declined. Evidence of this trend is primarily anecdotal:

- The President's Tort Policy Working Group noted: "Two major companies dropped out of the market in 1985, and by the end of the year only two companies were offering EIL coverage. Forty-seven companies were forced to close hazardous-waste management facilities for lack of EIL coverage" (Executive Office of the President, Tort Policy Working Group, 1986, 6).
- The two companies that were still writing EIL insurance policies by 1987 were the American Home Assurance Company and the Pollution Liability Insurance

Association (PLIA), but PLIA did not insure "chemical manufacturers, drug manufacturers or those who operate hazardous waste disposal sites" (Kunreuther, 1987, 182). The American International Group, with 110 affiliates, is the largest group writing EIL insurance, but the types of situations covered are limited (for example, underground storage tanks are not covered) and there are strict requirements for the areas that are covered.

- A survey of the National Association of Insurance Commissioners in December of 1986 found that 37 states had trouble with EIL coverage (Executive Office of the President, Tort Policy Working Group, 1987, 9).
- Kunreuther and Willingham (1986) observed that "between January 1984 and September 1985, approximately 70 percent of total EIL premiums written in the USA were taken off the market."

Sudden and accidental pollution coverage under CGLI has had similar setbacks as insurers have added pollution exclusion clauses. This behavior has in part been a reaction to recent court cases interpreting sudden and accidental pollution as gradual pollution episodes (Executive Office of the President, Tort Policy Working Group, 1986, 6).

Most significantly, treatment, storage, and disposal facilities were unable to meet the financial responsibility requirements under RCRA because of an inability to obtain insurance. As summarized by Katzman (1988, 76):

> Many waste-site operators were unable to meet the financial tests, so Congress twice extended the original deadline (January 1, 1985) for demonstrating responsibility. By the end of 1986, two-thirds of waste facilities had failed to obtain insurance, and no progress has been made in establishing financial responsibility requirements for generators. [Articles from the Bureau of National Affairs, Inc., *Environmental Reporter* are cited.]

Interestingly enough, some insurance specialists have noted that in spite of the apparent need for coverage generated by legislative financial responsibility requirements, some insurers claim that firms are not taking advantage of the insurance coverage that does exist. This has been attributed by some insurers to a combination of the low level of coverage relative to the extent of liability that can be experienced in catastrophic events, the failure of government to enforce financial responsibility requirements for fear of driving small firms from the hazardous waste management industry, and the availability of the Chapter 11 or bankruptcy mechanism as a solution to very large liability claims.[22]

Complex Conditions for Compensation: The Case of Workers' Compensation

Workers' compensation exemplifies how insurance can become limited by the complexity of conditions for receiving compensation.[23] Workers' compensation is a form of insurance that compensates workers for exposure to and injuries from risks related to employment. Compensation under such a program is usually given in the form of cash payments or medical coverage. Both federal and state governments administer workers' compensation programs. The two programs differ in the basis for or tests of liability or coverage. Federal workers' compensation is based on negligence, while state programs are a no-fault system (Curtin, 1988). Under state systems the employer carries the insurance and awards it to the injured worker regardless of whose fault the injury was. Meier (1985) notes that by 1948 workers' compensation programs existed in every state, covering about 85% of the nation's workers.

An advantage of workers' compensation for risk management is that since it is an expense that employers bear, the system provides an incentive for employers to reduce workplace risks (Baram, 1982, 78).

Meier (1985) outlines the following weaknesses in workers' compensation as a means of protecting workers from occupational risks:

- The system pays low benefits.
- It takes a long time to obtain the benefits after an accident.
- It does not prevent injuries from occurring; it only reduces the consequences.
- It is difficult to apply to occupational health problems, though extensions to this area are possible. As a result, few claims exist for health problems. Baram (1982, 80) points out that "less than 3 percent of all claims awarded by states are for death or disability due to an occupational disease."

Havender (1982, 53) has added other limitations to this list:

- The kinds of damages that a worker can claim are limited to direct physical harm and future disability and associated lost earnings, but not pain and suffering.
- The courts tend not to accept the statistical likelihood of harm as evidence of impairment and as the basis for a claim. An example of this is the likelihood of a worker developing cancer from a condition in the workplace.

Alternative Insurance Mechanisms

In spite of the general limitations of insurance, there are a number of alternative insurance mechanisms that are potentially applicable to health and safety risks. Many of these have, in fact, arisen in response to the shortcomings of traditional mechanisms. They include, for example, the design of innovative forms of insurance; caps or limits on or exemptions from liability created through legislation; insurance mechanisms such as insurance guarantees, pools, self-insurance, or captive insurance; the use of past insurance policies to cover present and future damages; increasing reserve ratios; risk reduction; the development of schemes to tier different forms of insurance to cover alternative scenarios related to cost and uncertainty; and innovative administrative arrangements to expedite insurance claims.

Innovative Forms of Insurance

The most common of these is EIL insurance, but victims' compensation and changes in conventional CGLI policies are also emerging mechanisms.

EIL insurance. According to Kunreuther's review of the history of environmental insurance (1987, 181), prior to major environmental catastrophes in the late 1960s (Torrey Canyon and the Santa Barbara Oil spill) gradual pollution episodes had been covered under CGLI. When CGLI no longer covered this type of event, other forms of insurance were developed. One of the earliest alternative mechanisms to conventional insurance was EIL, discussed briefly above. Environmental impairments can result from catastrophic or sudden accidents such as spills and explosions, or alternatively, from leakages and seepages of chemicals over a long period of time. In the past, EIL has been reserved for gradual pollution episodes (Executive Office of the President, Tort Policy Working Group, 1986, 6, 55). Klaus (1987) argues that in the past these events have been difficult to cover through conventional insurance mechanisms because the total damages are hard to quantify in monetary terms, the effects may not occur immediately after the incident, historical data on claims as a basis for establishing premiums are lacking, and the nature of the settlements and the characterization of losses are often highly political and emotional (Klaus, 1987).

According to Klaus (1987, 449), EIL insurance began around 1974. In spite of its innovations and attempts to fill a vacuum, it has experienced considerable impediments. Only a few policies have been issued, primarily since there have been few subscribers. There is a lack of reinsurance capacity, and a small amount of existing insurance capacity. Klaus points out that by the mid-1980s, a few national pools existed, but their combined capacity was not sufficient to meet the needs of even a single risk. Klaus (1987, 451) identified the following existing capacities (in U.S. dollars):

- Italy: $3 million
- France: $3 million (estimated)
- The Netherlands: $1.5 million
- Switzerland: $3 million

Some legislation requires environmental impairment liability insurance as one means of making sure that the sources of risk bear the financial responsibilities of such risk.

Victims' compensation. The growing number of injuries and fatalities from accidents involving toxic chemicals has led to proposals for compensating victims for such losses. Such compensation would come from insurance, rather than through individual court settlements. Attention to victims' compensation has arisen from proposed amendments to CERCLA that would compensate victims exposed to chemicals from abandoned hazardous waste sites and from large claims such as the $180 million settlement in the Agent Orange case. Initially CERCLA specifically did not provide for victims' compensation, but rather for liability in the form of remedial action (Hinds, 1982, 24).

One of the major obstacles to a broadly based system of victim compensation has been the problem of establishing causation. That is, where multiple causes of a given injury were considered possible, compensation was often denied. Apparently, some court decisions have been allowing victims to be compensated without showing unique causation, as long as it can be shown that the alleged risk contributed substantially to the injury (Fiksel, 1984).

Changes in conventional CGLI policies. As discussed above, a new type of policy, a claims-made policy, was developed to address many of the shortcomings of the previous occurrence-based policies. The pollution liability insurance developed by the insurance industry was based upon a claims-made policy in 1981 (Kunreuther, 1987, 181).

Limits on Liability or Claims

A few programs specify statute-based limits for liability, expressed mainly in terms of dollar limits on claims or awards. These limits are distinct from caps placed directly in insurance policies. It is argued that a statutory limitation in part circumvents the delays in litigation associated with coming up with a dollar value for a claim and the uncertainties surrounding establishment of causation and levels of compensation for harm. Those in favor of such limits argue that many uncertainties and the costs associated with addressing them are circumvented. Those opposing statutory limits argue that the limits can underestimate the true costs of harm, denying victims adequate compensation. Only a few examples of the use of the liability limit mechanism currently exist for activities that involve chemical risks per se.

Nuclear power plants. The most well-known example of a liability limit is the Price-Anderson Act of 1957 (amended in 1966, 1975, and 1988) for nuclear power plants. The act provides no-fault coverage and works by means of indemnification provisions for licensees and their contractors. When originally passed, the original limit on liability was $560 million for each accident in commercial plants. This number originated as the sum of private coverage plus what was assumed to be reasonable government coverage in the amount of $500 million (Wood, 1984, 105). In 1975 the limit was increased to $695 million. $160 million in mandatory insurance coverage was paid by private insurers, and the rest was paid by plant operators; if there were an accident at a reactor, each of the 107 operating facilities would be assessed $5 million. When the Nuclear Regulatory Commission licensed new commercial reactors, the amount of the limit would go up by $5 million per reactor, and the governmental share would be phased out at a certain point. A secondary insurance policy of $30 million is also required. For contractors under the Department of Energy, the liability ceiling was $500 million, paid entirely by the government, which remained unchanged by the 1975 amendments. Congressional action would be required for any additional coverage for an accident, but Congress did not have the authority to appropriate funds (U.S. General Accounting Office, June 1987). Under the act the maximum amount of insurance required for nuclear power plants is limited by what is available on the market. The act contains a provision for reevaluating the amount of the limit, and Congress has considered a number of bills over the years to change the limits. Both houses of Congress passed legislation to change the limit to $7.1 billion in 1988, with a substantial rise in the amount of the limit per plant. On August 20, 1988, the bill became P.L. 100–408, the Price-Anderson Amendments Act of 1988.

One major debate over the Price-Anderson Act centers on whether there should be a limit at all. A second issue concerns the number and kinds of costs a proposed limit should cover. To put the figure of $7.1 billion in the 1988 act in perspective, the U.S. General Accounting Office estimated that for 113 of the 119 plants operating in 1987, the financial implications of a catastrophic accident would be less than $6 billion (U.S. General Accounting Office, June 1987, 20). Even doubling the consequences of the accident results in 81 of the plants still falling below $6 billion (U.S. General Accounting Office, June 1987, 24). The original limitation on liability was considered too low to cover the various components of cost of a catastrophic accident and was ambiguous about evacuation costs (U.S. General Accounting Office, June 1987). This is especially true when an accident occurs and damages are not covered under the limit. Usually, there are no supplemental provisions to cover damages in cases like that other than relying on federal disaster assistance. Those arguing that the limits are too low base their experience on the accidents at Chernobyl and Three Mile Island and on numerous actions that have been taken by the Nuclear Regulatory Commission and the Department of Energy in response to safety problems at existing reactors. Those in favor of the limit argue that it avoids delays associated with tort litigation and it guarantees payment without requiring establishment of detailed causation (Huber, McCarthy, and Mills, 1985; Wald, 1987).

Hazardous waste sites. CERCLA placed a limit on liability for damages and response costs, except where the responsible party was acting willfully, was knowingly in violation of federal law, or was not cooperating with the National Contingency Plan (Hinds, 1982, 30). This limit is equal to the total response costs plus $50 million for facilities not related to transportation (CERCLA, Section 9607 (c)(1)(D)), or the full cost of response and damages if the action was willful misconduct or negligence or was a violation of standards or regulations, or if cooperation is not forthcoming (CERCLA, Section 9607 (c)(2)). For transportation facilities the limit is the greater of $300 per ton or $500,000; other amounts are set for other vehicles (CERCLA, Section 9607 (c)(1)(B,C)).

Other. Caps on liability from other areas lend experience and precedents for environmental risk areas. Liability limits or immunities from liability exist in several federal energy-related statutes, such as the Outer Continental Shelf Lands Act Amendments of 1978, the Deep Water Port Act of 1974, and the Trans-Alaska Pipeline Authorization Act of 1973 (Huber, McCarthy, and Mills, 1985, 2). Liability limitation for providers of facility parts (such as bolts and screws) under the Federal Acquisition Regulations is another example. This limitation has already been questioned by a Congressional Energy and Commerce Subcommittee on Oversight and Investigations in the context of the potential for nuclear accidents as well as accidents in other sectors of the economy, and the committee has recommended the removal of the limitation (U.S. Congress, Energy and Commerce Committee, 1988). States have also attempted to cap liability payments in medical malpractice cases as a means of reducing premiums; effects have been mixed (Huber, 1987, 35). A cap for medical malpractice was imposed in 1985 in New York State and was upheld in the State Court of Appeals. According to a recent *New York Times* article, this ruling was opposed by the insurers and praised by medical doctors (McFadden, 1988, B3).

Another approach to limited liability has been to limit the size of claims allowed under a variety of statutes. In reaction to the insurance crises of the early to middle 1980s, many states passed laws in 1986 that limited large claims in personal injury cases or made them more difficult for victims to obtain. Some of these claims were in the form of specific caps, while others established rule systems for the calculation of claims. These caps and the actions associated with them are itemized in Table 3.6, based on information provided by the National Conference on State Legislatures (NCSL). According to NCSL, state caps for noneconomic losses did not exist prior to 1986.

Immunity or Exemptions from Liability

Related to the limitation on the amount of liability is immunity or exemption from liability. In a move likely to have widespread consequences, the Supreme Court granted such immunity to government contractors in 1988 (Taylor, 1988). Contractors would be granted immunity from liability resulting from deaths and injuries due to a variety of causes. While this immunity particularly pertains to mechanical malfunctions and design defects, application to chemical sources is implied; in fact, the decision is believed to have affected the Agent Orange settlement. In 1988, Congress passed a bill exempting manufacturers from liability for punitive damages in cases involving their products if those products meet FDA regulations (Gilbert, 1988).

Table 3.6. **Alternative Liability Limit Actions Taken by Selected States Applicable to Toxic Torts, 1987–1988[a]**

State	Limit on How Much Plaintiffs Can Collect	Explanation of Action
Alabama	$250,000	Cap is for punitive damages for several liability only.
Alaska	500,000	Cap is for noneconomic damages; limits multiple claims against single defendant. (A defendant who is less than 50% at fault cannot be liable for claims equaling more than double the estimate of his damage.)
California	—	California abolished its cap for noneconomic damages.
Colorado	250,000	Cap is for noneconomic damages; it can be increased to $500,000 where a convincing case can be made. Liability is limited to defendant's share of damages.
Connecticut	—	If one defendant cannot pay his share, other responsible parties must pay.
Florida	450,000	This cap for noneconomic damages was overturned by the state legislature in 1987.
Hawaii	375,000	Cap applies only to physical pain and suffering, with no limit on other kinds of damages.
Idaho	400,000	Cap on noneconomic damages, adjusted for annual changes in wages.
Kansas	250,000	Cap is for pain and suffering, not for any other noneconomic damages.
Maryland	350,000	Cap on noneconomic damages.
Minnesota	400,000	Intangible losses such as "loss of consortium, emotional distress, or embarassment (not including pain and suffering)."
New Hampshire	875,000	Cap on noneconomic damages; punitive damages prohibited.
Oregon	500,000	Cap on noneconomic damages.
Texas	200,000	Punitive cap, equivalent to four times compensation of $200,000.
Virginia	350,000	Punitive cap only.
Washington	177,000– 493,000	A range for noneconomic damages based on an equation: "approximately 1/2 the average annual wage in state multiplied by the life expectancy."

[a]These data are drawn from figures and information provided in various documents by the National Conference of State Legislatures (NCSL), 1986–1988. The NCSL points out (Trolin, 1986) that five states have statutes that specifically prohibit placing limits on the amount of damages that can be recovered in personal injury cases, namely Arizona (ARS Constitution, Article 2, Part 32), Arkansas (Arkansas Constitution, Article 5, Part 32), Kentucky (Kentucky Constitution, Part 54), Pennsylvania (Pennsylvania Constitution, Article 3, Part 18), and Wyoming (Wyoming Constitution, Article 10, Part 4). Heller (1988, 3) further states that Virginia, North Dakota, Illinois, New Hampshire, and Texas have considered noneconomic caps to be unconstitutional.

Federal Insurance Guarantees

The federal or state government can underwrite insurance or provide insurance guarantees or subsidies. One example is flood insurance. Under the National Flood Insurance Program, homeowners in localities with approved flood protection mechanisms obtain highly subsidized insurance (Kunreuther et al., 1978). Also, the government provides relief for flood victims in the form of disaster assistance. Subsidized flood insurance has been criticized for encouraging development in floodplains, and as a result elaborate floodplain regulations and enforcement systems have had to accompany insurance provisions.

Another example of government-subsidized insurance is the Black Lung Benefits Program for coal miners who suffer from the effects of coal dust accumulation in the lungs. This program was set up under the Federal Coal Mine Health and Safety Act of 1969 and extended considerably under the Black Lung Benefits Act of 1972 and the Black Lung Benefits Reform Act of 1977.[24]

Risk Pooling

A popular and relatively new method of dealing with insurance limitations is risk pooling for risks not covered by conventional commercially available insurance. These options take several forms: self-insurance risk pools in the form of agreements among companies (not managed by a separate company), captive insurance organizations, and risk retention groups. These options differ from one another in degree of organization and formality. Self-insurance is the least formal, while the separate organizations set up to manage special insurance, such as captives or risk retention groups, are the most formal.

Self-insurance. Self-insurance involves the setting aside of corporate assets in the event that claims are made for which there is no insurance. It is considered to be one step above no explicit insurance at all. The disadvantage is that the reserves set aside for insurance purposes are fully taxable, unlike the reserves of a separate insurance company. Also, small firms with large exposure may not have the reserves to set aside for insurance (Executive Office of the President, Tort Policy Working Group, 1986, 58).

Captive insurance. One mechanism of risk pooling that has had a longer history than some of the others is captive insurance. Several companies can combine to pool risks by forming their own insurance company. This is a more formal arrangement than self-insurance or risk pooling by contract. Captive insurers have operated at a very high level of coverage, but are usually highly facility-specific. Examples of such arrangements for chemical and safety risks include the A. C. E. Insurance Company, set up in November 1985 for liability claims up to $150 million, but with a deductible of $100 million; and CASEX, for 15 chemical and petrochemical companies (Executive Office of the President, Tort Policy Working Group, 1986, 57). OIL is another captive. Set up for the oil industry, it operates at a coverage limit of about $250 million to $300 million. OIL will be covering the North Sea disaster that occurred in 1988, which will cause its reserves to dip substantially.

Risk retention groups. Risk retention groups are a kind of captive insurance but offer (at least initially) a lower level of coverage than captives. A federal law pertaining to risk retention, the Product Liability Risk Retention Act, was passed in 1981 and was amended by the Liability Risk Retention Act of 1986. The act allows the formation of insurance groups for the purpose of providing liability coverage that otherwise is unavailable or difficult to obtain. The groups are a type of self-insurance pool. A requirement of the act is that members of the group be engaged in a similar activity or (more importantly) share similar risks. Professional organizations or trade associations often provide natural organizational arrangements for such groups. Although registered under a single state, a risk retention group can do business nationwide. Certain restrictions apply regarding the kind of insurance these groups can offer. Also, while certain state regulations are preempted, other state laws must be followed. Risk retention groups are exempted from certain state laws pertaining to a similar type of insurance group, purchasing groups (Alexander and Alexander, Inc., 1987, 23–25, 32). The 1981 act was restricted primarily to products, whereas the 1986 act broadened the use of the groups to many different kinds of business activities and liabilities (Executive Office of the President, Tort Policy Working Group, 1986, 58). The risk retention concept was essentially approved by the Superfund Amendments and Reauthorization Act of 1986.

The Environmental Protection Insurance Corporation (EPIC), a risk retention group formed in July 1988 as an affiliate of Alexander and Alexander, started with an initial group of five participants in the chemical and related industries, and had about 60 companies considered active applicants, according to a spokesman for the organization. An offering was developed in the fall of 1988 to expand membership. EPIC offers $1 million coverage per incident and $2 million aggregate. As more capital comes in from membership expansion, the amount of coverage goes up. Another risk retention group aimed at chemical risks is Hypercept, a joint venture of an environmental firm and an insurance company, based in Monticello, New York. Hypercept was formed in 1985 and was licensed as a risk retention group in 1987. However, the effort was halted in July 1988 because of a lack of subscribers. Instead, Hypercept separated the writing of insurance from its capitalization in two different organizations.

Another early industry-owned organization that was similar to a risk retention group was set up by the commercial solid waste management industry. It was called Waste Insurance Liability Limited (WILL) and was based in Vermont. In order to join, companies had to meet certain financial requirements. It was limited to third-party claim coverage and covered both gradual and sudden pollution episodes. It had a relatively large coverage—$10 million per occurrence at a cost of $100,000 (along with a one-time contribution of $50,000) with smaller amounts available to smaller companies, and required a $1 million deductible per occurrence. The company required a minimum of 30 members with a total minimum capital of $17 million to operate.[25] This effort has not proceeded, apparently because of an inability to balance premiums and coverage (U.S. General Accounting Office, October 1987).

Thus, risk retention groups require a critical mass of subscribers in order to offer a certain level of coverage. On the other hand, the subscribers need a certain level of coverage to be induced to join. At the outset, it is a chicken-and-egg situation. One way of reducing the cost of offering the coverage (hence the amount of capital) is to reinsure part of it. The advantage of reinsurance is that the primary company maintains the full management system necessary to write the insurance (the full claims staff, policy issuing staff, accounting staff, etc.), while the reinsurer can operate with a much lower level of management.

Forcing Compensation of Claims Under Past Policies

Another approach is to expand the ability of past coverage to cover injuries, illnesses, and deaths that have had long latency periods (Steuber, 1987). According to Steuber, courts are favoring coverage for damages that occurred many years after an initiating cause, where the initiating cause occurred under an insurance policy that was active during that period. Coverage is still apparently valid, even though the policy may no longer be in force. Even if a policy believed to be operative at the time of the causative event cannot be found, a case can be made for its existence by means of showing policies before or after the time period in question, and its contents can be established by drawing on analogous or standard policies. Others argue against the use of past insurance policies for current liabilities, maintaining that such retroactive liability coverage was never intended by insurers (Kunreuther, 1986, 6). According to Steuber, the advantages for insureds of using older policies for coverage are that there is no pollution exclusion clause or deductibles, and the policies may cover companies preceding the one making the claim.

The issue of the coverage of future claims (as distinct from current claims) under past liability policies has been debated on a case-by-case basis. For example, an appeal is being filed against the $2.47 billion settlement in which the Aetna Casualty and Life Insurance Company was relieved of paying future claims related to the Dalkon Shield (Freudenheim, 1988b).

Increasing Reserve Ratios

The shortfall that insurers are projected to face in covering environmental health damages is estimated at about $62 billion between 1986 and 1989 (Kleindorfer, 1986, 22, citing Parker, 1986 and others). As a reaction to this, some insurers have opted to increase the conventional reserve ratio. Risk retention groups formed in the 1980s have been using a capital-to-coverage ratio (per incident) of approximately 5 or 10 to 1. This is a higher ratio than is typically used by the insurance industry. For individual firms, the insurance industry tends to adhere to a particular range of reserve ratios or ratio of premium to surplus. In 1985, the ratio of annual premiums to surplus was just under 4 (computed from Katzman 1988, 91). From 1985 to 1987, the premium-to-surplus ratio aggregated nationwide was reported as being under 2.0 for each of the three years (Insurance Information Institute, 1988).

Reliance on Other Financial Responsibility Mechanisms

Insurance is only one way in which firms can meet financial responsibility requirements of RCRA and other environmental laws. The U.S. General Accounting Office has cited other mechanisms, such as financial tests consisting of working capital or other assets and bond ratings, corporate guarantees, and various combinations of these with insurance (U.S. General Accounting Office, October 1987, 34). Several other mechanisms are discussed in subsequent sections.

Overall Risk Reduction

Sources of risk should be encouraged to reduce the risks associated with their activities in order to make these activities more attractive to insurers. Trade and professional associations are increasing their attention to risk reduction. In addition, newer initiatives are being undertaken by combinations of firms. For example, in 1988 12 companies initiated a joint environmental health and safety program called the Corporate Environmental Health & Safety Management Roundtable (*JAPCA*, 1988). A major effort is also under way to reduce risks via waste minimization (Baram, 1988; U.S. Congress, Office of Technology Assessment, 1986).

Tiering Mechanisms

As new forms of insurance have emerged, a number of mechanisms or strategies have been put forth that attempt to dovetail them with one another as well as with various control mechanisms in order to provide a comprehensive system of regulation and insurance coverage. Katzman (1988, 77) has suggested a tiering system for first party, third party, and worker damages and injuries, aimed at internalizing the cost of injuries. First party damages would be covered under conventional insurance policies. Tort law, statutory requirements, and user chargers complement one another in providing coverage for third party injuries. According to the relationships Katzman outlines, statutory requirements act prior to accidents as a preventive mechanism; user fees, while introduced prior to accidents, attempt to encourage less costly behavior and internalize the cost of damages before they happen; tort law is invoked after accidents to pay for damages. Finally, workers would be compensated under workers' compensation and through wage adjustments that theoretically could reflect risk as a basis for setting the wages.

Doherty, Kleindorfer, and Kunreuther (1988) have outlined a method of tiering different kinds of insurance to encourage risk reduction. Their system varies according to the amount of the loss and the degree of coverage provided. A first layer of coverage is provided through self-insurance, where both the amount of the loss and the coverage needed are small. Standard private insurance provides a second tier for larger losses and coverage, as long as the losses are relatively predictable so that premiums can be written. Alternatively, a risk pooling arrangement could be used in this second tier for higher amounts of coverage and loss. Finally, a third tier would be provided by the government for the very large losses. As one moves from tier one to tier three, the amount of risk spreading or the number of individuals that share the risk increases.

Innovative Arrangements for Payment of Claims

A major obstacle to the use of insurance for environmental health risks has been the absence of administrative mechanisms to handle claims, given the fact that claims and cases typically involve large numbers of people.

Administrative tribunals. One way of organizing compensation claims is through the use of administrative tribunals.[26] According to Soble (1977), who has developed applications for the concept of the administrative tribunal in the United States in the area of toxic substances, claims are diverted to an administrative tribunal rather than having them fall within the jurisdiction of the court system. Such tribunals overcome the limitations of workers' compensation in the area of occupational disease and cover class action suits for compensation of environmentally based diseases. There are a number of other advantages that Soble (1977) and Bardach and Kagan (1982c, 244) ascribe to the tribunal mechanism:

- It does not have to show absolute causality, only evidence of probable causality. For example, epidemiological evidence and other statistical evidence would be satisfactory as a basis for causality.
- Defendants would have the burden of proving a link between the disease and causes other than the cause argued by claimants. This would provide an incentive for chemical companies to keep track of adverse health impacts of their products more carefully.
- A tribunal has administrative advantages in providing resources to acquire evidence for claimants.

- The cases won by such a tribunal would be binding on future cases. They would also constitute strict liability on the part of the manufacturer.
- Damage amounts would not be fixed as is currently the case under workers' compensation.
- It would lower the overall costs of litigation, since it deals with many similar claims at one time and creates a body of information that can be used in future cases of a similar nature.

Trust funds. Trust funds can be used to dispense payments after an award to injured parties. Monies are paid to claimants by the administrators of the trust in much the same way as trust funds for other purposes operate. A trust fund has been proposed to distribute the $2.47 billion award in the Dalkon Shield case. The fund was proposed by the A. H. Robins Company, which manufactured the Dalkon Shield. The organization of that proposed trust fund consists of five independent trustees who distribute the monies to those claiming damages (Freudenheim, 1988a). A rather unique variant of the trust fund concept was proposed by Johns Manville for payments to its claimants. Johns Manville has proposed two trust funds, the "asbestos health trust" and the "property-damage trust fund." Under that plan, 50–80% of the stock from a new reorganization would go into the trust fund for asbestos victims, making the trust fund a dominant shareholder. Claims for DDT are also organized under a trust fund.

It is unclear how widespread the trust fund mechanism will become. The A. H. Robins Company and Johns Manville have organized trust funds through what are considered unusual uses of the bankruptcy laws. Both companies filed for bankruptcy under Chapter 11 (Manville in 1982 and Robins in 1985).[27] The insurers have tried to obtain immunity from further claims by contributing to the trust funds, which, it is claimed, looks as if the insurers are in Chapter 11 as well. Under the Manville plan, insurers contribute $700 million, as compared with $150 million in cash, $1.6 billion in 25-year bonds, and 20% of the profit each year after five years for the health fund and $125 million and residuals from the health fund for the property damage trust fund (Feder, 1988) and $425 million under the Robins plan. Vendors apparently want the same immunity. Both trust fund concepts have been appealed in a number of suits (Labaton, 1988).

The major source of dispute over this type of organization is when the claimants get paid (i.e., before or after stockholders). While the bankruptcy law does not allow stockholders to be paid until creditors are paid, there have been several deviations from this in proposed trust funds. In the Dalkon Shield case, the proposed trust fund would allow the stockholders to obtain a certain share subsequent to a takeover of the company by another company. The A. H. Robins company has assured claimants that a required $1.2 billion would be available even after the stockholders have been paid, but whether the $1.2 billion is sufficient is in dispute (Freudenheim, 1988b). A major appeal of the Robins trust fund has specifically addressed the issue of when claimants are paid relative to stockholders (Freudenheim, 1988b).

Another factor in the trust fund concept is the cost of administration. In the Robins plan for Dalkon Shield claimants, $200,000 was an initial contribution for administrative costs—figures of $30,000 per year plus $1000 per meeting have been quoted for the trustees (Freudenheim, 1988b).

Miscellaneous administrative mechanisms. Other forms of organizations exist for the management of claims and payments. In order to manage payments for asbestos claims, whose numbers are among the largest for environmental damages, an Industry Claim Facility was established. Sometimes law firms will directly manage claims. An example (from a product liability case rather than an environmental damage case) is the management of claims against Searle for an intrauterine device, the Copper-7.

Conditions Affecting Insurability of Health Risks

The crisis in insurance applicable to human health and safety risks from environmental causes is occurring against a backdrop of general problems in the insurance industry (many of which were highlighted in the earlier section on insurance limitations) as well as conditions unique to the environmental area.

Trends in Other Risk Areas

The crisis in insurance for chemical-related health risks may have largely been driven by developments in other areas requiring insurance coverage. One area is medical and other professional malpractice, business, and injury (non-health)–related product liability. In their study of trends in tort litigation, Hensler et al. (1987, 11) point out, using federal cases and California cases (as representative of state trends), that this category has grown "moderately in state courts and more dramatically in federal courts" since the mid-1970s. In the Hensler et al. study, trends in tort litigation are organized under three different headings: "routine personal injury torts," such as automobile cases; "high-stakes personal injury suits," including product liability, malpractice, and business injury cases; and "mass latent injury cases," including the chemical cases and others such as the Dalkon Shield (Hensler et al., 1987, 2–3).

While (according to the study) mass tort litigation has grown at a far greater rate and has greater potential for growth than any other area, the precedent of other areas such as medical malpractice may have influenced growth in the mass latent injury area. Also, within mass latent injury cases, cases pertaining to asbestos may have opened the way and been a driving force for similar litigation relating to other chemicals.

The effect that other areas may have on chemical risk cases applies to both the number of cases and the size of the awards. The Rand study shows that the size of awards for routine auto cases, non-health product liability cases, and high-stakes medical malpractice cases in California rose steeply between at least the mid-1970s and the early 1980s. The climb may have started even earlier. The increases estimated from that study range from 200% to 1000% over the 1960–1984 time period (Hensler et al., 1987, 16, 18).

Backlog in Toxic Tort Cases

According to Hensler et al. (1987, 10), since chemical health-related risk cases have come on the scene relatively recently, their explosive growth has been attributed to a backlog of cases. Thus, the explosion may be temporary.

Lack of Reinsurance Capability

There has been a decline in reinsurers willing to insure risks that primary insurers cannot bear (Parker, 1986, 16). Reinsurance is a critical part of the total insurance picture.

Competition for Insurance

It is ironic that a crisis in the availability, cost, and adequacy of insurance has occurred at a time when federal and state legislation has instituted financial responsibility requirements, including insurance, for hazardous activities. The viability of insurance depends on a certain minimum level of demand, and the demand has theoretically increased because of new legislative provisions, especially in the energy area. This large number of provisions, covering a potentially large number of entities, not only theoretically provides opportunities for coverage but creates competition for limited insurance as well.

The potential demand under RCRA for hazardous waste insurance could be quite large, given that by 1989 there were over 180,000 establishments classified under RCRA as generating hazardous wastes and requiring some sort of handling. While hazardous waste production for some of these generators is below the minimum for regulation under RCRA, they are not immune from liability. About 2200 establishments (some of which are generators and some of which are not) have submitted applications for storage, treatment, and disposal of hazardous wastes.[28] The potential for accidental releases is also quite large. A recent EPA study reported that over 11,000 accidental chemical releases occurred between 1980 and 1986 (Cummings-Saxton et al., 1988), and this figure excludes certain categories of accidents such as oil spills.

While the environmental statutes created some potential for growth in the demand for insurance, the failure of this demand to materialize is attributable at least in part to the following factors. First, it is stated under RCRA that having liability insurance is only one way in which a TSD facility can demonstrate adequate coverage. Other criteria include a financial test, corporate guarantees, and various combinations of these and insurance. Second, it has been argued that the financial responsibility requirements of the federal statutes have not been enforced rigorously. One reason given for the lack of enforcement has been government's fear that they will drive small firms out of the market. There is some evidence that these requirements may have done this already. Estimates of the number of facilities that could not meet the 1985 insurance requirement deadlines under RCRA range from 48 (Parker, 1986, 15) to 60 (Kleindorfer, 1986, 4). Kunreuther estimated the number at 10% of all regulated facilities (Kunreuther, 1986, 8). Katzman (1988, 76), quoting the Bureau of National Affairs, Inc., noted that the insurance requirement had not been met by two-thirds of the facilities regulated under RCRA.

Scarcity of Surplus Funds in the Insurance Industry

The viability of insurance depends on the existence of a certain amount of surplus funds. There has been a decline in industry surplus brought about by misestimation of the large number and size of claims in the area and the inconsistency in court judgments, which have made identification of risks difficult if not impossible (Parker, 1986, 16; Kleindorfer, 1986). New methods for risk pooling have also increased the reserve ratios to deal with this problem. The surplus problem may be turning around the industry as a whole—the Insurance Information Institute (1988) reported that policyholder surplus for 1986–1987 rose by 10.3% and for the year before by 24.9%. Policyholder surplus is considered an important insurance industry performance measure, since it measures net worth (U.S. Executive Office of the President, Tort Policy Working Group, March 1987, 24).

Summary

Conventional mechanisms of insurance have generally fallen short of dealing with environmental health risks. This occurs because the design of insurance mechanisms has not incorporated latency and unknown or uncertain causes of health effects. In addition, legal and judicial interpretations of what constitutes liability and how to assign liability have often been vague. Many innovative techniques have emerged to circumvent these limitations. Some of these techniques have been developed directly by government to manage risks, while others have been encouraged by government indirectly. While some have been borrowed from other areas, some have emerged because of environmental health risks. These devices are in the forefront of insurance design, and are very diverse. They include the development of new forms of insurance, liability caps, liability exemptions, federal insurance guarantees, risk pooling, application of past policies to current claims, increasing reserve ratios, financial responsibility requirements, overall risk reduction, tiering of different kinds of coverage, and innovative administrative mechanisms. These techniques are limited in how generally they can be applied and often require sophisticated managers for implementation.

Some of the conditions leading to the insurance crisis for environmental health risks originated outside of the environmental health risk area, while others are specific to the nature of environmental health risks. Those that are external to the health risk issue are the trends in automobile coverage, medical malpractice, and non-health–related product liability and their influence on chemical risks; competition for coverage from many sectors at once; and the low reinsurance capability and scarcity of surplus funds that the insurance industry is experiencing in general. Factors that are inherent in the environmental health risk issue include the backlog of claims for toxic torts and the fact that environmental health risks do not meet conventional insurance tests.

TAXATION

The Excise Tax

Taxation is a financial incentive system that can encourage risk reduction by redistributing income or inducing socially desirable behavior (for example, encouraging activities with low risk levels and discouraging activities with relatively high risk levels) (Breyer, 1982, 164). Taxes that eliminate, reduce, avoid, or avert environmental health risks associated with chemicals fall into a group of taxes called consumption taxes, and in particular, a subgroup of these called excise taxes (Pechman, 1987, 191). Pechman defines an excise tax as a tax "on the sale of a particular commodity or group of commodities," usually a function of the rate of use or cost of the product. While at the turn of the century this tax accounted for the majority of federal tax revenues, it has declined in importance relative to other forms of taxation imposed by the federal and state governments (Break, 1980, 10).[29] Pechman observes that in spite of the wide variety of taxes included in this category, this form of taxation is relatively uncommon in the United States. In 1985, excise taxes accounted for "14 percent of total federal, state and local tax revenues" (Pechman, 1987, 190).

Innovative Forms of Taxation Generated by Chemical Risk Concerns

Taxation can be effective as an incentive system to reduce risks or the adverse consequences of such risks being realized. Toward this end, innovative forms of taxation have been developed for potential application to chemical risks.

Taxes that have been and are being used either directly or indirectly for the purpose of reducing either the sources or effects of health and safety risks assume a number of different forms.[30] The most common types are (1) a tax on raw materials that are or are likely to become hazardous, (2) a tax on end products that can produce hazards during manufacture or use, and (3) a tax on process or waste treatment residuals, such as discharges of hazardous materials into the air or water.[31] Some examples of each of these taxes are discussed below.

Taxes on Raw Materials

The Comprehensive Environmental Response, Compensation, and Liability Act set up a Hazardous Substance Response Trust Fund of $1.6 billion to finance the cleanup of hazardous substances in the environment. The fund is partly financed by a "feedstock tax" on 42 substances identified under Section 221 of CERCLA. The primary purpose of the tax is to provide resources for hazardous waste site cleanups and the reduction of health risks after the fact, rather than to encourage risk-aversive behavior. This tax has been collected since June of 1981, and CERCLA contained a provision for renewal of the tax after 1985. SARA reauthorized this system and kept the feedstock tax intact. A feedstock tax is a tax on raw materials and crude oil petroleum products used as inputs into industrial production processes. A feedstock tax is defined as a tax that "is imposed at the beginning of the production process on the raw materials used to make the chemical products associated with hazardous substance generation" (U.S. EPA, December 1984a, 11). The tax is applied to both domestic and imported feedstocks. A major issue that EPA's Report to Congress addressed is the potential negative impact that the tax might have on the chemical industry. The EPA has justified the use of this tax economically based on its estimate that these taxes comprise less than 1% of the price of 32 of the chemicals; in only one case does the percentage exceed 3%. Some of EPA's conclusions in its most recent analysis of the impact of the tax on foreign trade are that: (1) while the dollar amount of the surplus that the U.S. maintains of the international trade in chemicals has been falling since 1980, in 1983 the U.S. share of chemical exports was higher (17%) than it had been in 10 years; (2) there are many other factors that overwhelm the impact of such a tax, such as crude oil price decontrols in the U.S., worldwide recession, exchange rate changes, and increases in foreign chemical production capacity. The tax's small negative impact underscores its purpose as an after-the-fact mechanism to obtain resources rather than to influence behavior. In fact, if firms using potentially hazardous raw materials did decide to change to raw materials not on the list, this would have an adverse effect on the ability of the Superfund to raise the funds required for cleanup.

Another form of a raw material tax is a tax on additives that may cause adverse health effects. An example of such a tax was the lead additive tax that New York City imposed on gasoline (one cent per gallon) in the early 1970s (Anderson et al., 1977, 58). In general, unless they are set at a high enough level to guide behavior in the direction of less risky materials, these taxes are somewhat far removed from risk reduction.

Taxes on End Products or Workplace Conditions

Taxes can also be placed on end products, ostensibly to reduce the risks associated with the production process for or use of the product. These taxes come in a wide variety of forms, aimed at working conditions, the health of consumers, and the impact of product disposal. Like taxes on raw materials, while they may provide the funds to ameliorate the immediate impacts of a hazardous condition, they may not necessarily have the effect of reducing the source of the risk.

Impact on worker health. One form of such a tax provides payments or other compensation for injuries to workers in the course of the production of a product, whether or not the product is hazardous. An example is the excise tax imposed on sales of coal to provide monies for the Black Lung Disability Trust Fund. The fund was set up under the Black Lung Benefits Reform Act of 1977. Affected miners receive payments from the fund, enabling them to purchase medical services and disability benefits (Baram, 1982, 87). Another example is the Japan Law for the Compensation of Pollution-Related Health Damage, passed in 1973. This broad-based law gives people injured from pollution financial payments for death- and injury-related expenses (but not pain, suffering, or property damage). The monies are supplied by charges levied on pollutant dischargers. They are distributed on the basis of whether the person exhibits disease symptoms associated with the discharges and proof of proximity to the discharges (Anderson et al., 1977, 49). In the mid-1970s, the concept of an "injury tax" was suggested as a means of implementing the objectives of OSHA. It was argued that such a tax would circumvent the need to inspect workplaces, since the tax would be tied to self-reporting or to workers' compensation claims (Nichols and Zeckhauser, 1977, 65). Along these same lines, another form of tax, the "exposure tax," has been suggested, and would be applied where emissions exceeded standards (Nichols and Zeckhauser, 1977, 67).

Impact on consumer health. Examples of taxes on commodities that are associated with negative health effects on consumers of a product rather than workers are cigarette taxes (also tobacco taxes) and liquor taxes. While these taxes are primarily invoked for the purpose of raising revenues, they could theoretically be used to alter consumption of the product. This assumes, however, a sensitivity of consumption to price, which may not in fact hold.

Impact of product performance on consumer health. Another example of a product tax is one based on product performance and the effect of that performance on human health. This tax is paid by manufacturers rather than consumers. One example of such a tax is the fuel economy tax imposed on automobiles, which came into effect in 1978. Manufacturers of automobiles pay a higher tax as the mileage per gallon of the automobile decreases (Pechman, 1987, 192). While its purpose is to promote energy conservation, in reducing the amount of fuel consumed, it presumably reduces air emissions as well.

Impact of product disposal. A disposal tax incorporates the cost of disposal of the product directly into its price. One type of tax has as its objective the provision of funds for the safe disposal of the product (assuming that arrangements are made to dispose of it properly). An example of this type of tax is the inclusion of disposal costs in the price of automobiles in Scandinavia at the time of purchase. Where automobile disposal is controlled, these funds can be channeled into disposal operations. Another type of tax attempts to provide a monetary incentive for safe disposal by directly linking it to a refund. One common form that has been tried and is currently in use in various ways throughout the country is the bottle tax. That tax is closely tied to recycling. The use of a disposal tax was recently advocated in connection with the disposal of nondegradable wastes, acting very much like a deposit system to encourage recycling (Windsor, 1988). Disposal taxes and deposit systems were explored extensively in connection with the disposal of bottles.[32]

Taxes on Residuals

Examples of residuals taxes include taxes on hazardous residuals regardless of how or whether they leave a site, effluent taxes for wastewater discharges, and emission taxes for air pollution discharges. These taxes can be based on the absolute amount of the discharge or on some net or marginal amount above some standard.

Hazardous residuals. Under CERCLA, one alternative to taxing chemical raw materials to provide funds for hazardous waste cleanup is to tax residuals via a "waste-end tax." A waste-end tax is a tax that is "imposed late in the production process on the generation, transportation, treatment, storage or disposal of hazardous wastes. A waste-end tax may be collected from either generators or managers of treatment, storage, and disposal facilities" (U.S. EPA, December 1984b, 11). It is usually based on the type and quantity of material generated.

Effluent charges for wastewater discharges. Effluent charges have been discussed for a long time as a means of achieving water quality goals. Kneese and Schultze point to the following advantages of effluent charges (which are "levied on each unit of pollution discharged into the watercourse"):

> First, the imposition of effluent charges encourages a pattern of waste management among different firms and municipalities that tends to minimize the costs of control for the river basin as a whole. Second, charges are more likely to be enforceable than is the chief alternative, the setting of effluent limits on individual firms and municipalities by a regulatory agency. Third, charges provide a continuing incentive to adopt improved technology as it comes along. (Kneese and Schultze, 1975, 87)

Furthermore, Kneese and Schultze point out that effluent charges allow polluters the flexibility of reducing risks in different parts of the production process. In that way, costs can be minimized at the level of the individual plant. Charges also provide a continuing incentive for firms to reduce risks below a standard, since effluent charges are paid on the basis of each unit of pollution.

The effluent charge has been analyzed for application in the Delaware River Basin and other river basins in the United States and Europe. In addition, an effluent charge system was conceptualized for the Reserve Mining Corporation discharges of taconite mine tailings at its Silver Bay plant on Lake Superior. Effluent charges were computed as a function of social cost of the discharge and annual discharges (Peterson, 1977). A number of such charges have been implemented in the form of fees for a service to remove the pollutants from the effluent based on the pollutants' strength.

Emission taxes. Several forms of emission taxes have been developed over the years as alternatives to the direct regulation of air emissions. Kneese and Schultze (1975) discuss a tax on sulfur oxides to encourage the use of low-sulfur fuels and a smog tax on automobiles to encourage reductions in automobile emissions through measures such as better automobile maintenance and the installation of pollution control devices.

Tax Write-offs, Exemptions, and Deferrals

Another set of financial incentives involving tax structures used as an alternative to regulation consists of income tax write-offs and deferrals, tax exemption of bonds, and sales tax exemptions for the financing of risk reduction techniques. This mechanism is different from those discussed above, since it reduces tax payments rather than requiring them. It has been most commonly applied to water pollution control facilities. Historically, many states have had provisions for sales tax exemptions on the purchase of pollution control equipment. Kneese and Schultze pointed out that the value of exemptions for industrial pollution control equipment stood at $1.8 billion in 1973, and exclusive of tax-exempt bank loans, was likely to be $2.9 billion by 1974 (Kneese and Schultze, 1975, 93). The Tax Reform Act of 1969 provided for accelerated depreciation of industrial waste treatment plant additions (Kneese and Schultze, 1975, 33). Kneese and Schultze argue that this mechanism introduces biases toward certain technological solutions to the pollution control problem, unlike taxation, which can allow more flexibility (Kneese and Schultze, 1975, 93).

Use of Taxation for Risk Management

The excise tax mechanism is traditional to government. Taxes such as those described above have been around in the United States practically since the country was founded. Pechman (1987) notes the existence of late 18th-century liquor and tobacco taxes resembling modern pollution and health taxes. Thus, the tradition of excise taxes in the country theoretically provided a framework for the emergence of today's specialized taxes, and for incentive-based systems for risk management and the discretion to invoke them. Taxes aimed at managing risks from chemicals now dominate the excise tax category.

Political institutions impose constraints on the implementation of these taxes. In spite of the interest in and conceptualization of taxes for managing environmental health risks, frequently they have not been implemented. Breyer (1982) points out that at the federal level a shift toward the use of taxation can be associated with a shift in the relative control of Congressional committees. Where taxation is used as a means of controlling risk, the control over an activity usually shifts from the committees that deal with substantive issues, such as energy, transportation, and environmental committees, to the finance and taxation committees. The finance committees tend to be less familiar with the substantive aspects of the risks than the more specialized committees, so that the tax mechanism imposed by the finance committees may not always correspond to the substantive needs of risk management.

Using the example of the emission fee, DeMuth succinctly underscores some of the political complexities at the federal level involved in implementing such a tax.

> Policies in the form of taxes and expenditures involve different political institutions than policies in the form of standards. An emission fee in the form of a flat national tax would have to be initiated by the House of Representatives and would have no success without the support of the Treasury Department. Such a tax would fall under the jurisdiction of several congressional committees, composed of representatives with interests and constituencies different from those which influenced the writing and administration of the present environmental laws (DeMuth, 1983, 269).

Structural requirements for taxation are not easily fulfilled by chemical risks. There are a number of structural requirements that a tax should meet if it is to be used as an incentive system applicable to the control of chemical risks. Some of the major requirements pertain to how rates are set and how the funds generated are used.

The following rules generally dictate how the levels of a tax or the tax rate needs to be set:[33]

- Tax levels should be set such that the more desired behavior (such as eliminating a hazard from a service or product) is less costly to pursue than the less desired behavior (retaining the hazards). This assumes that the taxpayer is interested in minimizing costs. Tax levels should reflect meaningful consumer choices and be cognizant of the fact that the consumer is not only weighing the relative risks of similar activities, but may be weighing other factors. For chemical risks, the backing for this kind of behavioral response is not well established.

- Taxes can be set at a level such that the marginal profits obtained from allowing the risks to continue unabated are redistributed to the consumers of the service or product.[34]
- Alternatively, taxes can be set at a level equal to the cost of risk reduction being accomplished by the government or some entity other than the generator of the hazard. (These are typically in the form of fees to recover costs of remediation.) In such a case, the tax rate should reflect the revenues needed to accomplish risk reduction.
- Tax levels should be set to reflect social costs as well as economic costs of risk reduction. A disadvantage often cited of certain forms of taxation is that social equity problems can arise due to the regressive nature of such taxes.[35]
- Tax levels need to be designed in a way such that they actually influence those producing risk to reduce it.
- The degree of variability in the tax rate should be understood, and should be a function of the risks to be reduced. The effects that such variability can have on the relative competitive advantage of products and services should be factored into the design of the tax as well. If the tax rate varies from a flat rate, variation can be along many lines. For example, in the area of chemical risks, the tax rate should be some function of the amount of chemical used and its toxicity. Breyer (1982, 278) gives an example of such a relationship in the Ruhr Valley, where charges per pound of pollutant discharged into the waterway are a function of the toxicity to fish.

Where the taxes are used to provide funds to reduce the consequences of a risk once it is realized rather than being used to encourage risk-aversive behavior, control should be exercised over the use of the tax funds generated. These tax funds should be directly funneled into achieving risk reduction goals rather than diverted to other functions unrelated to the risks. Where the purpose of taxation is to encourage risk-aversive behavior, however, control of the use of funds is not significant. Spill compensation funds that derive their revenues from taxing polluters are examples of funds where the monies are targeted to the purpose for which the taxes were collected in the first place. The targeting of funds in this way has not gone unchallenged; contributors or taxpayers often challenge the use of the monies.

MARKETABLE RIGHTS

Introduction: Marketable Rights Defined

A type of market-based incentive system for the regulation of environmental quality is the marketable rights concept. Marketable rights work in the following way.[36] A fixed amount of a particular right or activity is defined for an area in terms of risks to health or safety. An example of such a right is the amount of allowable pollutants that can be emitted into the air or into a waterway. This amount is divided into shares that are then bought and sold on the market. If the mechanism works properly, those activities that have to pay the most to avoid risks will buy up the rights. Those activities that have to pay little to avoid risks (including those whose activities are well in compliance with standards) will be sellers of those rights.

Thus, the components of such a system and the conditions for its implementation are (1) a geographically defined impact area, (2) a definable commodity or risk that is to be marketed, (3) a clear relationship between what is marketed and its impact on the geographic area, (4) the ability to translate the marketable components into shares, and (5) the existence of a market for the shares.

The marketable rights mechanism differs from taxation in a number of ways. Because of these differences it can be relatively more difficult for regulators to control the impacts of the mechanism. First, the marketplace determines the prices that are paid for the rights; thus, the prices vary according to the willingness of the firms to pay for those rights. Taxes as applied to health and safety, on the other hand, are often estimated on the basis of the cost to provide a service that reduces risk. Second, the market also determines the distribution of the rights within the geographic area for which the fixed amount of the right is defined. Third, unlike taxes, it is difficult to tell when the prices paid for the rights will be less than the costs of introducing risk reduction measures (Breyer, 1982, 172). Fourth, marketable rights are usually envisioned as being voluntary, whereas taxation is not.

Applications of Marketable Rights to Environmental Risks

The concept of marketable rights gained popularity during the 1970s and 1980s as an alternative to the command-and-control or regulatory approach to environmental management. A resurgence of interest in marketable rights occurred at the end of 1988 with the preparation of a bipartisan report, "Project 88" (U.S. Congress, December 1988), which recommended the use of a number of economic incentive systems as part of revisions to the clean air legislation. While its application and the debates surrounding its use have primarily centered on air quality, the marketable rights doctrine has been used in other environmental areas as well. Its extension into the area of environmental health risks (to the extent that they don't overlap with environmental risks in general) will depend on the extent to which the conditions for its use can be met.

As was the case with taxation, a wide variety of mechanisms are possible under the marketable rights system. Many new and innovative mechanisms have been generated by chemical risk problems. These are described below, along with their conditions of use and some examples.

Transfers of Development Rights (TDRs)

One of the first applications of the marketable rights concept was a general land use control mechanism called transfer of development rights. The TDR concept assumes that land development rights are subdividable into a series of rights that can be traded from one geographic area to another. Associated with the TDR concept is the "development rights bank." This enables land development rights to be stored (for example, for a piece of condemned land) until some future time when they are needed. TDRs are used to manage environmental risks indirectly by managing the location, type, and intensity of development. Developments in the use of TDRs should be watched carefully, in that they are likely to set precedents for the design and use of marketable rights in the environmental risk area.

Rose (1975, 6–7) points out that the roots of the TDR concept date back to colonial times. Recent precedents have been set in both Great Britain and the United States in the acquisition of land by condemnation or purchase in order to buy the development rights to the land for some future time or restrict future development. While this mechanism had wide applicability in urban planning and real estate development, it was also given considerable attention as a means for growth control and environmental planning (to encourage development to avoid environmentally sensitive areas) in the 1970s (Costonis and DeVoy, 1975). For example, the TDR concept has been embodied in some wetland and floodplain regulations: limited development may be allowed in a wetland or a floodplain if the developer promises to reconstitute the amount of the wetland or floodplain lost in some other area. The direct application of the TDR concept to environmental health and safety risks depends on the extent to which these health and safety risks can be related to development restrictions. In air and water pollution control these relationships have been established.

Air Emissions Trading: Bubbles, Nets, Offsets, and Banks

In air pollution control, the marketable rights concept has been applied to the trading of emissions within and among sources of air pollution. This concept is described in the emissions trading policy issued the U.S. EPA under the Clean Air Act (U.S. EPA, 1982a; 1986a). Emissions trading is primarily a voluntary policy that embodies four concepts: bubbles, netting, emission offsets, and emission reduction banking. As applied to stationary sources of pollution, the bubble policy allows the trading of emissions within an existing plant or among groups of existing plants as long as the total amount of emissions allowed remains unchanged (40 CFR 51.18, Appendix S). The first application of the bubble policy to an air emission permit was the permit granted to Naragansett Electric Company for two of its plants (Diemer and Eheart, 1988, 1005). In the area of stationary sources of air pollution, 40 bubbles were processed for volatile organic chemicals alone between December 1986 and

March 1989 (Creekmore, 1989). The bubble concept is also applied to mobile source emissions. One example of this is the bubble program for lead in gasoline, which lasted from 1985 to 1987. After 1987, the allowable lead level was only 0.1 g/gallon, a level considered too low for trading. Thousands of trades occurred under that program, and the price, estimated at about 0.2 cents/g when the program began, rose to 5.0 cents/g by the program's end (J. W. Caldwell, U.S. EPA, personal communication, 1989).

Netting refers to the trading of emissions from one part of a plant to another where a plant expansion or modification is involved, as long as the change in total emissions allowed from the resulting plant is insignificant. "Emission offsets" refers to the trading of emissions to allow a new source of air emissions to be constructed in an area, with the restriction that total emissions in the area remain within allowable limits. It does not allow emission trading among existing emission sources, however. Emission reduction banking is related to the other three policies, in that it allows emission reduction credits to be stored for use at another time. All of these concepts involve marketing the right to emit air pollutants as long as National Ambient Air Quality Standards are maintained. The units of emissions that are marketed are called "emission reduction credits."

An administrative mechanism that has been proposed to implement these trading policies is the transferable discharge permit (TDP). This approach allows discharge allotments under permit conditions to be traded among firms (Diemer and Eheart, 1988).

Wasteload Allocations for Water Pollution Control

In water pollution control, a concept analogous to a marketable right is used to achieve water quality standards under certain conditions. Under Section 303(d) of the Clean Water Act, states must calculate waste allocations for dischargers when a waterway will not meet water quality standards using the normal limitations on effluent discharges. Allowable concentrations of water pollutants for a given stream segment are translated into a total allowable pollution loading (called "Total Maximum Daily Load"). This loading is usually expressed as a weight of pollutants discharged over a given time period (for example, in pounds per day). This total loading is then allocated in portions to each of the activities (factories, residential buildings, institutional facilities, runoff over the land) discharging that type of pollution in the segment, leaving allowances for a margin of safety and seasonal variations. The allocations to "point sources" or individual buildings and discharge points are called individual "wasteload allocations" (WLAs).

The wasteload allocation process is similar to the marketable rights concept in that it establishes allocations that could be marketed or traded along a stream segment. The main difference is that the allocations are performed by government regulators (usually at the state level) rather than being developed and traded through the market system. As of the early 1980s, the EPA reported that practically all of the states had wasteload allocation programs for at least conventional pollutants (U.S. EPA, February 1984, 6). The transformation of this system into a marketable system of pollutant discharge rights was examined shortly after the passage of the Clean Water Act Amendments of 1972, which developed the system of wasteload allocations under an areawide water quality planning program. deLucia evaluated the effectiveness of a "Marketable Effluent Permit System" in achieving Clean Water Act goals. The system consisted of an initial allocation of discharges and permits by the government, after which point the holders of the permits could buy and sell them to obtain discharge rights (deLucia, 1974).

The bubble concept has been applied to a limited degree to water pollution control. Ingram and Mann (1984, 257) cite the 1982 wastewater discharge regulations issued in July 1982 for the iron and steel industry as an application of the intraplant bubble concept. There have also been some trades among plants to meet effluent limits as well as trades between point and nonpoint sources of water pollution (Levin, 1989, 128).

Factors Affecting Use of Marketable Rights

The marketable rights approach to managing risks has a number of advantages.[37] It is flexible in allowing a number of different routes by which a risk level can be attained. Its attractiveness lies in the fact that it reduces the overall costs to those engaged in activities involving environmental health risks while still attaining prescribed risk levels. There are, however, a number of conditions for its use, arising from the context of government, that have to be kept in mind for implementation and that can limit its usefulness. These factors are similar to those for taxation and include designing and costing the marketable rights.

Design and Costing Constraints

As was the case with taxation, the innovative forms of marketable rights have not been used extensively because they generally do not meet many of the criteria for use. From the experiences of previous marketable rights programs, it appears that in order for a marketable rights system to be an effective risk management tool it should consider the following factors:

- A demonstration needs to be made of the existence of cost incentives over conventional pollution reduction techniques. A number of arguments have been developed in favor of the marketable rights concept (McGartland, 1988, 35), but many of them assume conditions of perfect competition.
- The nature of the marketable components and their sources should be well-defined (for example, which pollutants or safety systems are marketable and

where they originate). A prerequisite to this is that the impacts of these pollutants should be clearly defined. It has been pointed out that where pollutants, their sources, and impacts are not well known, as is the case with "nonpoint source" water pollutants (e.g., pollutants running off land masses from a variety of unidentifiable sources), the marketable rights approach is difficult if not impossible to apply (Segerson, 1988).

- It should be clear whether whole permits or licenses are marketable or just certain conditions or limitations in these legal documents.
- The geographic scope of the market transfers should be clear, such as (1) defining a particular watershed or an airshed and (2) ensuring that geographic variability does not produce unintended adverse consequences.
- The time periods over which trades are designed to occur have varied widely. For example, trades for VOCs have occurred over a 24-hour period for cross-line averaging over similar operations (Creekmore, 1989, 5), while trades for lead in gasoline occur over a calendar quarter (J. W. Caldwell, U.S. EPA, personal communication, 1989). Banking can extend time periods still further. There are indications that trades over long periods of time can result in violations of standards (Creekmore, 1989, 6). If this is the case, then more policy guidelines may be needed in this area based on impact analyses.
- The method of trading and who is involved in the trading greatly influences the cost; for this reason, a cost analysis of alternative methods should be performed prior to the design of any given system. Transaction costs tend to rise as the number of markets for offsets goes up. Offsets traded in a market open to many different sources of emissions rather than on trading within a single industry require much more information about participants and prices; hence, transaction costs may be greater (McGartland, 1988, 37). Costs go up where participants have different dispersion characteristics and affect different receptors. Costs also vary according to whether the trading is done at one time or over different time periods (McGartland, 1988, 39–41).
- The role of regulatory agencies in overseeing the transfers should be defined, and the design of permits and permit conditions should be consistent with the transfer options.
- Provisions should be developed to avoid the formation of monopolies, such as limitations on the distribution of shares. This is needed in order to maintain a competitive market, which is considered necessary for rights to be distributed optimally and for bidding to be as truthful as possible (Diemer and Eheart, 1988, 1004).
- An administrative and monitoring mechanism should be included to ensure the continued effectiveness of the system. The administrative system should ensure the financial viability of the holders of marketable rights. Administrative arrangements should also ensure the ability of market shareholders to abide by the conditions of these rights and to adapt to changes in conditions over time that might force changes in the level or distribution of the rights. The means of financing the administrative costs of the system should be incorporated in the cost of the rights transfers.
- Where implementation plans exist for meeting ambient environmental quality levels, the plans must be adaptable to a marketable rights approach. For example, State Implementation Plans (SIPs) under the Clean Air Act are the basis

for determining the suitability of implementation approaches for National Ambient Air Quality Standards. These plans have incorporated a "Uniform SIP Rollback" strategy for air pollutants, which would not be consistent with a TDP concept. A comparative cost analysis for the two alternative strategies for acid deposition control found that the TDP approach would increase costs by less than 2%, compared to a 3% cost increase for the uniform SIP rollback approach. SIP revisions would be required anyway to implement the TDP approach, since the SIPs are not written generically to allow the most cost-effective pollution control approach (Diemer and Eheart, 1988, 1004). Areawide water quality management plans prepared under the Clean Water Act are written more generically, and conceivably would allow a permit transfer mechanism to implement the wasteload allocations. The relationship between trades and planning provisions should be spelled out in the plans.

- The design and relative cost-effectiveness of a marketable rights system is extremely sensitive to environmental regulations. First, systems will vary depending on whether they are forbidden to degrade existing environmental quality (the "nondegradation" requirements under the Clean Air Act and Clean Water Act). McGartland (1988, 35) compares the effect of alternative assumptions about degradation on outcomes of marketable discharge permit systems. Second, they will vary according to whether there are restrictions on participants (new vs existing dischargers, discharges from different sources, etc.), the number of substances involved in a trade, and the geographic area involved.

Administrative Requirements

The administration of marketable rights can be quite complex, especially since their distribution may have to be controlled across different political jurisdictions, and the formation of monopolies has to be avoided. Associated with administrative complexity is monitoring complexity because of the flexibility and uniqueness of the rights in each case. The variations in the arrangements from source to source in the case of wastewater discharges or stack to stack in the case of air emissions require close monitoring to ensure that the tradeoffs are being carried out according with agreements made.

The federal government plays a key role in the administration or overseeing of marketable rights, though actual transfers are designed by private brokers. A clear regulatory policy set by the government with respect to these rights is an important prerequisite to any administrative arrangement. While in the air quality area the EPA policy on trading has been finalized, some claim that the backlog in processing applications even after the policy was issued (applications were first processed in 1988) is indicative of continued uncertainty on the part of the government with regard to implementing the policy (Levin, 1989). Moreover, there is a question of who actually administers the rights and coordinates the tradeoffs. While this could conceivably occur on a contractual basis among individual firms, banking necessitates the need for a more permanent coordinative entity. The EPA, by virtue of its role in approving the calculation of total rights and the permits for redistributing those rights, indirectly administers the trading.

Equity Considerations

Using the marketable rights approach requires consideration of the very complex question of equity in distributing the rights among activities. Firms have challenged the allocation of pollution rights where two similar firms in different locations were allocated different pollutant discharge limits, and could continue to do so under the marketable rights approach.

Another aspect of the equity question that occasionally arises has to do with who gets the credits. In the case of chlorofluorocarbons (CFCs), for example, disputes arise over which importer along a complex chain of transfers from initial manufacturer through various distributors and brokers would receive the credits. The decision made in the case of CFCs was that the importer of record who certified that the product met U.S. standards usually would obtain the credits (M. Gibbs, ICF, Inc., personal communication, 1989).

Technical Requirements

The marketable rights mechanisms that have been adapted for air and water pollution control are directly applicable to environmental health risks that occur from such pollution. Both systems, to the extent that they are used at all, currently rely heavily on ambient environmental standards as a framework. These ambient water and air standards are based on health risks, among other things. In order to safeguard environmental health completely through the use of marketable rights, however, many more chemicals would have to be specified in the ambient standards. This could be done easily by relying on the full array of chemicals specified in discharge permits (New Source Performance Standards [NSPS] and National Emission Standards for Hazardous Air Pollutants [NESHAPs] under the Clean Air Act and National Pollutant Discharge Elimination System [NPDES] permit conditions under the Clean Water Act) or on health-based ambient standards that are being developed by many states.

In addition to standards, a key requirement of the trading process is the development of a baseline amount of a chemical from which credits can be calculated. This was identified as the major problem area experienced in the bubbles processed for volatile organic chemicals (Creekmore, 1989, 7).

MISCELLANEOUS FINANCIAL RESPONSIBILITY REQUIREMENTS

Government agencies have developed a number of mechanisms to recover the costs associated with preventing risks from being realized or lessening the impact of their realization. The mechanisms are often used in conjunction with direct regulation as well as indirectly as an incentive system by providing positive or negative reinforcement.

Financial responsibility or financial assurance requirements can take a number of different forms: contributions to special funds, such as trust funds, restoration funds, and compensatory funds; contributions to escrow accounts; letters of credit; surety and performance bonds; or tests of financial strength, including statements of assets, insurance, or revenue tests for municipalities.

None of these mechanisms is new. They have been used for years by private investors and government to ensure adequate performance by industry. The government appears to have largely relied on conventional financial mechanisms in designing both direct regulatory programs and incentive systems to encourage and ensure risk reduction in the future. As a result of governmental requirements, the formal use of and reference to both funds and tests of financial viability (temporary and long-term) have experienced a precipitous increase with the rise of environmental health risk concerns. Monitoring, enforcement, and administrative considerations will continue to influence the viability of these mechanisms.

Special Funds

Trust funds are a general means of setting aside money for a particular future purpose. Restoration funds and spill compensation funds are two examples of types of trust funds used in risk management. Baram defines the restoration fund in the following way:

> A restoration fund is a trust fund created to ensure that funds will be available to rectify problems caused by a specified occurrence. Upon the happening of the specified occurrence, losses incurred by the obligee are compensated automatically by proceeds from the restoration fund. (Baram, 1982, 96)

The Hazardous Substance Response Fund under CERCLA is an example of a restoration fund. The chemical feedstock tax discussed above finances 86% of the fund. The rest is financed by cost recovery from government cleanup of privately owned sites, fines, general revenues, and other sources (U.S. EPA, December 1984b, 1-1).

Many states now have spill compensation funds maintained by contributions or mandatory payments from spill-prone industries. The New Jersey Oil Spill Compensation fund was set up in 1977 under the Spill Compensation and Control Act. The fund is financed by a tax levied on oil companies and other major facilities. Oil companies are taxed at $0.01 per barrel, but if outstanding claims against the fund exist, the tax can increase to $0.04 per barrel (New Jersey Statutes Annotated, 58:20–23.11.).

Under the Trans-Alaska Pipeline Act of 1973, a fund financed by industry was set up requiring that in the event of an oil spill, shippers are to pay the first

$14 million in damages, after which the fund would pay $86 million on a no-fault basis, with a cap of $100 million on cleanup costs (Wald, 1989).

Trust funds usually have the following requirements or specifications:

- methods for estimating the amount of money that should be placed into the fund, and procedures for updating cost estimates over time if conditions change. As applied to hazardous waste treatment facilities, the cost estimates pertain to implementing facility closure and postclosure procedures.
- a specified "pay-in" period (when the first payments and total payments should be made into the fund). Arguments have been made in favor of early pay-in periods on the basis that financial responsibility may weaken over time.
- method of making payments to the fund and providing assurances of such payments, such as taxation, surety bonds, letters of credit, and insurance
- specification of how funds are managed (e.g., whether funds can be invested) and who manages the fund
- specification of how priorities are established for fund pay-outs, especially in cases where there are competing claims that could exceed the amount in the fund

Escrow Payments

Baram summarizes the current concept of an escrow in the following way:

> An escrow is a written instrument that . . . imposes a legal obligation and is deposited together with a specified amount of funds with a non-party to the contract (or other business transaction) in question. The written instrument and funds so deposited are held by the third party until the full performance of an agreed-upon condition or the happening of a specified event, such as an accident. Upon fulfillment of the condition or happening of the event, the deposited instrument and funds are given to the person indemnified. (Baram, 1982, 96)

Bonds/Letters of Credit

Baram has defined a "bond" in the context of risk management to mean a performance bond "which guarantees the performance of a contract." More specifically:

> A performance bond is an obligation in writing binding the signatories to pay a certain sum upon the happening of a particular event, such as a failure to fulfill the terms of a contract. It usually consists of the obligation or promise to pay a specified sum to a person named and at a stated time; the condition(s) for payment or occurrence(s) necessary for payment, if any; and the signed commitment by the obligor to honor the bond agreement. (Baram, 1982, 95)

Performance bonds are often required prior to construction activities even under nonhazardous circumstances, simply to ensure that the contracted work gets done. An early example of the use of performance bonds to avoid environmental damages occurred in the area of strip mining. Firms engaged in strip mining were required under federal strip mining legislation to restore the land after the mining was completed (Bardach and Kagan, 1982c, 246). In certain court cases involving hazardous waste sites, bonds have been required to be posted to prevent financial insolvency.[38]

Letters of credit are another device used to ensure that the source of a potential hazard will provide resources to avert the risks from that hazard. One example of its use was in the cleanup of dioxin in Newark, New Jersey from an abandoned Agent Orange manufacturing plant. The State of New Jersey required the company to file a $16 million letter of credit to cover some of the cleanup costs.

Tests of Financial Strength/Solvency

Some requirements for financial soundness are found directly in legislation pertaining to the control of hazardous activities. RCRA has a provision that requires owner/operators of treatment, storage, and disposal facilities to show a certain amount of insurance coverage per site. For owners or operators of TSDFs, this is $1 million per occurrence for sudden and accidental pollution incidents and an annual aggregate of at least $2 million. For surface impoundments, landfills, and land treatment facilities, an additional coverage for gradual pollution incidents is required of $3 million per occurrence and $6 million annual aggregate. In lieu of insurance coverage, assets can be used to demonstrate financial solvency.

Some transportation acts require that owners and operators of motor vehicles be at least partially financially responsible for accidents. The Motor Carrier Safety Act of 1980 requires minimum levels of financial responsibility for trucks. The Bus Regulatory Reform Act of 1982 includes minimum levels of financial responsibility for motor carriers (Congressional Quarterly, Inc., 1982).

In the case of corporations, financial strength is usually defined in terms of corporate assets or insurance. In the area of hazardous wastes, generators of hazardous wastes and/or storage and treatment facilities are required to carry several million dollars of insurance or to have assets covering that amount.

GRANTS AND LOANS

The federal government funds state and local governments and private industry on a regular basis to administer or implement programs aimed at reducing environmental health risks. Government agencies exercise control over activities through grants and loans in a variety of ways, and these methods have to some extent been adapted to environmental health risks. First, general formulas for the allocation of grant and loan monies are based on criteria that usually relate to need (and could also relate to health and safety). Second, conditions relating to the behavior or performance of the activity are imposed as a prerequisite for obtaining a particular grant. Third, in the course of monitoring and auditing the allocation or use of grant funds, the federal government can revoke or reduce funds where violations of the conditions or ineffectiveness in meeting the conditions occur. These grants can and often do function as a management tool to control environmental health risks by introducing risk reduction objectives into the criteria for allocation formulas, placing these objectives as conditions for an individual entity receiving the grant, and exercising the power to revoke or reduce funds if conditions are not met.

The ongoing federal grant and loan programs relevant to chemical risks exist primarily in the areas of infrastructure planning and construction monies; overall resource and environmental protection planning activities; and research, demonstration, and monitoring and surveillance. Federal grant allocations (as distinct from the monies agencies use directly for projects that benefit local government) go primarily to government entities; grants go directly to facility owners and operators only on a limited basis. Categorical and formula grants predominate, which tends to limit the discretion of grant recipients. Grant conditions are quite commonly used to control chemical risks, primarily for wastewater treatment plants.

Who Recipients Are: Facility Owners or Government

Grants can be categorized according to the type of recipient. In addition, Break (1980, 74) offers the categories of how the recipient uses funds, method of allocation of funds to recipients, and how much participation the grantor exercises.

Two major categories of federal funds that have functioned as alternatives to regulating risk are (1) grants directly to the owners and operators of facilities that are the source of risks or (2) grants to state and local governments to operate management programs that reduce risks ("grants-in-aid"). Grants-in-aid have been defined as "money payments furnished by a higher to a lower level of government to be used for specified purposes, and subject to conditions spelled out in law or administrative regulation" (Reagan and Sanzone, 1981, 54, cited in Gordon, 1986, 161). Grants directly to facility owners/operators are primarily for facility design, construction, and operation. They are not commonly used. Grants-in-aid are typically for program development and administration, except where governmental agencies also assume the function of facility operation and maintenance. They have been very popular, but declines in overall funding have limited their usefulness. One of the largest grant programs aimed at controlling environmental risks is for the construction of wastewater treatment plants under Section 106 of the federal Clean Water Act. Since 1972, over $25 billion has been awarded under the program.

The effectiveness of the granting mechanisms has from time to time been compromised by federal governmental action with regard to federal funds. For example, Nixon impounded Clean Water Act monies in the early 1970s (Schick, 1980, 45–6). Allocation of Superfund and Clean Water Bill monies was delayed during the 1980s under the Reagan administration. In the early 1980s, the federal government, after delegating considerable responsibility to the states to run environmental programs, cut their funds significantly. According to Davies (1984, 150), "the Congressional Budget Office reported that, in constant 1982 dollars, state water grants declined 53 percent between 1981 and the 1984 budget request. The comparable cut in air grants was 33 percent. Similar, although somewhat less drastic, cuts were made in other assistance programs for state environmental functions."

Some laws in the 1980s introduced new state and local grant programs, e.g., in water supply. In the area of water supply, past federal government investments have concentrated on planning, the construction of major storage systems, and (to a lesser extent) sporadic grants for small water systems. General water resources planning is financed on a regionwide and often multi-state basis through special purpose legislation and through portions of programs funded under the Clean Water Act. The construction of water supply storage systems usually occurs in connection with dams and energy, recreation, or flood control projects and is almost entirely under the jurisdiction of the Army Corps of Engineers and the Bureau of Reclamation in the Department of Interior, whose investments have totaled hundreds of millions of dollars over the years. The Corps is funded through special congressional appropriations. Small systems and rural systems had been receiving grants and loans through the Farmers Home Administration (at levels between $400 million and $1 billion since 1926), the Economic Development Administration, the Department of Housing and Urban Development, and the Appalachian Regional Commission (U.S. Congress, Congressional Budget Office, 1983, 127–129). Until recently, none of these monies were specifically or directly aimed at the quality of the supply, prevention of contamination, and reduction of risks that pollutants in drinking water posed. A study of groundwater contamination nationwide in the early 1980s concluded that "currently no Federal program has earmarked funds specifically for the protection of groundwater quality. In addition, funding for programs that have supported groundwater-related activities has been reduced or eliminated" (citing water quality and solid waste planning activities) (U.S. Congress, Office of Technology Assessment, October 1984, Vol. I, 11). The report did point out that groundwater protection can be covered under state planning grants under the Clean Water Act. Since that time, the Wellhead Protection Program under the SDWA of 1986 was developed, providing assistance for the development of plans to prevent contamination of well recharge areas. The amounts allocated to the grants ranged between $20 million and $35 million per year.

Types of Grants

Categorical Grants

Break (1980) categorized grants according to how grant recipients use the funds. The categories include unrestricted grants, general (with limited restrictions) grants, block grants, and categorical grants (by function). According to Gordon, most grants are categorical grants, which usually allow less discretion for the recipient and more discretion for the granting agency. They are usually accompanied by extensive regulations outlining conditions for the use of the money (Gordon, 1986, 162). The government has relied on categorical grants more than any other type.

Formula Grants

Break also categorized grants based on how funds are allocated: by formula (and there are several ways in which formulas can be designed) and by competition among potential recipients for project grants. These two kinds of grants can be subcategories of categorical grants. Grants that use formulas afford less discretion to recipients than project grants.[39]

The tendency in federal financing is to use formula grants rather than project grants. Maass (1983, 69) suggests two reasons for this policy. First, the greater equity likely to be afforded by the formula grants furthers Congress's attempt to be equitable and prevents or limits executive discretion with respect to special interest groups. Second, since formula grants generalize programs, they help Congress avoid "pork barrel" projects and "logrolling"—the focus is shifted to formula allocations rather than projects, though formulas are often justified in terms of projects. The impact of formula grants is more general and widespread. It "universalizes the particular" (Mayhew, 1974).

Use of Grant Conditions for Risk Management

Chemical risks are controlled under the wastewater treatment plant construction grants program through extensive use of grant conditions. First, approval of the siting and design of the facility is contingent on preparation of an environmental impact statement (EIS). In the EIS, the environmental, social, and economic impacts (including health impacts) of the facility are described and analyzed for the proposed facility and its alternatives. The Council on Environmental Quality (CEQ), which is the clearinghouse for EISs, reported that through 1984, EPA had filed hundreds of statements for all of its activities, but the majority are typically for wastewater treatment plants. In 1985, 10 EISs were filed for wastewater treatment plants (Council on Environmental Quality, 1986, 174). In 1986 two were filed, reflecting a temporary gap in funding prior to the passage of the Clean Water Act amendments (Council on Environmental

Quality, 1987, 246). Second, risks are controlled through the grant program by means of imposing standards for the quality of the wastewater that will be discharged into natural waterways from the facility. These standards are usually equivalent to the effluent standards set under the NPDES (the program regulating the discharge once the facility is complete). Presumably, they are developed taking health risks into account.

Grant conditions are also used as a sanction (that is, a grant can be rescinded if risk reduction measures are not implemented). Grants to states under the Federal Highway Safety Act of 1978 and its subsequent amendments are one example of the use of grant conditions as a sanction. Sanctions are aimed at reducing the risks of automobile travel from one source: speeding. Speeding is defined as exceeding a specified speed limit. Receipt of federal highway grants is conditioned on states ensuring that a certain percentage of traffic obeys the speed limit. If a percentage of traffic greater than that specified (currently 50%) exceeds the speed limit, sanctions (loss of up to 10% of highway aid) are imposed (National Research Council, 1984, 151).

Another example of the use of grant conditions as an alternative to regulating risk exists in the area of transit safety. The Urban Mass Transportation Act has a provision to allow the federal government to withhold funding to operators and owners of mass transportation systems where safety hazards are discovered.

Trends in the Use of Grants and Loans for Risk Management

General Popularity

The use of grants and loans as an incentive for reducing environmental health risks has been dominated by the governmental context in which it operates. Gordon (1986, 161) has pointed out in a synopsis of federal grant programs that there has been tremendous growth in the number of federal grant programs, which could spill over into the area of chemical risks. Separate federal grant authorizations increased from 45 in 1961 to 400 in 1969, 500 by 1978, and 580 by mid-1981, followed by a downturn during the Reagan administration. There has also been significant growth in the total dollar size of all programs: from $7 billion in 1960 to $24 billion by 1970, $49.8 billion by 1975, and $91 billion by 1980.

In the three federal grant and loan categories that tend to influence environmental health risks—infrastructure, planning, and research and demonstration—the trends have been somewhat different. Tarr has pointed out that the federal share of infrastructure construction has steadily declined since its peak during World War II (Tarr, 1984, 23). In contrast, planning grants steadily increased during the 1960s and 1970s; however, funding for continued planning diminished considerably during the 1980s, with the exception of a few new programs, such as wellhead protection.

Discretion

Most federal grants are categorical grants. These give a lot of discretion to the administering agency (since they can write the regulations for the program) but little to the governing body receiving the grant. Block grants and General Revenue Sharing grants (begun under the Nixon administration) afforded recipient governments much more discretion. They only accounted for a small portion of total grants. Eventually, General Revenue Sharing was phased out, and only a few block grant programs remain (Gordon, 1986, 161–168). Thus, the use of the categorical grant mechanism for chemical risk management follows a prevailing trend.

CONCLUSIONS

Incentives vary as to whether they emerged specifically for chemical risks or have been around a long time as traditional processes of government. As a general statement, the incentive systems are familiar procedures, but their application to chemical risks has led to the development of innovative extensions of traditional mechanisms. However, these innovations arose largely from the environmental protection movement, not from issues of environmental health risk. Furthermore, while these innovations exist, their use on any large scale has been minimized by court cases and the lack of hard evidence regarding their overall effect on risky activities (Melnick, 1983).

Environmental health risks have ridden the tide of a nationwide movement toward greater number and size of claims for injury. There is reason to believe that claims for medical malpractice and non-health–based product liability were actually the forerunners of similar claims in the risk area. At any rate, these health risk claims at least started out as being no different from those occurring in other sectors. Similarly, the response of the insurance industry to risk appears to have been no different than its response in other areas. The use of other incentive systems such as taxation and grants and loans to reduce risks also drew from a preexisting framework.

Where the environmental health risk area has diverged from other forms of injury is in the degree to which it has extended existing mechanisms. In insurance, application of a variety of mechanisms to overcome limitations, such as caps, risk pooling, and other devices, has been significant in the risk area and has provided an example to other injury areas. In finance, innovative taxation mechanisms have been conceptualized for risk, though the obstacles to implementation are sizable.

In other financial areas, environmental health risks have drawn on traditional mechanisms already in use (taxation, marketable rights, and grants and loans), but their application to the chemical risk area has produced considerable variations in the kinds of mechanisms used. Because of a basic lack of fit between chemical risks and these incentive mechanisms, the absence of clear support for their use, and lack of efficiency testing methods, they have not been used to any great extent.

NOTES

1. The descriptions of common law doctrines and their relative advantages and disadvantages contained in this section are drawn primarily from Baram (1982) and Breyer (1982), except where noted otherwise.
2. Initially, tort law rested on two theories—deterrence and compensation. However, with the growing popularity of a no-fault basis for litigation (see strict liability discussion), it is believed that tort law is moving more toward compensation than deterrence. (For a discussion of this, see Executive Office of the President, Tort Policy Working Group, 1986, 30–31). Nevertheless, even if compensation is becoming more prominent, compensation and deterrence are not mutually exclusive, since high levels of compensation can act as a deterrent to risky activities.
3. Baram (1982, 6) points out further that these doctrines are not exclusively associated with the common law. In fact, they can be defined in statutory law as well. See Baram (1982, 28) regarding statute-based strict liability.
4. Except where indicated otherwise, the sections on each of the doctrines are drawn primarily from Baram (1982).
5. According to Katzman (1985, 24), a plaintiff can only show by inference that the defendant was negligent.
6. For the historical development of this concept, see Silver (1986).
7. This section is drawn from Hinds (1982, 2–4), except where indicated otherwise.
8. The Supreme Court case of *Boyle* v. *United Technologies*, 108 S.Ct. 2510 (1988), was referred to by Justice Antonin Scalia as federal common law. In *City of Milwaukee* v. *Illinois*, 101 S.Ct. 1784 (1981), the Supreme Court applied the federal common law doctrine for interstate water pollution, though this was later weakened on appeal as a sole basis for establishing liability pertaining to hazardous wastes (Hinds, 1982, 2, 11, 13). Earlier, the federal common law of nuisance was used in a few air and water pollution cases involving interstate pollution (Hinds, 1982, 3).
9. For example, only remedial relief can be obtained under RCRA and CERCLA, not damages (Hinds, 1982, 15). Furthermore, Hinds points out that this may in fact be a "catch-22" situation. While federal common law could be invoked in a given court case for damages, federal common law is presumably displaced by statutory law. Under common law, the conditions for obtaining damage payments vary as discussed above. Under federal nuisance law, irreparable damage has to be demonstrated (Hinds, 1982, 18).
10. These have commonly been identified as regulatory alternatives. See for example, Baram (1982) and Mitnick (1980).
11. Insurance was first applied to property damages, and was gradually extended to threats to life and health. Covello and Mumpower (1985, 109) review the use of insurance to manage risks beginning in ancient times. According to Covello and Mumpower, insurance was used in Mesopotamian civilization earlier than 3000

B.C.; the Code of Hammurabai institutionalized insurance (bottomry) in 1950 B.C. and included risk premiums for maritime contracts based on the likelihood that vessels would be lost and loans would not be paid; in Roman times, health and life insurance existed; and in 17th-century Europe, numerous kinds of insurance existed for casualties.

12. See 40 CFR, Part 387, administered by the Bureau of Motor Carriers.

13. Insurance does not necessarily circumvent the need for accurate risk assessment information, which can be time-consuming and costly to obtain (Ferreira, 1982, 268); risk assessments are traditionally conducted as a basis for insurance.

14. See, for example, *Summit Associates, Inc.* v. *Liberty Mutual Fire Insurance Company* 229 N.J. Super. 56, 550 A.2d. 1235 (1988). This case ruled that the pollution exclusion, which did not allow certain risks to be covered, should not dictate the extent of coverage, since it was not in the interests of good public policy (Doherty, Kleindorfer, and Kunreuther, 1988, 16).

15. Because of the extent of the debates over these issues and the fact that they influence (but are somewhat tangential to) environmental health insurance, coverage of the issues pertaining to insurance industry changes and judicial and legal changes is beyond the scope of this book.

16. The court case cited by the U.S. General Accounting Office (November 1986, 7) as upholding these policies is *Keene Corp.* v. *Insurance Co. of North America*, 667 F.2d 1034 (D.C.Cir. 1981), *cert. denied*, 445 U.S. 1007, *rehearing denied*, 456 U.S. 951 (1982).

17. One difference between the two kinds of policies is that where claims-made policies are issued over successive years, the current policy covers claims filed for incidents covered in previous years (for a retroactive period) but only if a claims-made policy was in force at that time; where occurrence-based policies are issued over successive years, the policy issued during the year the incident occurred is the one that covers claims for that incident. The two kinds of policies will only provide equivalent coverage if the claims-made policy is uninterrupted (U.S. General Accounting Office, November 1986, 13).

18. The following discussion on exclusions is drawn from U.S. General Accounting Office (November 1986, 16–18).

19. This section on the discovery rule is drawn from Dore (1988) except where indicated otherwise.

20. Dore (1988, 12–4) points out that New York State passed such a statute in 1986, allowing a one-year delay in filing claims for five substances: asbestos, diethylstilbestrol (DES), tungsten carbide, chlordane, and polyvinyl chloride. This was to benefit claimants whose cases had been dropped because of statutes of limitation.

21. This summary is drawn from the *New York Times* (December 20, 1988).

22. These views have been expressed by Clayton Cook of Hypercept, a risk retention group that is a joint venture of Aralie, Co. and Rhulen Insurance Co. (Monticello, New York).

23. Except where indicated otherwise, this section is summarized from Meier (1985, 205–6).

24. A few other federally based subsidy programs aimed at insurance exist for the purpose of protecting property rather than health and safety. One example is the

Urban Property Protection and Reinsurance Act of 1964. There are also subsidy programs that are not insurance-based, such as the financial assistance given to residences and businesses for relocation where condemnation is involved.

25. For a description of this program see Parker (1986, 17).

26. This description is drawn from Bardach and Kagan (1982c, 244–246) and Soble (1977).

27. Filing for bankruptcy is one way that polluters can avoid meeting cleanup requirements. Katzman (1988, 81) cites the court case *Ohio* v. *Kovacs*, 469 U.S. 274 (1985), in which the Supreme Court ruled that if a firm filed for bankruptcy it could avoid toxic waste cleanup costs.

28. This estimate is from the U.S. Environmental Protection Agency, Office of Solid Waste and Emergency Response, computerized database for the RCRA program, April 1989.

29. Break (1980, 10) notes that sales and excise taxes as a percentage of total general revenue declined between 1948 and 1978 from 20% to 10% for federal revenues and from 34% to 18% for state revenues while remaining virtually unchanged for local revenues at 3%.

30. Taxes in this category are a subgroup of the category of excise taxes.

31. Quite a number of typologies exist for these kinds of taxes. See, for example, Anderson et al. (1977).

32. For a discussion of the pros and cons of a bottle tax on plastic containers in New York City under Local Law No. 43 (1971), see *The Society of the Plastics Industry, Inc.* v. *The City of New York and Richard Lewisohn*, 68 Misc.2d 366, 326 N.Y.S.2d 788 (N.Y.Cnty 1971).

33. These are drawn primarily from Breyer (1982) and Baram (1982), except where indicated otherwise.

34. Inferred from Breyer (1982, 167).

35. Freeman (1989) recently refuted this criticism, using the example of a gasoline tax. The argument used against gasoline taxes is that the poor drive the larger, older cars that consume more gasoline, and thus they would be taxed to a greater extent than more well-off social groups. Freeman found that while the gasoline tax as a percentage of income does increase at lower income levels, as a percentage of expenditures (a proxy for a better measure of real income) it does not.

36. This introductory description of the marketable rights doctrine is drawn from Breyer (1982, 171).

37. These advantages have been pointed out by Breyer (1982) and others.

38. Noted in Hinds (1982, 22) who bases this on several cases involving the Hooker Chemical waste sites in New York State.

39. Gordon (1986, 163) has categorized grants according to degree of discretion.

PART II

Risk Management in the Federal System

Part I focused on the substantive aspects of managing environmental health risks, encompassing the content of the statutory base, nonstatutory common law, and incentive-based approaches to risk management. This sets the stage for an exploration of how these management tools for environmental health risks from chemicals have become institutionalized within the federal government.

The federal government's executive, legislative, and judicial branches, established by the Constitution of the United States, all play a major role in risk management. The executive branch consists of the Office of the President and its line agencies and offices. Independent commissions and agencies comprise still another dimension of executive agencies associated with the executive branch.* These agencies interpret, administer, and enforce risk-based laws through formal and informal rulemaking. The executive branch conducts oversight and budgetary review over all its line agencies as well. In the legislative branch, over a dozen committees of Congress and their subcommittees play a role in formulating risk policy by developing legislation related to societal risk and through oversight. Furthermore, Congress controls the budgets of administrative agencies through authorizations, by reviewing Presidential budget submissions, and by making appropriations and hence is a major controlling influence over the resource base with which agencies can pursue risk management. Congress can also exercise authority over executive agencies using its veto

---

*Independent agencies such as the EPA and independent commissions such as CPSC and the NRC are more independent of the executive branch than departments, but agency administrators and commissioners are appointed by the President. They are discussed here along with the executive branch as executive agencies (Office of the Federal Register, 1988).

power to override an agency action, such as agency rulemaking, where such authority is placed directly in legislation. Finally, the judicial branch plays a considerable role in risk management through its system of courts, which rule on both administrative procedure and questions of fact arising from risk management decisions made by federal executive departments, agencies, and commissions, along with the private sector.

While these governmental units and commissions are dealt with in separate chapters, the strong linkages among them must not be overlooked. These interrelationships are the focus of Chapter 7. In fact, a number of overlapping functions blur some of the distinctions between the three branches of government, especially in the area of risk management. Molton and Ricci (1985, 2) point to a number of examples of this. Congress assumed certain powers of the executive branch in the saccharin case when it overturned a ban imposed by an executive agency, the Food and Drug Administration. Congress has assumed powers of the judiciary when it holds quasi-judiciary oversight hearings. The executive branch, through the Office of the President, assumes legislative functions when it issues Executive Orders. The executive agencies perform a legislative function in rulemaking and quasi-judicial functions when they hold administrative hearings. Finally, the judicial branch performs legislative functions in altering laws as the result of court rulings. The courts also perform administrative functions in ruling on agency procedures that interpret and implement laws. For example, according to Melnick, the courts disagreed with the EPA on certain interpretations of the Clean Air Act. The court ruled that EPA could not allow "significant deterioration" of existing air quality and interpreted air emission requirements more restrictively than the EPA (Melnick, 1983, 11).

In order to understand governmental processes in Congress, executive agencies, and the judiciary, two of the three analytical perspectives that were presented in the introductory chapter are emphasized: (1) the degree of bureaucratic organization and behavior and its characterization in terms of discretionary action, and (2) the way that the three branches react to their own performance or the way that an evaluation of outputs is used as feedback to produce system changes. The chapters on the executive agencies and the judiciary touch on the third concept presented in the introduction, intensity and direction of action, in describing the rate of agency formation, budgetary trends, and court caseloads. The intensity and direction of Congressional action with regard to legislation was dealt with extensively in Chapter 2.

CHAPTER 4

Congress

INTRODUCTION

Congress, along with the courts and the President, is a major formulator of public policy (Mazmanian and Sabatier, 1983, 6). From as early as the 1950s through the 1970s, Congress dominated other branches of government in formulating certain risk-related policies and programs, playing a leadership role in developing clean air and water and health research programs (Ripley and Franklin, 1987, 219).

While Congress also engages in policy implementation through its use of hearings and its general power of oversight, this role is primarily supportive and indirect. According to Mazmanian and Sabatier (1983, 6), there are too many political reasons why members of Congress tend not to emphasize implementation: they are reluctant to jeopardize longstanding relationships with government agencies and prefer to act by building constituencies rather than through public confrontation. Some exceptions to this concept of Congress as a benign implementor in the environmental risk area do exist, however. For example, in 1982 Congress voted down EPA's request for a five-year extension in the deadlines for meeting "best available control technology" requirements for wastewater treatment plants under the Clean Water Act.

In the area of environmental health risk policy, Congress exercises its authority through its traditional roles of lawmaking, budgetary reviews and appropriations, and oversight. These mechanisms have been institutionalized through a complex system of committees, subcommittees, and procedures for the enactment of legislation that influence the way risk policy is formulated.

The first part of this chapter deals with the degree of bureaucratization and discretion in (1) the way Congress is organized to formulate and otherwise respond to risk policy (focusing primarily on the committee system) and (2) the procedures Congress uses to enact legislation (focusing on voting and various veto authorities of the members of Congress outside of the committee system). The second part of the chapter addresses how Congress evaluates agency and other governmental performance or outcomes, using its oversight function to make future adjustments in the risk management system.

THE CONGRESSIONAL DECISIONMAKING
SYSTEM FOR RISK MANAGEMENT

Over the past two decades there have been a number of distinct trends and developments in Congressional action with respect to risk management. First, two kinds of committees—legislative and appropriations—have had a considerable influence on the level of resources allocated to risk programs and the allocation of authority to administrative agencies in the risk area. In order to exercise its functions, including enactment of legislation and oversight, Congress has largely been dividing up chemical risk issues among a diverse set of existing legislative and appropriations committees and their subcommittees. This is partly driven by the nature of chemical risk, which involves many risk agents, affects many different activities, and is a relatively new issue area. Second, Congress has relied heavily on its traditional function of the enactment of legislation and on the voting mechanism that is part of the enactment process in formulating risk policy. These functions are undertaken directly by members of Congress outside of the committee system. The outcome of this process was discussed in Chapter 2 in terms of the rate and quantity of lawmaking in the risk area.

The Organization of the Committee System

Diversity of Committees Engaged in Risk Issues

Congressional committees are generally grouped into three functional categories: governmental operations committees, legislative or oversight committees, and appropriations and other budgeting committees (Ripley and Franklin, 1987). These committees are permanent or "standing" committees that change only when major reorganization occurs. Legislative committees conduct much of the factfinding and oversight activities for Congress. Appropriations committees review the budgets. Other special committees carry out particular investigations or are joint committees (including ad hoc conference committees) between both houses of the Congress (Office of the Federal Register, 1988, 29).

Table 4.1 lists the major Congressional committees and subcommittees responsible for legislation and oversight in connection with five federal agencies with major risk-related responsibilities. Sixteen standing committees in both houses of Congress potentially can cover environmental health issues. While this appears to be a relatively large number of committees for a given issue area, it is less than half of the total number of standing committees (38) in existence in Congress. Ten out of 22 in the House and 6 out of 16 in the Senate are involved with risk issues pertaining to the five agencies.

These committees have a long tradition within Congress; in fact, many have been in existence for well over a decade. As indicated in Table 4.2, many of these committees were formed in the 19th century, though a number of reorganizations of the committees occurred later, and subcommittees tend to be of a more recent vintage. The appropriations committees of the two houses of Congress that deal with risk management issues for the five major risk management agencies are listed in Table 4.3. When the number of subcommittees within both legislative and appropriations committees is taken into account, it can be seen that a large potential exists for dividing up an issue among different entities.[1]

Decentralization of Decisionmaking

The strength of the Congressional committees in legislative decisionmaking has often been underscored. It has been argued that the substance of Congressional activity and the heart of decisionmaking in Congress lies at the committee level (Melnick, 1985, 655; Ripley and Franklin, 1987, 5). Congressional committees, together with the administrative agency officials with whom they interact and nongovernmental groups that exert influence on Congress, have been characterized as comprising a subgovernment that shapes policy in general. This characterization also applies to risk policy in particular. Ripley and Franklin (1987, 8, 10) define subgovernments as "clusters of individuals that effectively make most of the routine decisions in a given substantive area of policy," adding that "participation in subgovernments offers the most pervasive and effective channel for interest-group impact on policy and program decisions."

The committee system was reorganized in the House in 1974 and in the Senate in 1977 (Ripley and Franklin, 1987, 225; Maass 1983, 54). As mentioned earlier, the House of Representatives has a total of 22 standing committees and the Senate has 16 (Office of the Federal Register, 1988, 29). In addition to the standing committees, the subcommittees of these standing committees have typically numbered well over 200. In the 97th Congress there were 238 subcommittees with four joint committees accounting for another six subcommittees (Congressional Quarterly, Inc., 1982, 454). In the 99th Congress there were 200 subcommittees, in addition to four joint committees and seven select committees (Barke, 1986, 33).

Issues pertaining to chemical risks are distributed within the committee system in Congress in a highly decentralized manner because of the large number of subcommittees and the decisionmaking power afforded to them. In the 1970s, according to various scholars (Melnick, 1985; Maass, 1983; Ripley and Franklin, 1987), there was a proliferation of these subcommittees and a growth in their authority and size as well. This trend magnified the decentralization of Congress into committees (often at the expense of the power of the committees themselves). The courts played an important role in this decentralization process (Melnick, 1985, 654–655). The importance of the subcommittees is reflected in the fact that they are where most of the work of Congress occurs. During the 95th Congress, about 80% of the work group meetings occurred in subcommittees (Davidson and Olesznek, 1981, 117). In further support of the power of

Table 4.1. Standing Committees of the U.S. Congress Responsible for Legislation and Oversight with Respect to Selected Federal Agencies Engaged in Risk Management

Agency	Committees and Subcommittees of the House of Representatives	Committees and Subcommittees of the Senate
Consumer Product Safety Commission (CPSC)	Government Operations • Commerce, Consumer and Monetary Affairs Energy and Commerce • Health and the Environment • Telecommunications, Consumer Protection, and Finance	Commerce, Science and Transportation • Consumer Protection
Environmental Protection Agency (EPA)	Government Operations • Environment, Energy, and Natural Resources Agriculture Energy and Commerce • Health and Environment Merchant Marine and Fisheries Public Works and Transportation Science and Technology • Natural Resources, Agriculture, Research, and Environment Small Business • Energy, Environment, and Safety Issues Affecting Small Business	Environment and Public Works • Environmental Pollution Select Small Business Commerce, Science, and Transportation Agriculture, Nutrition, and Forestry
Food and Drug Administration (FDA)	Energy and Commerce • Health and Environment • Oversight and Investigations Science and Technology • Investigations and Oversight	Labor and Human Resources

Table 4.1, continued

Agency	Committees and Subcommittees of the House of Representatives	Committees and Subcommittees of the Senate
Nuclear Regulatory Commission (NRC)	Interstate and Foreign Commerce • Energy and Power Science and Technology • Energy Research and Production • Natural Resources and Environment Interior and Insular Affairs • Energy and Environment	Environment and Public Works • Nuclear Regulatory Governmental Affairs • Energy, Nuclear Proliferation and Federal Services
Occupational Safety and Health Administration (OSHA)	Education and Labor • Health and Safety Small Business • Energy, Environment, and Safety Issues Affecting Small Business	Labor and Human Resources Small Business

Source: Summarized from the Congressional Quarterly Inc. (1986). *Federal Regulatory Directory.* (Washington, DC: Congressional Quarterly, Inc.).

Table 4.2.   Standing Committees of the U.S. Congress and Dates of Formation

Current Name	Previous Name	Date Formed
House of Representatives		
Agriculture	–	1820
Education and Labor	–	1867
Energy and Commerce	Interstate and Foreign Commerce	1795
Government Operations	Expenditures in Executive Departments	1816
Interior and Insular Affairs	Public Lands	1805
Merchant Marine and Fisheries	–	1887
Public Works and Transportation	Public Buildings and Grounds	1837
Science and Technology	Science, Technology and Astronautics	1958
Small Business	Select Small Business	1942
Senate		
Agriculture, Nutrition, and Forestry	Agriculture and Forestry (Agriculture)	1825
Commerce, Science, and Transportation	Commerce (Commerce and Manufacturers)	1816
Energy and Natural Resources	Public Lands and Survey	1816
Environment and Public Works	Public Works (Public Buildings and Grounds)	1837
Government Affairs	District of Columbia	1816
Labor and Human Resources	Labor and Public Welfare (Education and Labor)	1869

Source: Congressional Quarterly, Inc. (1982). *Guide to Congress,* 2nd ed. (Washington, DC: Congressional Quarterly, Inc.) p. 366. George Goodwin (1970). *The Little Legislatures* (Amherst, MA: University of Massachusetts Press).

the subcommittee, it has been observed that "much of their [the subcommittees'] work—both on the House and the Senate side—is routinely endorsed by the full committee without further review" (Congressional Quarterly, Inc., 1982, 456).

The power of the subcommittees is not invulnerable. One way in which the power of committees can be overridden is if the whole House of Representatives, the Speaker of the House, or a subset of the House membership (by petition) dislodges a bill from a committee that refuses to report it. Although it happens rarely, this can be done where a bill is critical in settling a particular case. The House can also exert influence over committee decisions by amending the bills the committees develop under certain conditions (Maass, 1983, 86–88). At any rate, it is in the context of the growing importance of the committee and subcommittee system within the Congress that environmental health risk policy reached its apex.

Table 4.3. Subcommittees of the Congressional Appropriations Committees
           Responsible for Appropriations for Selected Federal Agencies

| Federal Agency | Name of Subcommittees | |
	House of Representatives	Senate
U.S. Environmental Protection Agency	HUD[a]	HUD
Consumer Product Safety Commission	HUD	HUD
Occupational Safety and Health Administration	Labor–HHS[b]	Labor–HHS, Education
Nuclear Regulatory Commission	Energy and Water	Energy and Water
Food and Drug Administration	Rule Development	Agriculture

Source: Appropriations Committees of the House of Representatives and the Senate, 1988.
[a]HUD: Housing and Urban Development.
[b]HHS: Health and Human Services.

The existence of subcommittees has to a large extent been a function of legislated procedures and the style of committee leadership. Some legislation increased the number of subcommittees. The 1974 Committee Reform Amendments required that each committee have at least four subcommittees (Congressional Quarterly, Inc., 1982).

Prior to these amendments, committees varied considerably in the number, power, and coverage of their subcommittees. Rather than giving committee leaders more control over the subcommittees, this legislation enhanced decentralization (Maass, 1983, 96). Another requirement was that committees exceeding 15 members had to establish separate oversight subcommittees (Foreman, 1984, 196), a requirement that had the potential for increasing the number of subcommittees still further. Other requirements attempted to address the problem of the proliferation of subcommittees. Maass (1983, 97) reported that in 1981, House Democrats proposed that the number of House subcommittees be limited to a maximum of eight (or to the existing number, whichever was smaller) to prevent such proliferation.

The congressional subcommittee structure led to a decentralization of decisionmaking. One ramification of this was that many subcommittees were given the authority to make decisions about legislation at the expense of either the committees or Congress as a whole. At the committee level, this practice was actually encouraged by a procedure of multiple referrals in the Senate (allowed through unanimous consent motions) and joint referrals in the House (practiced since 1975) (Davidson, 1981, 120–121), where issues were new or had unclear or multiple jurisdictions.

The significance of the fragmentation is that the evaluation of issues can be duplicated on the one hand, and overspecialization can occur on the other hand. Overspecialization can result in some issues being missed, which Barke (1986, 35) has referred to as "underlapping" jurisdiction. Foreman (1984, 196–197) has provided numerous examples of duplication from the 97th Congress in which the subcommittee structure enhanced the proliferation and decentralization of hearings on air pollution, in spite of the fact that there was a lead agency on air pollution at that time (the Health and Environment Subcommittee of the Committee on Energy and Commerce).

Attempts to consolidate subcommittees in order to integrate Congressional policies have often met with opposition, because the supporters of the subcommittees argue that they promote checks and balances and more thorough review of issues. Approaches other than reduction in the number of committees as a way of dealing with the fragmentation of the committee system included the use of ad hoc and joint committees. These approaches were considered to be somewhat more successful than the attempts at reducing the number of committees or subcommittees directly (Sundquist, 1981, 431–433). It has been argued that in spite of the large number of subcommittees, several forces operate that restrict the authority and discretion of subcommittees, acting to constrain decentralization of Congressional decisionmaking. First, the subcommittees do not have separate appropriations (they depend on the full committees for these resources) and second, the subcommittees rely on the full committees to define their jurisdiction (Davidson, 1981, 117) and to vote on bills before they go to the floor of Congress.

The following are several representative examples of how a large number of committees and subcommittees have dealt with environmental health risk issues, leading at times to dissatisfaction on the part of administrative agencies:

- During the 97th Congress, the lead committee held 30 days of hearings on various aspects of air pollution, but there were 18 additional days of hearings on specific aspects of air pollution by four other subcommittees (Foreman, 1984, 196–197).
- Douglas Costle, an EPA Administrator during the Carter administration, complained that his agency had to report to 44 separate congressional committees or subcommittees (Foreman, 1984, 197).
- The Congress performed its investigations of the Ann Gorsuch administration in EPA and EPA's performance under Superfund through its committee and subcommittee structure. In early 1983, investigations were under way in the House by five subcommittees, and attempts by Thomas P. O'Neill, Jr. of Massachusetts to consolidate them failed (Kenski and Kenski, 1984, 110).
- With regard to the FDA's involvement with Congressional subcommittees, Brickman, Jasanoff, and Ilgen (1985, 45) observe that " . . . in 1978 the Food and Drug Administration took part in fifty-one hearings before twenty-four different congressional committees and subcommittees . . ." which were in addition to GAO studies and other investigations by the Congress, and "in the same year, GAO issued eight investigative reports on issues affecting FDA, and the agency responded to 4,463 written inquiries from Congress."
- The proliferation and fragmentation of these subcommittees and competition among them has been considered to be a contributor to Congress's failure to identify growing problems in the management of the country's nuclear weapons plants (Butterfield, 1988).

• Quite a number of committees dealt with the accident at Bhopal and its after-
math. A Congressional Research Service study noted that:

> The House Committees that have conducted most of the hearings
> about Bhopal and related issues include the Energy and Commerce Com-
> mittee, which has focused on environmental laws regulating toxic
> chemical manufacture, transportation, disposal, and emissions into the
> atmosphere; the Education and Labor Committee, which has focused
> on occupational safety and health issues; and the Foreign Affairs Com-
> mittee, which has focused on export controls for American products and
> technology. In the Senate, the Environment and Public Works Commit-
> tee has focused on hazardous materials inventories and emergency plan-
> ning in case of an accident. (Aidala, 1985, 9)

• Kessler (1984, 1039) has pointed out that any new legislation to resolve some
of the deficiencies in FDCA would have to be dealt with by committees with
overlapping jurisdictions—the Agriculture Committee and the Energy and Commerce Committee in the House of Representatives and the Agriculture Committee and Labor and Human Resources Committee in the Senate.

This pattern of fragmentation that characterizes the system of committees on
new risk issues seems to be a norm within Congress, regardless of the type of
issue. This observation was underscored by Weiss (1989, 412–413) in the
context of the needs of policy analysis, which requires a more comprehensive
outlook: "When a new issue erupts into prominence . . . the existing committee
structure finds a dozen or so committees and subcommittees with claims on
some piece of the action."

Influence of Committees on Risk Issues

Legislative committees. Legislative committees affect risk policy through
both formal and informal decisionmaking. They use formal decisionmaking to
design legislation, to conduct administrative oversight, and to authorize programs in legislation. In these capacities legislative committees define their agendas by acting both on their own initiative, i.e., through discretionary authority
and in response to charges from the congressional leadership. This exercise of
discretion dates back to the early 19th century when committees were first
established (Maass, 1983, 36).

Legislative committees have exercised discretion in taking action in a number
of areas involving toxic substances where agencies have been unable or reluctant to do so. When EPA had only passed a few rules under Section 6(a) of
TSCA to identify and place prohibitions on chemicals considered too risky to
allow on the market, the Congressional Interstate and Foreign Commerce Committee's Subcommittee on Oversight and Investigations held numerous hearings
on the issue (Stever, 1988, 2–26, footnote 49). When EPA refused to act under

TSCA in the area of asbestos removal in schools, Congress passed a new piece of legislation, the Asbestos Hazard Emergency Response Act of 1986, which formalized the process (Stever, 1988, 2–52.1). Congress reacted to barriers to EPA's promulgating final drinking water standards by passing the Safe Drinking Water Act Amendments of 1986, which put the standards-setting process on a strict timetable.

Legislative committees and conference committees charged with designing legislation have made heavy use of the mechanism of committee reports accompanying legislation to extend and clarify the meaning of the legislation they design, especially where controversial legislation has been referred to a conference committee. These committee reports are often referenced by administering agencies and the courts as a way of determining legislative intent. An example of use of the committee report to strengthen congressional intent for a risk policy where the intent was not stated directly in the legislation pertained to the relationship between sulfur dioxide emissions and acid rain. In 1977, the House Subcommittee on Health and the Environment deliberately left out a requirement for gas scrubbing from a proposed version of the Clean Air Act Amendments of 1977, but put it into the committee report "in order to slant the legislative history in favor of mandatory scrubbing" (Foreman, 1984, 189; Ackerman and Hassler, 1981, 29–30).

In addition, legislative committees have engaged in information gathering and dissemination, e.g., through the use of hearings in the course of administrative oversight functions. The hearing process also can act as a forum for airing opinions and grievances by special interest groups and the general public on public issues. Hearings held on the FDA proposal to ban saccharin in March of 1977 reinforced public opinion, which ultimately led to a delay in the ban for 18 months pending further studies. Such publicity and investigative activity often bypasses many of the delays of the legislative function (Maass, 1983, 12; Barke, 1986, 24). In the case of the saccharin ban, however, Congress ultimately instituted the ban by enacting a law, the Saccharin Study and Labeling Act (P.L. 95–203).

Legislative committees authorize programs in legislation, while appropriations committees appropriate the money so that the programs can be carried out. According to Maass (1983), legislative committees can at their discretion limit the power of the appropriations function by exercising a number of different, often conflicting controls. Over the years the use of these controls for program authorization has increased the influence of legislative committees relative to appropriations committees and the executive branch. Some of the ways in which this is accomplished follow.

1. They can control the procedures and standards that administrators follow, thereby influencing administrative discretion in the use of funds.[2]
2. They can determine whether or not a time limit (and hence periodic reauthorization) is to be allowed for programs. Annual authorizations are required for

many U.S. EPA programs, for example.[3] As Schick (1980, 171–172) points out, through annual authorizations and the consequent annual reviews, legislative committees can exercise oversight functions. This often occurs at the expense of administrative efficiency. The move has been toward the imposition of annual authorizations on many agencies.

3. They can decide whether or not to place monetary limits on authorizations (the fewer limits placed, the less need there is for periodic reauthorizations).

4. Finally, they can determine governmental organization, work force, and implementation of programs (Maass, 1983, 120).

Informal decisionmaking is used to a lesser extent than formal decisionmaking. There have been a few notable examples of legislative committees managing environmental risk by exercising informal, direct linkages to both agency staffs and the courts. Such behavior has typically bypassed the Presidency and Congress as a whole. The few examples that exist of informal decisionmaking in the risk area are striking. An important example, described by Maass (1983, 114–115), involved interpretations of and potential amendments to the 1970 Clean Air Act. One issue he discusses concerned the legitimacy of using certain approaches to disperse air pollutants, such as tall stacks, as alternatives to pollution control devices for the reduction of pollutant concentrations. In this case the Senate Environmental and Public Works Committee, under the leadership of Edmund S. Muskie (Democrat), controlled the information on the President's bill to support the use of such dispersion techniques by failing to bring it before the entire Congress. Another air pollution control policy area involved the legitimacy of setting standards, stricter than national ambient air quality standards, in the form of "prevention of significant deterioration" (PSD) requirements that would restrict the extent to which air quality could be degraded below air quality standards in areas meeting those standards. Muskie's committee convinced EPA not to resist the PSD standard and to prevent a new Presidential amendment to eliminate it. Thus, in both cases, the Senate committee was the initiator of these informal arrangements with EPA staff. According to Maass (1983, 114–116), informal strategies such as these were practiced by that committee through two presidencies.

Thus, Congress has influenced risk policy more commonly by exercising its formal decisionmaking powers, but with a few striking examples of the use of informal authority.

Appropriations committees. Budgetary decisions are one of the major means for Congressional control over how and whether risk policies can be implemented. It has been argued that Congress has used its budgetary authority more than its authority over the substance of programs to react to Reagan administration environmental policies (Kenski and Kenski, 1984, 99). While a comprehensive review of budgetary decisionmaking is beyond the scope of this book,[4] a few observations on how major expenditure recommendations of the appropriations committees influence or reflect risk policy are noteworthy, particularly in analyzing the role of Congress during the Reagan administration years.

The revenue committees within the U.S. Congress are the Appropriations Committees, the House Ways and Means Committee, and the Senate Finance Committee. The Appropriations Committees are responsible for spending or appropriating funds that have been authorized in legislation by the legislative committees, and the House Ways and Means Committee and the Senate Finance Committee are responsible for revenue generation through their jurisdiction over tax legislation (Schick, 1980, 30). As is the case with legislative committees, the appropriations committees work through a complex system of subcommittees, listed in Table 4.3.

The budget process often begins with recommendations by the President. The President's Office of Management and Budget plays a key role in developing a fiscal program and shaping the budgetary recommendations to Congress, in line with budgetary control and regulatory reform policies (Office of the Federal Register, 1988, 89). These budgetary recommendations are reviewed and subject to a vote by the committee members. The budget that the President submits is broken down by agency. The agency budgets are reviewed separately by a particular subcommittee of the appropriations committee in the House of Representatives and a similar set of subcommittees within the appropriations committee in the Senate.

The House provides a more in-depth review than the Senate. The House subcommittee, in fact, is considered to be highly specialized in the area of the particular agency whose budgets it reviews (Maass, 1983, 132). The subcommittee review can either cut or promote program spending. Finally, the votes taken by the Senate and the House can approve, disapprove, or change the recommendations of the appropriations committees.

As in the case of legislative committees, the appropriations committees exercise control over spending in a variety of ways. These are, according to Maass, (1) specifying the purpose and level of spending for appropriated funds, (2) placing conditions on spending (by attaching riders to appropriations bills), and (3) using hearings, reports, and debates (not required by statute). The use of these controls allows the appropriations committees considerable discretion and provides alternative mechanisms to restrict the power of legislative committees. The legislative committees also have access to controls that allow them to influence the domain of the appropriations committees (Maass, 1983, 136–141; 146).

The Reagan administration's approach to budgetary appropriations in the environmental area was to cut the programs drastically. Briefly, Congress's response was largely to go along with these cuts but minimize the degree to which they were made. For example, Congress was able to undermine the Reagan administration's cost-cutting philosophy by means of an appropriations bill disallowing a cut of 20% in the grants to states (Kenski and Kenski, 1984, 99). Congress also used its appropriations authority to place powerful conditions on the regulatory authority of administrative agencies. For example, during the 1970s, Congress used appropriations riders to place conditions on OSHA regulations (Davidson, 1981, 127).

While decisions of the appropriations committees are rarely challenged, the voting process in the full House or Senate can dramatically alter their budgetary recommendations. A striking example of this occurred in the chemical risk area in 1983. At that time, Congress attempted to reverse the cuts in the EPA budget that had been typical of executive policy during the preceding few years. In 1983, the Reagan administration approved recommendations to increase EPA's budget. These recommendations were voted by the Senate. The House, however, voted a dramatic increase over what its own appropriation committee had recommended: "The House . . on June 2, in a highly unusual move, . . . chose to overrule its own Appropriations Committee 'and voted to give EPA $353 million more than the administration had requested and $222 million more than the committee sought' . . . The vote was 200–167 . . . " (Kenski and Kenski, 1984, 115–116; Mosher, 1983, 1344). These actions still did not restore budgetary levels to what they had been before the cuts.

However, there are a number of limitations to the scope and authority of the appropriations committees (Schick, 1980). First, the appropriations committees have jurisdiction over only a portion of the federal budget. Certain "backdoor spending" categories are excluded from their domains. Second, they are responsible primarily for appropriating funds rather than authorizing the outlay of funds on an annual basis, which is the responsibility of the legislative committees and ultimately appears in the statutes. Furthermore, the actions of these committees with respect to risk issues have to be viewed in the context of the budgetary politics and the constraints ensuing from the conflict between authorizations (the function of legislative committees) and appropriations. Since the 1974 Congressional reorganization, another layer of committees, the budget committees, review the work of the appropriations subcommittees.

As in the case of legislative committees, appropriations committees can exercise considerable discretionary authority over the executive branch. The mechanisms by which this authority is exercised and the pattern of authority typify many of the actions of government outside of the risk area as well as within it.

Shaping of Risk Policy Through
Legislation and Lawmaking Procedures

Congress uses the committee system described above as a major vehicle for shaping legislation. Briefly, the process of enacting legislation works in the following way: Bills or resolutions are introduced into the House, and are assigned to an appropriate House committee where they are voted on. Successful bills are passed on to the full House, and if the vote is favorable in the House they are referred to the Senate, where a similar process of committee review and voting by the full Senate occurs. Where the two houses disagree, a conference committee is set up to work out a compromise which is then resubmitted to both Houses for approval. Approved bills are then presented to the President for approval or veto (Office of the Federal Register, 1988, 31).

The way in which the structure of the committee system influences risk policy through this process of legislative enactment was discussed in the previous section. Below are highlighted a number of components of and variations on this process that are usually implemented directly by the members of Congress. These are patterns of Congressional voting as an indication of the way Congress exercises its authority through voting on legislation, ways in which the members of Congress can get bills out of committee, the role of the Presidential veto and ways Congress can bypass it, and finally, the legislative veto.

Voting Patterns of Members of Congress

Members of Congress ultimately influence risk policy by means of their voting decisions on bills that come before them during each Congressional session. A review of votes on environmental health risk issues in the House of Representatives between 1981 and 1988 revealed that about 52 issues (listed in the *Congressional Quarterly*) were formally voted on during that period (Zimmerman, 1989). Most of the issues that arose before Congress pertained to the renewal of legislation. During that period, major revisions of FIFRA (1983), RCRA (HSWA of 1984), CERCLA (SARA of 1986), the SDWA (1985), and the CWA (1987) occurred, as well as the passage of specialized, separately titled amendments targeted to specific problems, such as the Asbestos Hazard Emergency Response Act (an amendment to TSCA), the Ocean Dumping Ban Act (an amendment to the Marine Protection, Research, and Sanctuaries Act), and the Lead Contamination Control Act (an amendment to SDWA). In addition, votes were taken on agency appropriations and administrative proceedings, the most important of which pertained to Ann Gorsuch's leadership of EPA. Analysis of voting on environmental health risk issues related to chemicals for four congressional sessions across the 435 districts in the House of Representatives revealed the pattern shown in Table 4.4.[5]

These figures appear to show that a majority of the votes on risk-related issues in the House of Representatives was positive or pro-environmental, regardless of Congressional session. This trend toward pro-environmental voting also appears to increase over the years between 1981 and 1988.

A number of factors are associated with voting patterns on environmental issues. An earlier analysis of the voting behavior of individual members during the 97th Congress, based on League of Women Voters data, found that in general, "ideology, party affiliation, region, and constituency all are important factors affecting environmental support on floor votes in both the House and Senate" (Kenski and Kenski, 1984, 114–115):

- About half of the votes supported environmental legislation. The percentage was higher in the House than in the Senate.

Table 4.4.   Voting Patterns on Chemical-Related Environmental Health Risk Issues

Congressional Session	Year	Number of Environmental Health Issues	Percentage of Votes Pro-environmental (House of Representatives)
97th	1981–1982	14	62.0
98th	1983–1984	14	70.1
99th	1985–1986	16	76.8
100th	1987–May 1988	8	78.6

- Partisan behavior was evident: Democrats had double the percentage of members voting for environmental issues than Republicans, regardless of which house they were in.
- Regional differences were evident: strength of vote for the environment appeared in the following order—east, midwest, west, and south.

Congressional Tactics to Influence Legislation Prior to the Vote

Members of Congress can influence legislation in many ways prior to the formal floor vote and even prior to its introduction into committees. These strategies include influencing (1) the referral of a bill to a particular committee, (2) when and how a bill comes to the floor, (3) the extent to which changes will be allowed to a bill once it comes to the floor, and (4) delays in the voting process.

Even before a bill is introduced into a committee, the members of Congress can influence which committee will control it. House and Senate rules govern the allocation of bills; however, if bills are new or have overlapping jurisdictions, there is some discretion in the way the bill is assigned to a committee (Congressional Quarterly, Inc., 1982, 465). Recognizing this, a bill can even be written in such a way as to steer it to a certain committee. It has been argued that this discretion is an outcome of "a failure of Congress to restructure the committee system to meet changing developments and national problems" (Congressional Quarterly, Inc., 1982, 456).

The timing of a bill and the way it comes to the floor affect how much time the members have to debate it. There are a number of ways Congress can bring a bill to the floor if a committee refuses to do so (Congressional Quarterly, Inc., 1982, 414, 416; Maass, 1983). Where a rule is required as a prerequisite to bringing a bill before the members of Congress and the rule is not passed, then members of the House can use discharge procedures and members of the Senate can use the mechanism of Unanimous Consent. These mechanisms have rarely been used. By far the most common of the mechanisms to dislodge a bill is suspension of the rule requirement by the House Rules Committee, which requires a two-thirds vote. This is often used for more controversial legislation. In the 96th Congress (1979–1981) bills pertaining to environmental research and the delay of the saccharin ban in particular were brought to the floor of the House as a result of the suspension of the rules (Maass, 1983, 78).

Where bills require a rule to be passed prior to their introduction for a vote, rules can be used to influence the amount of debate that occurs. Rules can be passed suspending the passage of amendments to the bill, specifying how much time members can debate it, and preventing objections from being voiced (Congressional Quarterly, Inc., 1982, 414).

Finally, the members of Congress can delay the vote though (1) a Senate filibuster and (2) a House quorum call combined with attempts to keep legislators away so no quorum is reached. There are also speeding-up tactics to shorten procedures, e.g., through the use of suspension of rules, a strategy used in one instance to bypass environmental opposition to a water project (Congressional Quarterly, Inc., 1982, 422).

The Presidential Veto

According to Sundquist, Congress's authority to enact legislation is often limited by cumbersome procedures and limitations imposed by the Presidential veto (Sundquist, 1981, 317). Congress can bypass a Presidential veto by writing amendments to authorizing legislation, as an alternative to writing new legislation that could also be vetoed by the President (Sundquist, 1981, 358). While schedules for reauthorization are often written into legislation, Congress has taken considerable advantage of the amendment route in the risk policy area. A Presidential veto was applied, however, to the Clean Water Act bill in the mid-1980s.

The Legislative Veto

One major vehicle Congress used to exercise control over the authority of the executive branch of government has been the legislative veto. As reviewed by Fisher (1988, 224), under the veto provision Congress would delegate certain responsibilities to agencies with the condition that Congress had a certain amount of time (90 days) after a delegated action was taken by an agency to approve or disapprove it without further Presidential intervention. If neither house of Congress acted, the rule would become a law within 60 days. Thus, the mechanism allowed both the agencies and Congress considerable discretionary authority. Through this mechanism, Congress could make mid-course corrections in agency rules, regulations, and procedures, while the agencies could also pursue changes without the risk of presidential intervention. In addition to influencing agency rulemaking, another purpose of the legislative veto was to influence the content of legislation. Strickler and Williams (1986, 286) ascribe several other purposes to the legislative veto, such as acting as a means for compromise when the details of a piece of legislation cannot be agreed on. Congress could carry out the legislative veto either through the actions of a single house of Congress, of both houses, or by committee (Fisher, 1988, 224). Thus, the legislative veto, while not applicable to the enactment of legislation per se, can be a major factor in lawmaking.

The veto was first introduced in 1932, but only began to be used extensively during the 1970s. The authority for legislative vetoes was written into laws (especially after 1973); by 1979, some 113 laws had provisions for the veto (Sundquist, 1981, 345). Veto provisions were not restricted only to the statutes, and could be attached to authorization and appropriation bills (Strickler and Williams, 1986, 286). The number of veto provisions attached to bills, some of which were signed into law, was far greater. On June 23, 1983, the one-house legislative veto was struck down by the Supreme Court in *Immigration and Naturalization Service* v. *Chadha*, 462 U.S. 919 (1983). It was argued that Congress could only act by changing legislation directly by passing a law, rather than through a veto of an action allowed in legislation. Other lower courts have also ruled on the legislative veto (Strickler and Williams, 1986, 289, 298, footnote 4). The case pertained to the constitutionality of a one-house veto only, however. At the time of the 1983 court ruling on legislative vetoes, more than 200 statutes had provisions for the veto, including the one-house veto (O'Brien, 1985, 3). A third of the vetoes had been incorporated into legislation between 1976 and 1981 alone (Congressional Quarterly, Inc., 1982, 440). After 1983, the legislative veto was replaced by the joint resolution veto. This reinstated the viability of a veto, but only where it is agreed on by both houses. Since 1983, the joint resolution veto has become quite popular, according to Fisher (1988, 225). Fisher notes that between June 23, 1983 and October 18, 1986 (the end of the 99th Congress), 102 new vetoes were incorporated into 24 different statutes. Most of these emphasize the authority of the appropriations committees rather than legislative committees. Unlike the legislative veto, the joint resolution veto is subject to a Presidential veto (Strickler and Williams, 1986, 287).

It is difficult to determine how veto power enabled Congress to influence agency procedures. One study of the effect of veto power on interactions between Congress and agencies found that the veto in general did not change existing relations between Congress and the agencies. Where it did, it tended to influence the independent agencies more than the cabinet-level departments (Strickler and Williams, 1986, 285).

At least on paper, the legislative veto was visible in legislation dealing with risk policy. Prior to the *Chadha* case in 1983, the House of Representatives attached veto provisions to bills pertaining to water and air pollution control, pesticides, toxic substances regulation, and programs run by the EPA and the CPSC (Sundquist, 1981, 351, footnote 18). The veto was written into Article 305 of CERCLA for any rules and regulations issued under Subchapter I. After the *Chadha* case, the veto provision on an agency rule or regulation exercised by both houses of Congress concurrently was incorporated into the 1988 amendments to the Federal Insecticide, Fungicide, and Rodenticide Act (P.L. 100-532, 7, U.S.C.A. Article 136w). It is also contained in the indemnification and limitations on liability provisions of the Atomic Energy Act in connection with compensation where damages from a nuclear incident are estimated to exceed aggregate public liability covered under the act (42 U.S.C.A. Article 2210).

POLICY ADJUSTMENTS THROUGH PERFORMANCE EVALUATIONS: THE ROLE OF OVERSIGHT

One of the main responsibilities of Congress is to evaluate the organizational programs, structure, design, and leadership of the government through its oversight function. Through oversight as well as through legislation, Congress can effect changes in administrative processes. Oversight encompasses the "after-the-fact" functions of review, evaluation, investigation, and fact-finding; the findings of these processes are made available to the public (Sundquist, 1981). Legally, oversight is advisory only and emphasizes information-gathering rather than the actions that can result from the information (Sundquist, 1981, 326):

> Oversight encompasses the efforts of the Congress to find out, after a law is passed, what happens as a consequence, and its purpose as stated in the rule [Rule X(b)(1) of 1974] is to enable the Congress to determine the need for modifying existing law or enacting new law. In practice, some review and study activities have only a remote legislative intent; they may be undertaken to influence the way in which officials administer the law, or their object may be to expose misconduct. But it is useful to limit the term oversight to information-gathering and use other terms to denote the wide range of actions that may be undertaken on the basis of the information: otherwise much of the legislative process itself is encompassed under the heading of oversight.

In spite of the advisory nature of oversight, its influence has been pervasive. Congressional oversight is highly variable over time and in terms of its depth and breadth (Ripley and Franklin, 1987, 220–224; Sundquist, 1981, 328). The degree of oversight and its influence on risk policy is largely determined by Congress's priorities in exercising its function as overseer or watchdog. The mechanisms that Congress uses for oversight are highly variable as well and involve virtually all Congressional powers: hearings, administrative investigations, and review of personnel decisions.

According to Sundquist (1981, 328, 333), oversight activity rose substantially during the 1970s, regardless of the indicator used. It involved the activities of special offices, committees, and individual members of Congress as well and was accompanied by an increase in the size of the congressional staff.

According to Ripley and Franklin's review (1987, 221) of the work of Ogul (1981), oversight tended to occur more in the presence of distrust of bureaucratic style, motivation, or competence; a crisis environment; major changes in policy including expenditures of money; greater resources to conduct investigations; and legislative mandates. Furthermore, they point out that several periods characterize the last category—legislative mandates for congressional oversight. These included the Legislative Reorganization Act of 1946, the Intergovernmental Cooperation Act of 1968, the Legislative Reorganization Act of 1970, the 1974 Congressional Budget and Impoundment Control Act (Ripley and Franklin, 1987, 224), and a variety of bills introduced in 1979 to improve the oversight process and make it part of individual pieces of legislation (Sundquist, 1981, 340–341). The period following the Watergate investigations was marked by especially heavy oversight activity by Congress. The reforms of 1974 were far-reaching and account for the growth and complexity in the legislation that governed risk policy in its early years.

Oversight for Organizational Change

Agency Size and Programs

Congressional action produces changes in the size and operation of government institutions. As Sundquist (1981, 316) has pointed out, it is easier for Congress to control the size of the bureaucracy than to make it more accountable or sensitive to public issues, because the latter requires collective rather than individual actions of members of Congress. Through its legislative functions, budgetary authority, and hearing processes, Congress provides the jurisdictional authority and resources that create and structure organizations and programs (Ripley and Franklin, 1987, 65). According to Ripley and Franklin, Congress's concern with government structure and organization is primarily focused on the department or commission level, rather than lower subunits of government. In addition, they point out that Congress pays more attention to the independent regulatory commissions, since they are arms of Congress.

Agency Leadership

A major use of Congressional oversight is to evaluate and thereby influence leadership through the review of executive personnel and appointments. The impact of this oversight authority was clearly felt in the area of risk policy implementation during the Congressional review of the EPA's role in the Superfund program under the administration of Ann Gorsuch. Gorsuch was appointed by the President as EPA Administrator in the early 1980s, with Congress's approval. During her administration, EPA policies were aimed at easing hazardous waste disposal regulations, but Gorsuch rescinded a number of these under the pressure of Congress. As a result of constant battles with Congress and the indictment of one of her officials, Gorsuch resigned in 1983 (Congressional

Quarterly, Inc., 1986, 114). In spite of the fact that Congress was able to win in this situation, it has been pointed out that this only occurred after considerable damage had been done (Kenski and Kenski, 1984, 101). Subsequent GAO reports and summaries of the EPA's accomplishments in hazardous waste site cleanups during the Gorsuch administration and beyond showed the pervasiveness of problems in the implementation of hazardous waste site policies.

Congress's influence over governmental leadership extended to other agencies involved in risk issues. Similar pressures by Congress were used in investigations of OSHA's operation under Thorne Auchter, and the Department of Interior's operation under James Watt. In addition, in 1981, the President wanted to eliminate the Consumer Product Safety Commission or reorganize it within another agency. Congress did not support this, and it did not come to pass (Foreman, 1984, 205).

Oversight Through Program Evaluation

Congress performs oversight by means of a variety of evaluation mechanisms. These mechanisms include studies by specialized Congressional offices, hearings (both special and routine), and periodic reports to Congress from executive agencies, which are reviewed by members of Congress.

Six special offices exist within Congress that perform evaluation functions by conducting special studies, in addition to legislative and budgeting support functions (Office of the Federal Register, 1988, 42–61). Of these, the GAO, formed in 1921, and the OTA, formed in 1972, play a large role in risk policy relative to other special offices in Congress in conducting studies and evaluations of programs, agencies, and other matters of interest to Congress.[6] These two offices conduct program evaluations on a more comprehensive basis than those provided by the oversight hearing records of the special committees. The findings of these studies often result in changes in program design and administration and are the foundation for new legislation.

The GAO's studies that evaluate risk-related programs tend to be narrowed to specific programs and a relatively short time period. The GAO conducts program evaluations and program audits as well. The effect of the GAO investigative reporting on Congressional and other governmental decisionmaking has not been well studied, so little is known about it (Maass, 1983, 207). Only a few studies have studied the output of the GAO, and a number of these are histories (Mosher, 1979). By the end of the 1970s, Mosher's compilation of GAO issue areas (1979, 338) indicated that risk issues as a group constituted under 10% of the GAO resources during fiscal year 1978. Environmental protection accounted for 2.4% and consumer and worker protection, 1.3%. While other risk issues might have been aggregated under other categories, such as health, science and technology, and energy, they were unlikely to bring the total resource allocation above 10%. In recent years, the GAO reports issued on environmental risk issues have been more prominent and are clearly part of the organization's mission.

The OTA is usually charged with studies of a broader nature, dealing with impacts of technology on society. Examples of some of the major reports dealing with health risk issues during the 1980s included hazardous waste management, high-level radioactive waste management, hazardous material transportation, ocean incineration, evaluation of the Superfund program, and special reports pertaining to incidents such as Love Canal. In a number of instances the OTA studies were instrumental in altering the direction of governmental policy in the risk area.

CONCLUSIONS

Traditional Processes

Many characteristics of the structure and functions of Congress that have markedly influenced risk policy are independent of the risk issue and were in place long before it emerged. As a general statement, Congress has exercised its traditional role as formulator of public policy in the risk area as it has in other areas of policy. The committee and subcommittee structure of Congress that has played such a pervasive role in formulating risk policy existed prior to the attention to risks. Congressional committees actively developed legislation and exercised oversight functions in the risk area. They issued subpoenas, overrode agency decisions through vetoes of agency programs and procedures, reviewed executive appointments through confirmation hearings, and controlled agency appropriations. These are all traditional powers of Congress that were in place prior to the emergence of the risk issue and were applied to risk problems.

The decentralization, diffusion, and competition that existed within the committee and subcommittee structure of Congress affected how the risk policy issue was treated, just as it affected other issues. This influence was not unique to risk policy, and the situation occurred in spite of the numerous legislative attempts by Congress to reform its structure to reduce the large number of committees. This decentralized and often fragmented committee structure gave rise to the partitioning of a given issue among subcommittees. While numerous examples exist of the negative effects of this situation, the overall effect of fragmentation is difficult to ascertain without considerably more investigation. Rather than being a negative factor, division among committees can theoretically have the advantage of promoting both a broader perspective on an issue and the use of more checks and balances within government.

The voting process is the ultimate means by which members of Congress influence risk policy through the enactment of legislation. The large volume of legislation passed in the risk area and the predominance of pro-environmental votes are testimonies of the use of the voting power of Congress to promote environmental health risk policies. It was pointed out that there are many mechanisms that members of Congress can use to influence bills prior to the vote. Several of these mechanisms, which are traditional to the way Congress operates, were applied to risk issues.

Initiatives

Congress has taken considerable initiative, often exercising unusual discretion and aggressiveness within traditional areas. In spite of the fragmentation the committee structure imposed on the passage of new laws, numerous examples exist of committee initiatives. Also, highly visible oversight hearings have been held where agencies failed to act or acted in a manner inconsistent with the intent of legislation. Legislative committees took these initiatives in areas such as asbestos removal in schools and the issuance of rules for restricting or banning the manufacture of risky chemicals. Appropriations committees took an aggressive stand in stemming the cuts that the Reagan administration had made in EPA's budget during a time when its mission was expanding. Congress actively used and extended its oversight function to dislodge controversial and ineffective leaders at EPA and other agencies. It judiciously applied its primary function of the enactment of legislation and was largely responsible for the precipitous rise in legislation covering chemical risks where a void had existed. The legislative veto was written into a number of pieces of environmental legislation; even after its form was altered in 1983, it continued to be used.

Thus, the mainstream of Congress's structure and activities was not altered by the emergence of the risk issue during the 1970s and 1980s. The traditional system that had dominated Congressional decisionmaking even before the risk issue emerged influenced the risk issue as it had other issues before it. At the level of fine detail, however, in areas that involved extending and applying the authority that it had, Congress displayed major initiatives within the traditional system.

NOTES

1. A study of the role of Congress in risk assessment reviewed many of the Congressional oversight committees that deal with over two dozen risk-related statutes (J. H. Wiggins Co., 1981).
2. One major example is the use of entitlements. By specifying a rate of spending rather than a total amount, Congress can allow somewhat open-ended expenditures of monies.
3. No time limits existed on programs prior to 1960. At that time, limits were imposed primarily through appropriations. Many programs now carry a requirement for an

annual reauthorization so that no appropriations are allowed before the program is reauthorized by legislation and signed by the President (Maass, 1983, 121).

4. For a review of the federal budgetary process, particularly as it pertains to the congressional role, see Schick (1980) and Maass (1983). The summary is drawn from these two sources.

5. This analysis did not include votes on environmental issues not directly related to human health, such as animal protection legislation.

6. Other offices conduct studies as well, but their relationship to environmental risk policies has been less well covered. The Congressional Budget Office specifically focuses on economic impacts in its studies. The Congressional Research Service and the Library of Congress have produced research reports related to risk issues at the initiative of committees.

CHAPTER 5

Executive Departments and Independent Agencies

INTRODUCTION

Administrative agencies implement the laws and policies developed by Congress and the President. In addition, they interpret and change the law through administrative rulemaking procedures and administrative decisions that execute those rules with discretionary or nondiscretionary action. New laws and frequent changes in existing laws, the demands of incentive-based systems for risk management, and the challenges of Congressional and judicial activity have created a growing administrative burden for federal risk management agencies. The ramifications of this burden can be evaluated in terms of the criteria presented in the introductory chapter for evaluating governmental activity in the risk area against the backdrop of traditional government. These criteria were expressed in terms of intensity and direction of actions, degree of bureaucratization, and the process whereby outcomes are used as the basis for future adjustments in the system. These three characteristics are applied specifically to executive agencies in the following way:

1. Trends in organizational capacity are examined in terms of the rate of agency formation and the magnitude of and change in budgetary resources.
2. The degree and form of bureaucratization are examined primarily in terms of agency discretionary authority in the use and structuring of the risk assessment function, the use of advisory committees, and the conduct of the standards-setting and rulemaking functions.
3. Outcomes of agency activities are examined in terms of how federal agencies involved in risk management are using program outcomes or evaluating performance to make future refinements within their agencies.

Before applying these three areas of evaluation in more detail, a brief overview is presented of the activities and organization of the five major federal agencies involved in risk management that are the focus of this chapter.

MAJOR AGENCIES INVOLVED IN RISK MANAGEMENT[1]

Five federal agencies and their affiliated organizations play major roles in the management of environmental health risks. These agencies are the EPA, the CPSC, the FDA within the Department of Health and Human Services (DHHS), OSHA within the Department of Labor, and the Nuclear Regulatory Commission (NRC). While this chapter concentrates on the executive agencies, the role of the President is, of course, a pervasive and influential one and is central in setting the parameters under which these agencies operate. The role of the President will be highlighted as it arises in connection with particular risk issues faced by the agencies.

The organizational and administrative form of an agency reflects its power within the federal bureaucracy. Federal organizations responsible for safety, health, and environmental protection differ from one another in the form of organization they have. The forms that these agencies assume include cabinet-level departments, independent agencies, administrations within departments, or independent commissions. Two of the five agencies covered here are commissions—the Nuclear Regulatory Commission and the Consumer Product Safety Commission. The other agencies are either offices within larger cabinet departments or independent agencies.

Mitnick (1980, 66–71) notes that considerable structural differences exist between commissions on the one hand and agencies or departments on the other hand, no matter what their functions are. By virtue of these differences, commissions and agencies have different advantages and disadvantages. Commissions have multiple leaders, whereas an agency has a single head. Multiple leaders provide greater executive input into decisions; however, they can diffuse leadership. Single agency heads concentrate leadership; if structured properly, such an agency can obtain a variety of inputs into decisionmaking from subunit directors. Commissioners are typically appointed, whereas the heads of bureaus often come up through the civil service system and presumably are less subject to political pressure. Commissions enjoy a certain independence of decisionmaking that agencies do not. Some disadvantages of commissions are that salaries that may be set by a legislature are more difficult to change; thus, commissioners may be lured away by better paying jobs. The fixed number of commissioners is not readily adaptable to changes in the workload, and commissioners may react by simplifying regulatory procedures.

Table 5.1 shows that a large number of administrative entities in the executive branch are involved in governmental decisionmaking about risks. The five federal agencies stand out prominently in the area of risk management, on the basis of the scope of their responsibilities, the size of their budgets, and the attention they have received as visible players in the risk area.

Table 5.1.   Federal Agencies in the Executive Branch Engaged in Risk Management

Agency	Date Established
Department of Agriculture	
• Forest Service	1905
Department of Commerce	
• National Oceanic and Atmospheric Administration	1970
Department of Defense	
• Army Corps of Engineers	1824
Department of Energy	
• Energy Information Administration	1977
• Federal Energy Regulatory Commission	1977
Department of Health and Human Services, Public Health Service	
• Agency for Toxic Substances and Disease Registry	1983
• Centers for Disease Control	1973
• Food and Drug Administration	1930/1968
• National Institutes of Health	various
Department of Interior	
• Fish and Wildlife Service	1871/1956
• Geological Survey	1879
Department of Justice	
• Land and Natural Resources Division	1913
Department of Labor	
• Occupational Health and Safety Administration	1970/1973
• Mine Safety and Health Administration	1973/1977
Department of Transportation	
• Coast Guard	1915
• Federal Aviation Administration	1958/1967
• Federal Highway Administration	1967
• Federal Railroad Administration	1967
• National Highway Traffic Safety Administration	1970
• Research and Special Programs Administration Materials Transportation Bureau	1975
• Urban Mass Transportation Administration	1968
Environmental Protection Agency	1971
Federal Emergency Management Agency	1979
Consumer Product Safety Commission	1973
National Transportation Safety Board	1966/1975
Nuclear Regulatory Commission	1975
Council on Environmental Quality	1969
Office of Science and Technology Policy	1976

Source: Dates are drawn from Office of the Federal Register (1988).

Scope of Responsibilities

Chemical health risks comprise a major portion of the concerns of the five agencies. Most of the organizations singled out have substantial (in some cases exclusive) responsibilities over the risks associated with the products or services they regulate or manage. From the perspective of market share, the agencies are prominent by virtue of the size of the industries regulated. Meier (1985, 82, 100, 102) has indicated, for example, that the FDA regulates a $30-billion-a-year industry. The CPSC establishes safety standards for some 15,000 consumer products resulting in accidents that cost some $10 billion per year for emergency room expenses (U.S. Consumer Product Safety Commission, 1986, 1). OSHA is responsible for 70 million employees in over 5 million establishments (Congressional Quarterly, Inc., 1983, 406). The chemical and related industries and waste disposal operations that the EPA regulates and the nuclear industry that the NRC regulates (which represented 113 operating plants and 10 with construction permits by mid-1989 [U.S. Nuclear Regulatory Commission, Office of Public Affairs, July 31, 1989, personal communication]) are at least equally sizable.

Budget

A second factor that makes these agencies prominent is the magnitude of their budgets in the area of regulation of environmental protection and health. The budget allocations and personnel for each agency having risk-related responsibilities are given for 1988 in Table 5.2. The budget and personnel of these agencies represent a significant portion of the nation's regulatory budget. Of the five agencies, the EPA commands the largest budget, and the CPSC has the smallest, being funded at under 1% of the EPA budget and under 10% of the FDA budget. Trends in agency budgets over time are described later in this chapter.

Attention in Legislation and Nationwide Studies Pertaining to Risk

Finally, the agencies selected here have achieved national attention and recognition in the area of public risk management. Considerable attention has been paid to their risk management role in current or proposed legislation or in national studies of risk management. For example, a National Academy of Sciences (NAS) study on risk management in 1983 (National Research Council, 1983) singled out four of the agencies—EPA, FDA, OSHA, and CPSC—as having major roles in regulating the potential risks associated with environmental contaminants.

Table 5.2.   Budget Allocations and Personnel for Selected Federal Agencies Having Risk-Related Functions, 1988

	1988 Budget Appropriations (in millions of dollars)	Personnel
CPSC	32.7	459
EPA	4,968.4	11,127
FDA	483.1	7,032
NRC	392.8	3,136
OSHRCª	5.9	90
DOL		
MSHA	160.2	2,352
OSHA	235.5	2,702

Source: Executive Office of the President, Office of Management and Budget. *Budget of the United States Government*, Fiscal Year 1988.
ªOccupational Safety and Health Review Commission.

Agency Profiles

U.S. Environmental Protection Agency

The Environmental Protection Agency was formed as an independent agency in 1970 under Executive Reorganization Plan No. 3. The EPA oversees most of the federal laws that govern the environment and in this capacity regulates governmental and private sector activities that adversely affect the environment. It performs these responsibilities either directly or by delegating regulatory and grants programs to state and local governments (Congressional Quarterly, Inc., 1986).

Major changes in federal environmental laws were made in 1972 after Earth Day, expanding the agency's responsibilities. By the mid-1980s, EPA administered nine major statutes, handling risks associated with air pollutants, general water quality, drinking water quality, noise, toxic and hazardous substances including radiation and pesticides, and solid and hazardous wastes. Some of EPA's functions relating to environmental risks are as follows:

- It develops environmental criteria and standards for air quality and radiation, water quality, drinking water, and solid waste management.
- It administers regulatory programs in the form of permits, licenses, registrations, environmental reviews, and funding programs either directly or through the states. Examples of some larger programs are the National Pollutant Discharge Elimination System for wastewater discharges under the CWA, reviews of new sources of air pollution under the CAA, permits for hazardous waste disposal facilities under RCRA, and pesticide registrations under FIFRA.
- It has environmental review authority over the activities of several other federal

agencies. For example, it reviews environmental impact statements prepared by all federal agencies under the National Environmental Policy Act.

- It provides research, technology transfer, and guidance to program administrators and to those regulated (which is a role that the federal government has played in the environmental area prior to the formation of the EPA).

At its headquarters offices in Washington, DC, the EPA is organized both functionally and categorically. The organization of the headquarters office is shown in Figure 5.1. At headquarters, almost a dozen assistant administrators or directors responsible for functional or categorical areas report to the Administrator of EPA. The functional categories include enforcement and compliance, policy, planning and evaluation, research and development, and various program administration functions. The categorical areas are water, solid waste and emergency response, air and radiation, pesticides, and toxic substances. The agency is divided further into 10 geographic units or regional offices, which are each headed by regional administrators who also report to the Administrator in Washington.

The agency was formed from functions originally located in 15 federal agencies in three departments (Agriculture; Health, Education, and Welfare; and Interior), totaling some 6000 employees (Marcus, 1980, 275). For example, the water quality functions alone were drawn from the Department of the Interior and the Corps of Engineers. Pesticides management functions were drawn from the Departments of Agriculture and the Interior, as well as the Food and Drug Administration (Meier, 1985, 145).

Consumer Product Safety Commission[2]

The Consumer Product Safety Commission was established in 1973. In the area of chemical risks, it administers the Consumer Product Safety Act and assumes similar functions under the Flammable Fabrics Act, the Hazardous Substances Act, and the Poison Prevention Packaging Act. According to its own policy statement and the goals of the acts it administers, the major purpose of the commission is to guard the public against unreasonable risks of injury associated with consumer products. In that capacity, some of the major functions of the commission are as follows (Congressional Quarterly, Inc., 1986, 82):

- It is directed to issue voluntary standards (and mandatory standards where voluntary standards prove to be insufficient) aimed at consumer protection. Acting in an emergency capacity, it can set imminent hazard standards.
- It can establish priorities among the products it regulates.
- It can ban products it considers deleterious to consumer health and safety.
- It performs research and information dissemination functions.

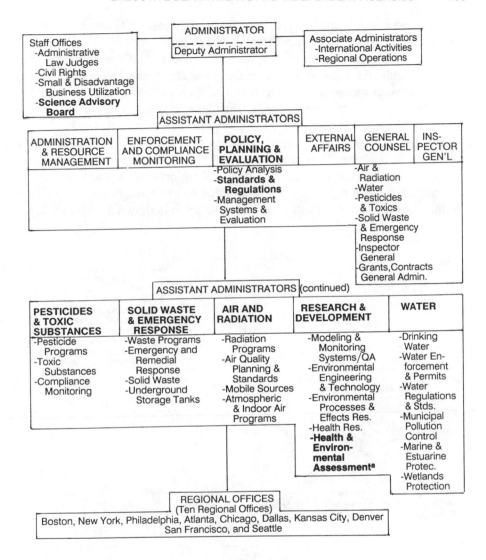

Figure 5.1.   Organization of the U.S. Environmental Protection Agency. *Source:* Drawn from the U.S. EPA Organizational Chart in the Headquarter Telephone Directory (Summer 1989) and "Who's Who in EPA and Their Phone Numbers." This chart reflects EPA organization under Lee M. Thomas.

Notes: Bold titles have been added and indicate those offices with very direct risk management functions. Subcategories are included only for those offices where significant risk management functions are undertaken. [a]The Office of Health and Environmental Assessment within the Office of Research and Development is the key office that actually conducts risk assessments. It contains the Exposure Assessment Group; the Carcinogen Assessment Branch and the Carcinogen Assessment Statistics and Epidemiology Branch (formerly the Carcinogen Assessment Group) within the Human Health Assessment Group; and the Genetic Toxicology Assessment Branch and Reproductive and Developmental Toxicology Branch (formerly comprising the Reproductive Effects Assessment Group) also within the Human Health Assessment Group. It also contains the Risk Assessment Forum (U.S. EPA, June 1989).

The commission's emphasis, at least in the early part of its history, had been on injury and safety rather than chemical health risks per se.

The commission's structure is typical of an independent governmental commission, with five commissioners, including a chairman. The commissioners are Presidential appointees subject to Senate approval. While the chairman has considerable administrative authority, the CPSC as a unit can only act with a majority vote of the commissioners (National Research Council, 1983, 42). The commissioners also have approval authority over various appointments to the commission (LaPorte, Gies, and Baum, 1975, 1094). The commissioners play a role analogous to that of the executive of a cabinet-level agency. In order to maintain bipartisan representation, only three commissioners can be from one political party. The organization of the commission is shown in Figure 5.2.

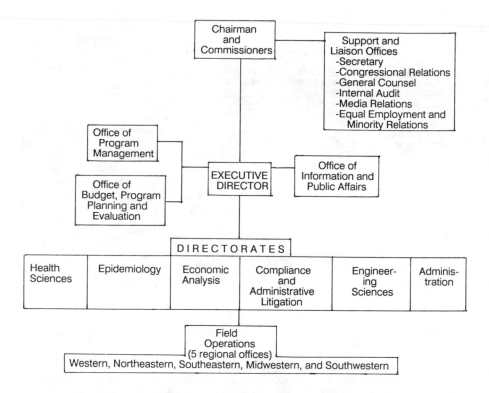

Figure 5.2.   Organization of the Consumer Product Safety Commission. *Source:* Summarized from U.S. Consumer Product Safety Commission (1986, vi).

Department of Health and Human Services:
Food and Drug Administration

The Food and Drug Administration is located within the Public Health Service of the Department of Health and Human Services. It is one of several agencies within that department that manage health risks. The other agencies within DHHS with risk management related functions are described briefly below:

- The Centers for Disease Control were established in 1973 for the control and prevention of disease. They administer programs for disease prevention, provide expertise and technical advice, and perform emergency response functions.
- The Agency for Toxic Substances and Disease Registry (ATSDR) was established in 1983 under CERCLA to protect the public and workers from risks from toxic substances. The agency acts as a clearinghouse for information and expertise, develops and analyzes scientific and technical information and procedures, has been charged with conducting health risk assessments at hazardous waste sites designated under Superfund, and has an emergency response function.
- The National Institutes of Health (including the National Cancer Institute) conduct and finance research and training for health protection.

The Food and Drug Administration's chemical risk management function primarily covers contaminants in food and drugs. It also is concerned with the risks to public health from the use of cosmetics and medical devices. It shares some of its functions in food safety with the Food Safety and Inspection Service (FSIS) established within the Department of Agriculture in 1981, which regulates meats and poultry primarily by means of plant inspections. It shares other functions with the Department of Commerce, which regulates fish and shellfish, and state governments, which regulate dairy products (Meier, 1985, 88).

FDA's predecessor agencies date from at least 1907 when the Food and Drug Act of 1906 was passed. It was given its current name under the Agriculture Appropriation Act of 1931 when its functions with respect to foods and pesticide control were carried out in the Department of Agriculture (U.S. General Accounting Office, April 1987, 21). The functions of the FDA were transferred to other agencies several times. In 1953, it was incorporated into the Department of Health, Education, and Welfare; in 1979, it was incorporated into its current location, the DHHS. Each transfer generally enlarged its functions. (Congressional Quarterly, Inc., 1983, 1986).

The FDA is an administration within a department and is thus further removed organizationally from the Executive Office of the President than the EPA or some of the independent commissions. The organization of the FDA is shown in Figure 5.3.

Some of the major functions performed by the FDA that directly or indirectly

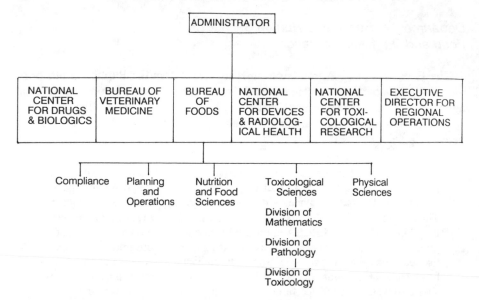

Figure 5.3. Organization of the Food and Drug Administration of the Department of Health and Human Services. *Source:* National Research Council (1983, 100). *Note:* This chart does not represent the entire organization of the FDA. It emphasizes the organization and placement of the Bureau of Foods, which has had one of the major responsibilities for risk assessment within the agency.

involve risk assessment are as follows (Congressional Quarterly, Inc., 1986, 332–334):

- It inspects food processing and other manufacturing establishments under its jurisdiction, such as manufacturing facilities for drugs and medical devices.
- It registers drug manufacturing and processing facilities, commercial processors of canned foods with a low acid content, manufacturing establishments producing medical devices, and manufacturers of electronic products emitting radiation.
- It licenses and conducts premarket testing programs for new drugs, biological products, and medical devices.
- It issues and administers labeling requirements.
- It establishes regulations, including standards, for safety, levels of quality, or additives in the products it regulates, and applies these standards in the exercise of its regulatory functions of inspection and registration. These standards include food standards called "action levels," guidelines for good manufacturing practice, and new drug regulations.
- It enforces regulations by issuing warning or regulatory letters, recalling products, issuing injunctions (usually to stop continued manufacture and distribution), issuing citations informing violators of impending criminal action, seizing existing shipments, prosecuting violators through the U.S. Attorney General's office, or imposing fines for certain infractions.

Department of Labor:
Occupational Safety and Health Administration

OSHA was established in 1970 to administer the Occupational Safety and Health Act (OSH Act). Under that act, the agency is charged with maintaining the health and safety of employers and employees in the workplace environment, with the exception of the federal government workplace, which is not covered under OSHA.[3] In order to accomplish this mission the agency conducts the following functions:

- It develops and enforces health and safety standards and programs in four areas: general industry, agriculture, maritime, and construction activities. There are three kinds of standards: interim standards, emergency temporary standards (OSH Act, Section 6(c)), and permanent standards (OSH Act, Section 6(b)). Interim standards are Threshold Limit Values ("consensus standards"), set by the American Conference of Governmental Industrial Hygienists (ACGIH) prior to the creation of OSHA and updated annually, that were adopted by OSHA in its early years. Emergency standards address imminent danger from new hazards, and permanent standards are developed six months after emergency standards are set (McCaffrey, 1982, 73).
- It mandates recordkeeping of injuries and illnesses by business establishments.
- It conducts inspections of individual workplaces to ascertain compliance with the standards.
- It enforces its regulations by issuing injunctions where dangerous conditions are found.

During the 1970s the agency was criticized for its inability to enforce a large number of standards. It was also criticized for emphasizing areas considered trivial and for emphasizing worker safety to a much greater extent than worker health (Congressional Quarterly, Inc., 1983, 398). During the late 1970s, the regulatory reform movement affected OSHA just as it did other agencies. As a result, OSHA reduced and simplified the number of standards it was promulgating and enforcing. It also eased requirements for recordkeeping and reduced the number of plant inspections it conducted.

Like the FDA, OSHA is an administration located within a larger department, the Department of Labor. It is one of about a half a dozen or so major units reporting directly to the Secretary of Labor. As such, it theoretically has less authority than the Administrator of EPA or a commissioner. The organization of OSHA is shown in Figure 5.4.

Nuclear Regulatory Commission

The five-member Nuclear Regulatory Commission was established by the Energy Reorganization Act of 1974 and Executive Order 11834 of 1975. The NRC regulates both nuclear materials and power plants. The same legislation

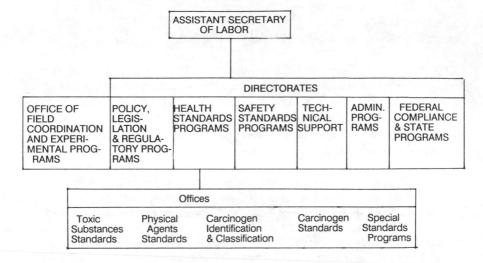

Figure 5.4. Organization of the Occupational Safety and Health Administration within the Department of Labor. *Source:* National Research Council (1983, 95). *Note:* This chart does not represent the entire organization of OSHA. It emphasizes the organization and placement of the Directorate of Health Standards and Programs, which has had one of the major responsibilities for risk assessment within the agency.

that created the NRC created the Energy Research and Development Administration, which oversees nuclear development. Prior to the act (and since 1946), these two functions, regulation and development of nuclear energy, were conducted by the same agency, the Atomic Energy Commission (AEC).

The regulatory functions of the NRC are as follows:

- It licenses, regulates, and inspects the construction and operation of nuclear power plants as well as facilities for research, demonstrations, and power generation.
- It regulates all aspects of nuclear materials handling and the export of certain materials.
- It develops standards, makes safety determinations, and promulgates other regulations pertaining to nuclear materials and facilities and their operations.

The organization of the NRC is shown in Figure 5.5.

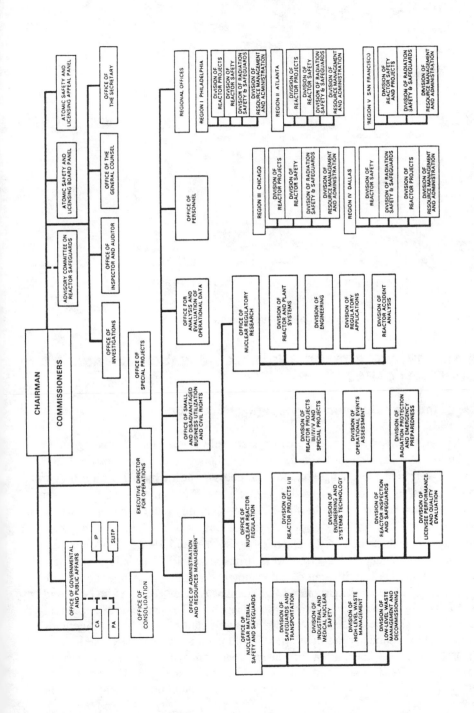

Figure 5.5. Organization of the Nuclear Regulatory Commission. *Source:* U.S. Nuclear Regulatory Commission (1988).

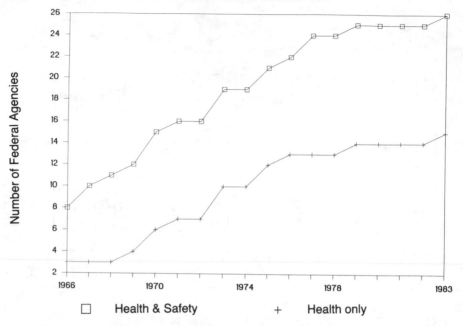

Figure 5.6. Cumulative agency growth: health and safety.

TRENDS IN THE SIZE OF THE RISK BUREAUCRACY AND ITS BUDGETARY CAPACITY

Size of the Risk Bureaucracy: Agency Formation

By 1988, over two dozen agencies were actively involved in the management of health and safety risks at the federal level. Many of these agencies are listed in Table 5.1, and the change in their numbers is portrayed as a cumulative growth curve in Figure 5.6. As that figure shows, the cumulative growth in these agencies has been steady since the 1970s. This pattern can in part be attributed to the growth in risk-oriented legislation, occurring at about the same time, that was described in Chapter 2. Since the mid-1970s, however, the addition of new agencies has come to a virtual standstill. The only major new additions of agencies in the area of environmental health risks since the middle 1970s were the Federal Emergency Management Agency (FEMA) in 1979 and the ATSDR in 1983. Thus, agency growth in the 1970s appeared to parallel growth in legislation, while in the 1980s no significant number of new agencies were added to administer the many new laws that continued to be passed.

The influence of legislation in earlier years on agency growth has been underscored in several studies. Brickman, Jasanoff, and Ilgen (1985, 80) note that the parallel growth of laws and agencies is an outcome of what has been observed as Congress's way of delegating one law to one agency, or "the U.S. penchant for entrusting new legislation to new agencies."[4] O'Brien (1986a, 35) also observes that regulatory agencies in this area tend to be "single-purpose" agencies. Important exceptions to this notion of single-purpose agencies do exist, however. For example, the EPA administers some nine statutes, not just one, and several of the other agencies typically administer more than one statute. Nevertheless, from a larger perspective, general subject areas such as worker health, food, and the general environment do tend to be concentrated in the hands of separate agencies.

The key question addressed here is whether this growth in agencies was unusual when compared with levels of activity in agency formation in government in general. While data are difficult to obtain to answer this question definitively, several studies of the growth of government bureaucracies do provide a good basis for comparison. Lowi recounts the history of governmental agencies in general from the turn of the century to the middle of the century, and observes a growth rate similar to the one occurring in the environmental health agencies (Lowi, 1969, 31). Lowi underscores this trend in terms of the large per capita dollar outlay for governmental administration, total dollar outlay for administration and activities, and personnel levels by governments at all levels. An even stronger argument for this growth is presented by Kaufman as he traces a precipitous (though not necessarily steady) growth in the cumulative curve for all federal governmental organizations since the late 18th century (Kaufman, 1976, 48–49). He attributes this to the fact that the rate of organizational creation and the endurance of organizations far exceed the death rate.[5] Finally, MacIntyre provides a nice link between the growth of agencies in general and the growth in laws:

> One of the more profound developments of this century has been an enormous expansion in the power of government bureaucracy. The proximate cause of this growth has been the congressional inclination for writing more laws—laws which have typically been vague and ambiguous. (MacIntyre, 1986, 67)

In summary, the rise in the number of risk organizations has been precipitous in the past decade. While it is sizable, when viewed against century-long trends for federal governmental agencies in general it is not unusual and can be explained in terms of the way government organizations appear to proliferate over time and traditionally respond to legislation.

Capacity for Risk Management: Resource Allocation and Its Growth

Budgetary factors influence agency decisionmaking and the extent to which agencies can control their actions, often in inconsistent ways. The existence of extra resources provides what has been called "organizational slack" (Cyert and March, 1963). Such slack or flexibility, according to Mitnick, enhances organizational discretion—for personal or public interest ends or for the interests of regulated parties (Mitnick, 1980, 149). In reality, resources never seem to be enough to cover agency needs because of "the legislature's tendency to grant many more statutory duties than it provides resources to carry them out" (MacIntyre, 1986, 77). Resource constraints can seriously impair an agency's ability to carry out its mandate but can in some ways also enhance agency discretion. It has been argued that budgetary constraints allow agencies to set priorities, as long as funds are not at such a low level as to inhibit an agency's ability to carry out legislative mandates.

Organizational resources are not always expressed directly in terms of funds. Meier (1985, 15) has characterized organizational resources, particularly those allocated to regulatory agencies, as falling generally within five categories. These are (1) expertise or the development and storage of knowledge; (2) cohesion, or the united pursuit of goals by agency personnel; (3) legislative authority, as reflected in legislative goals, delegation agreements, sanctions, and procedures; (4) political salience, or the importance of the regulatory area as perceived by the public; and (5) the quality and goals of leadership.

Looking at the dollar budgetary allocation for the major programs and agencies involved in risk management, some clear trends are apparent, in terms of both program area and agency. Some dramatic changes in the federal budget between the 1970s and the 1980s are discussed below.

Budgetary Trends in Risk-Related Program Areas

Budgets for individual federal agencies do not distinguish line items for risk assessment or risk management from other functions. This hinders direct identification of resources for risk management. In lieu of direct figures, budget items can be used for a few isolated programs primarily in the regulatory area whose functions are closely identified with risk management. These programs include the development of standards and enforcement in environmental protection, environmental health, and occupational health.

The *Congressional Quarterly* estimated that as a gross trend over the decade from 1970 to 1980, regulatory agency expenditures (including personnel costs) for health, safety, and environmental protection alone increased 10-fold, from $0.5 billion to over $5 billion per year (unadjusted for inflation). Since that time, however, the annual increase over the 1980 level has rarely exceeded 10% (Congressional Quarterly, Inc., 1983, 4). More detailed and subsequent budgetary figures both for outlays by the federal government directly and for federal outlays to state and local government show the same trend: increases through the 1970s, followed by declines in the early 1980s. Some reversal of the downward trend appeared toward the end of the 1980s, but earlier budgetary levels were not restored. This trend is true regardless of the expenditure category.

Some of these budgetary trends, expressed in 1982 dollars,[6] are illustrated in Figures 5.7 through 5.10. Direct federal outlays for two categories, (1) pollution control and abatement and (2) consumer and occupational health and

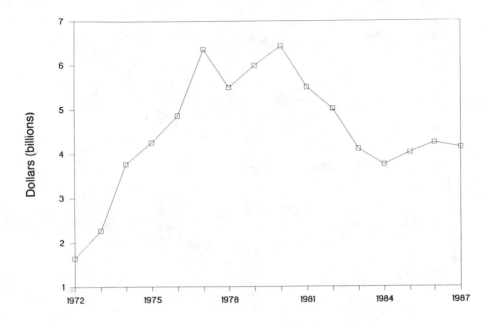

Figure 5.7.   Annual federal budget outlays for pollution control and abatement (billions of 1982 dollars).

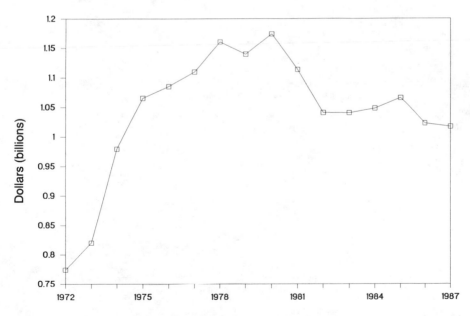

Figure 5.8.   Annual federal budget outlays for consumer and occupational safety and health (billions of 1982 dollars).

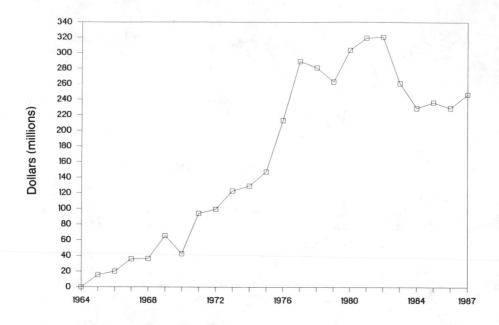

Figure 5.9. Annual federal budget outlays to state and local governments for pollution abatement, control, and compliance (millions of 1982 dollars).

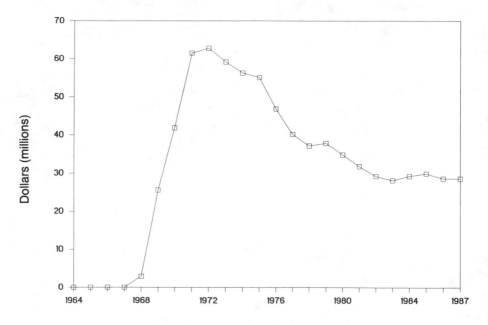

Figure 5.10. Annual federal budget outlays to state and local governments for food safety and inspection (millions of 1982 dollars).

safety, show similar trends. Pollution control and abatement outlays (shown in Figure 5.7) peak in 1980 with declines through 1984 and a reversal to approximately the 1982 level by 1987. Direct federal outlays for consumer and occupational health and safety (shown in Figure 5.8), also peak in 1980 and decline through 1982, after which time the outlays level off.

Figures 5.9 and 5.10 show federal outlays to state and local governments in 1982 dollars for a variety of programs that involve risk management. These figures are significant because a number of programs involving risk management have been delegated to state and local governments for implementation, such as management of the construction grants program for wastewater treatment plants, air emission control programs, and wastewater discharge programs. Comparing direct federal outlays with federal outlays to state and local governments addresses an argument often put forth that federal outlays are decreasing in a given program area because outlays to other levels of government are increasing. The federal outlays for state and local governments are grouped into the categories of (1) abatement, control, and compliance (2) food safety and inspection, and (3) occupational and mine safety. When the effects of inflation are taken into account, the trends in the abatement, control, and compliance area show declines after the early 1980s, with slight increases appearing after the mid-1980s. In the food safety and inspection area, however, the declines in 1982 dollars are precipitous from 1972 on. Thus, the declines in the federal budget in areas related to risk management are not explained in terms of allocations to state and local government, since these budgetary outlays do not increase as the federal outlays decrease.

Budgetary Trends by Agency

EPA. The allocation of resources in general for environmental protection in the 1970s reflected the growing concerns of the environmental movement. These increases were followed by budgetary declines, reflecting a decreased emphasis on these programs due to inflation, regulatory reform movements that placed economic constraints on environmental policy, increased delegation of environmental programs to state agencies, and changes in the political climate. The budgetary trend for the EPA is shown in Figure 5.11 from 1980 through 1987. While the declines through 1981 are reversed to some extent, the 1987 level never attained earlier levels, when adjusted for inflation.

CPSC. The CPSC has never been able to obtain the funding levels that have been authorized by Congress. In the first five years of its operation, it is estimated that the CPSC was only appropriated about two-thirds of what was authorized (Bick and Kasperson, 1978, 34). After that, it suffered severe budget cuts that reduced the size of its staff, which has recently numbered between 500 and 600 and is responsible for some 15,000 consumer products. Figure 5.12 shows the precipitous declines in the commission's budget since 1980, most of

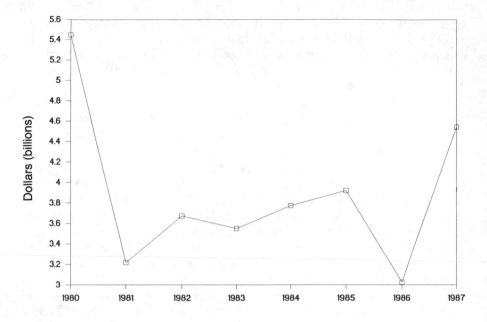

Figure 5.11.   Annual budget for the EPA (billions of 1982 dollars).

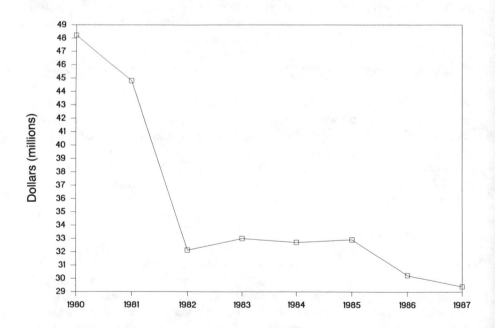

Figure 5.12.   Annual budget for the CPSC (millions of 1982 dollars).

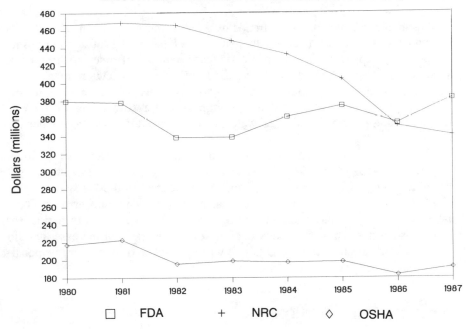

Figure 5.13. Annual budget: FDA, NRC, and OSHA (millions of 1982 dollars).

which occurred between 1980 and 1982. The trend is especially pronounced after figures are adjusted for inflation.

FDA. The FDA budget has on average remained relatively stable between 1980 and 1987, showing slight cyclical fluctuations in budget levels. The trend is shown in Figure 5.13.

OSHA. OSHA's funding over the 1980 to 1987 period has, like FDA's, been relatively stable. (See Figure 5.13.) Thompson shows that the level of funding for one of OSHA's functions, standards-setting, which is a major risk-related function, was usually well under 10% of the agency's total appropriations annually between 1973 and 1980. In addition, he points out that the amount allocated to standards-setting remained constant over that period regardless of the increase in the level of activity in the standards-setting area (Thompson, 1982, 209). As a result of limitations on the size of the inspection force, the agency has been forced to establish priorities for workplace inspections. For example, in 1982, 1200 inspectors were available to cover some 70 million employees in over five million establishments (Congressional Quarterly, Inc., 1983, 406).

NRC. Budgetary levels for NRC show steady declines from 1980 through 1986, with the beginning of a reversal appearing in 1987. Trends in these levels are shown in Figure 5.13.

To summarize, the budgetary trends by agency and program show increases through the 1970s and declines thereafter until the mid-1980s, when some recovery was apparent. These are trends that a number of other governmental sectors felt as well. Attempts by Congress to reinstate these cuts in many cases did not bring back the amounts to previous levels. The only exceptions to this were major expansions in legislated programs, such as allocations to SDWA, the Superfund program, and wastewater treatment plant construction under the CWA.

The general thrust of oversight reports and commentary on the federal budget for the management of risks from chemicals is that the dollar amounts are not sufficient to cover the scope of responsibilities defined in the legislation. This is especially true in light of the fact that many programs appear to be very labor-intensive in areas such as enforcement and gathering information for standards-setting. Thus, budgetary declines or at least rates of increase that did not equal earlier rates occurred at a time when the scope of regulatory responsibility was increasing.

DISCRETIONARY AUTHORITY

The latitude or degree of choice allowed to agencies in conducting their activities is often referred to as discretionary authority. The concept of organizational discretion is widespread in the literature on organization theory and management and was covered partly in Chapter 1. Much of the literature on discretion arises from studies of both public and private organizations. Barnard (1938) referred to the concept of the "zone of indifference." Simon (1976) adapted this concept, referring to a "zone of acceptance"; it was developed further by Thompson (1967). An early source of the use of the terms discretion (as autonomy) and prescription in the field of political theory is attributed to Dahl and Lindblom (1953). The concept has also been applied extensively in the literature on regulation as central to explaining administrative behavior and regulatory outcomes. For example, Kagan (1978) provides an application of the concept of discretion to regulation. Greenwood (1984) links the form of discretion to the state of knowledge in environmental, health, and safety decisions. Greenwood has also identified aspects of administrative discretion that differ from those identified above, namely discretion in "interpreting statutory language," "balancing conflicting values," "determining priorities," and "answering scientific and engineering questions" (Greenwood, 1984, 5–6; Shapiro, 1984, 94).

Administrative discretion and its determinants provide an important framework to understand how administrative and organizational developments in risk policy compare with those in other governmental operations. Chapter 4 emphasized the ways in which Congress controlled the boundaries of discretion allowed to administrative agencies through the enactment of legislation, its oversight function, and its authority over appropriations. The content of the legislation itself, described in Chapter 2, reflected the substance of many of these Congressional actions. The executive office placed severe constraints on agency discretion in environmental health regulation in the form of executive orders. The most significant orders were Executive Order 12044 under the Carter administration and Executive Order 12291 under the Reagan administration (Kraft, 1984). In addition, the role of the Presidential Office of Management and Budget increased executive oversight of agency operations.

Administrative discretion and its ramifications for risk management are addressed in terms of four dimensions:

1. Discretionary authority is reflected in the design of the organization of the agencies. In particular, each of the agencies can exercise discretion in locating the risk management function within the organization structure. This location often determines how much access to and influence on the chief executive these functions can have. It sets organizational priorities for key decisionmakers and technical specialists, who control and generate the information that goes into risk assessments. Organizational location, at least theoretically, reflects the degree of power, responsibility, and authority that a function has within an organization.

2. The agencies can to a large extent exercise discretion in the use of in-house and external advisory committees, which expand their technical resource base. This discretion is exercised within certain limits. The formation of advisory committees and the development of their procedures are not entirely discretionary, since they are often subject to the requirements of the Federal Advisory Committee Act and statutory prescriptions for the role of advisory bodies. Nevertheless, the extent to which an agency uses an advisory committee and its findings is discretionary.

3. Agencies vary in the latitude they have in setting standards, which includes choice of chemicals to regulate, setting of levels for them, and scheduling these activities.

4. Agencies vary in the discretion they exercise in the rulemaking process, particularly in the way each agency can deviate from the formal procedures for rulemaking outlined in the Administrative Procedure Act (APA). According to Davis (1972), informal rulemaking is the most common form of governmental rulemaking.

Discretionary authority is a quality that agencies in general desire to possess. It has been argued in a number of reviews of legislation that agencies will tend to maximize discretion and resist prescriptive mechanisms of analysis in order to maximize their power and authority (DeLong, 1983, 5). The extent to which agencies resist prescribed procedures is exemplified by EPA's reluctance to give up its case-by-case negotiation approach to setting wastewater effluent limits in favor of the development of standards and particularly the granting of waivers on a class-by-class basis only. Congress's emphasis on reducing agency discretion is an outgrowth of its fear that management agencies will be "captured" by the very interests they are to regulate (Wilson, 1980; Kagan, 1978).

The degree of agency discretion is defined and bounded by legislative mandates, the norms of administrative procedure (including the tradition of delegation), and resources. The ground rules for discretion are set forth in legislation either through direct references to agency authority or by omission. Chapter 2 pointed out that many agencies typically operate under vague statutory mandates. Furthermore, in designing these statutes Congress sets the parameters for agency discretion. Whether or not these actions on the part of Congress represent a departure from what had been typical of government in the past depends on what time period is considered. If one looks at the past two decades, according to Greenwood (1984, 4; citing Ackerman and Hassler, 1981), the passage of the major environmental, health, and safety statutes "expanded the federal government's authority and responsibility to intervene in the private sector for the purpose of protecting the natural environment and health and safety of Americans."

As defined by administrative procedure, discretion is linked to the process of delegation of authority from one branch of government to another and from one level of government to another. There is a long and pervasive tradition of discretion in government as applied to delegation.[7] Greenwood (1984) cites arguments showing that the height of such delegation actually occurred around the period of the New Deal and has been declining since then. Mitnick, in contrast, sees delegation as more of a continuum in the direction of an increase over time. Tracing the history of Congressional delegation to agencies back to the 19th century, Mitnick observes: "Thus, Congress at first tried to administer the task [of maritime affairs] itself through committees of its members. It turned to delegation, i.e., the creation of agents to perform an act or task heretofore performed by the delegating party," when it found a variety of ills such as impossible work loads, inefficiencies, high turnover, lack of expertise, etc. (Mitnick, 1980, 23). Furthermore, the delegation of discretionary authority from other branches of the federal government to the administrative agencies was paralleled by a delegation of authority from federal administrative agencies to state and local government (Davies, 1984, 151). This occurred in many areas and was extensive during the early years of the Reagan administration, as the CEQ points out:

> . . . by the end of 1982 state governments had been delegated enforcement responsibilities for over 95 percent of applicable National Emissions Standards for Hazardous Air Pollutants, and over 90 percent of applicable New Source Performance Standards, up from 64 percent at the beginning of the year. In addition, of the 60 state and local agencies eligible to grant and enforce new source permits, 48 had been delegated full or partial authorities by the end of 1982, up from 26 at the end of 1981. (Council on Environmental Quality, 1983, 75)

Another aspect of administrative procedure as a factor in agency discretion is the way the judiciary and Congress shape agency discretion. It will be shown in the next chapter that the judiciary, along with Congress, also affects agency discretion in the environmental risk area as it does in other areas through the judicial review of agency procedures and substantive decisions. In fact, when the actions of Congress and the judiciary are viewed together, opposing trends emerge, both within each of the two entities and between them, with respect to their philosophies toward agency discretion. On the one hand, Congress passed many laws that enhanced discretion by being vague or ambiguous. The responsible agencies were forced to engage in considerable rulemaking to compensate for such statutory vagueness. On the other hand, Congress also prescribed a good deal in the area of types of chemicals to be regulated. The judiciary reinforced Congressional constraints on agency discretion in some of its case reviews under the legislation. It did so by conducting a *de novo* review of agency decisionmaking and fact-finding.[8]

Agency Discretion in Organizing the Risk Assessment Functions

The creation of an administrative entity charged with risk assessment and risk management within government is not prescribed in any of the legislation discussed in Chapter 2. The creation of such entities as well as their location within government is purely a discretionary act. Over the past couple of decades, there has been a dramatic rise in the exercise of such discretion in terms of the number of entities charged with risk management functions that have appeared within government. The importance, power, and priority of the risk management and assessment unit relative to other agency responsibilities is reflected in its location in the organization. One way of characterizing organizational location is in terms of distance, or number of echelons, to the top of the organization, i.e., to a commissioner or chief administrator.[9] Over the last half century, most environmental functions that have dealt with health risks have continually risen to the top of the structure of both public and private organizations. External factors often influence the priority and hence the location of a given function in an organization. The location of the risk assessment function is influenced by the relative strength of legislative directives to consider activities that involve risk.

One issue that has been controversial in connection with the location of risk management functions pertains to how close the two functions of risk assessment and regulation should be organizationally. A major study by the National Research Council (1983) on federal risk management and risk assessment functions considered a proposal to join the two activities in the same area of a given organization.[10] The advantage of such an organizational arrangement would be to improve coordination between the two functions. However, the NRC implied that within a given organization the functions should be separate. Separating the two functions in different organizations was not recommended in the end. It was argued that the complete separation of the risk assessment function from the regulatory function would hinder the responsiveness of regulatory agencies to risk assessment (National Research Council, 1983, 6, 152).

Greenwood (1984, 114–120) has discussed the pros and cons of alternative locations for the risk assessment function in connection with EPA and OSHA, with particular attention to the relationship between location and competence. He argues that one advantage of separating the two functions and concentrating risk assessment in a separate agency is that it is easier to recruit skilled people in a single area, such as risk assessment. Also, he argues, by separating the two functions, the scientific aspects of risk assessment will not be influenced by policy considerations. In contrast, many have argued that the science of risk assessment cannot, in fact, be isolated from policy (Greenwood, 1984, 115; Latin, 1988).

EPA

EPA manages environmental health risks in a number of different ways. First, it manages the application of risk assessment methods for the development of environmental and health criteria and standards. Some examples of this are the use of risk assessments for the preparation of criteria documents that are the basis for National Ambient Air Quality Standards (NAAQS) under the Clean Air Act, for criteria documents for Water Quality Criteria under the Clean Water Act, and for the establishment of Maximum Contaminant Levels for drinking water contaminants as part of the National Primary Drinking Water Regulations under the Safe Drinking Water Act.

Second, EPA manages the use of risk assessments to regulate specific sources of risk on a case-by-case basis under regulatory programs involving permits, licenses, or registrations for facility construction or operation. For example, a risk assessment was conducted in the preparation of a permit for the ocean disposal of sewage sludge from New York City under the Marine Research, Protection, and Sanctuaries Act, a law that is also administered by EPA. In this case the relative risk of disease from the various chemicals in the sludge under alternative treatment techniques was evaluated via risk assessment. Risk assessment has also been used to develop evaluation criteria for alternatives to ocean disposal of sewage sludge. Under the Safe Drinking Water Act's Wellhead Protection Program, EPA has been developing guidelines incorporating risk assessment for water purveyors, so they can determine the seriousness of contamination in wellfields. Another example is the use of risk assessments for preparing environmental impact statements under the National Environmental Policy Act, especially where public health effects are a prominent concern. Such assessments are routinely prepared for impact statements involving any hazardous or toxic substances during construction or for disposal. Risk assessments also accompany every remedial investigation/feasibility study prepared for Superfund sites.

The EPA has tried to use its discretionary authority in designing its organization to give some prominence and priority to risk issues. The development and application of risk assessment methods tends to be centralized in the agency's headquarters in Washington rather than decentralized in the regional offices.

Within its headquarters office, EPA has centralized the major responsibilities for risk assessment within the Office of Health and Environmental Assessment. The subgroups within this office that conduct various aspects of risk assessment are organized by stage in risk assessment. They include the Exposure Assessment Group, the Carcinogen Assessment Branch and Carcinogen Assessment Statistics and Epidemiology Branch (formerly the Carcinogen Assessment Group) within the Human Health Assessment Group, and the Genetic Toxicology Assessment Branch and Reproductive and Developmental Toxicology Branch (formerly comprising the Reproductive Effects Assessment Group), also within the Human Health Assessment Group. A major coordinative function is executed by the Risk Assessment Forum, which was formed in 1984 after the National Research Council study was performed. According to EPA, the forum brings together agency risk assessment experts in a formal process to gain consensus on the methods and guidelines the agency publishes. Organizationally, the forum is located within the OHEA, but the forum's staff reports to the agency's Risk Assessment Council. The council reports directly to the Administrator of the Office of Research and Development. The agency publishes a variety of guidance material in the area of risk assessment, including health assessment and criteria documents on a chemical-by-chemical basis. The agency also has made available a computerized database of chemical health risks. The activities at headquarters are supplemented by assessment work at other centers of the agency, namely the Environmental Criteria and Assessment Offices in Cincinnati and in Research Triangle Park, North Carolina (U.S. EPA, June 1989).

Within the headquarters, however, the risk assessment function is also dispersed throughout a number of offices, partly as a result of the large number of offices within EPA that use risk assessment. The extent of this dispersion is reflected in the data on the location of these offices given in Table 5.3. The directors of each of these offices report directly to an assistant administrator, who in turn reports to the EPA Administrator. Thus, for the most part, the risk assessment function is at least three echelons down from the Administrator in the headquarters office. As Greenwood (1984, 120) points out, within a given office, such as the Office of Air Quality and Planning Standards, risk assessment activities can be subdivided even further among the various divisions. This moves them even further from the top of the organization.

CPSC

Risk assessments and related activities within the CPSC are conducted by the Directorate for Health Sciences, which reports directly to the Executive Director of the commission. Between 1981 and 1989, almost two dozen risk assessments were prepared by this directorate (M. Cohen, CPSC, September 8, 1989, per-

Table 5.3. Organizational Subunits Conducting Risk Assessments Within EPA

Office	Legislation	Program for Which Assessments Are Conducted
Assistant Administrator for Water		
Office of Drinking Water	Safe Drinking Water Act	National Primary Drinking Water Standards
Office of Ground Water Protection	Safe Drinking Water Act	Underground Injection Control Program; Wellhead Protection Program
Office of Water Regulations and Standards	Clean Water Act	Effluent guidelines; water quality guidelines
Office of Water Enforcement and Permits	Clean Water Act	NPDES permit program
Assistant Administrator for Solid Waste and Emergency Response		
Office of Emergency and Remedial Response	Superfund	Remedial investigation/feasibility studies
Office of the Assistant Administrator for Air and Radiation		
Office of Air Quality Planning and Standards	Clean Air Act	National Ambient Air Quality Standards; Criteria Documents
Office of Radiation Programs	Atomic Energy Act	Radiation standards
Office of the Assistant Administrator for Pesticides and Toxic Substances		
Office of Pesticide Programs	Federal Insecticide, Fungicide, and Rodenticide Act	Pesticide standards
Office of Toxic Substances	Toxic Substances Control Act	Premanufacture Notifications
Office of the Assistant Administrator for Research and Development		
Office of Health and Environmental Assessment	Various laws	Risk assessment procedures

sonal communication). The Office of Program Management, which also reports to the Executive Director, is responsible for the development of the commission's standards as well as those recommended by others via petition. It also sets requirements and rules and conducts impact analyses for various products (Congressional Quarterly, Inc., 1986). In this capacity it plays a major role within the commission in risk management. This office is divided into programs along the lines of types of hazards and types of assessments. Other entities provide technical support functions for risk management, namely the Directorates of Economic Analysis and Engineering Sciences.

One of the commission's activities reflecting considerable discretionary authority in risk management has been the Chemical Hazards Program, administered by the Directorate for Health Sciences since 1981. This program attempts to set priorities by identifying and evaluating chemicals that pose potential chronic health hazards to consumers. All information on identified chemicals, including risk assessments from other agencies, is incorporated into the program as the basis for recommendations on the further use of the chemicals. The evaluation of chemicals involves a systematic screening procedure (Menza, 1983). A major component of the Chemical Hazards Program is the System for Tracking the Inventory of Chemicals (STIC). It is used by the CPSC as an internal management tool. Chemicals for further evaluation are chosen from the inventory and on the basis of information from the inventory. Chemicals are suggested for incorporation in the inventory from various EPA reports (such as those prepared under Section 8(e) of TSCA), trade publications, and consumer complaints. In its early years (as of June 1983), there were 951 chemicals in the inventory; by April 1989 the number had increased to 1034 (W. P. Menza, Directorate of Health Sciences, CPSC, September 8, 1989, personal communication).

The status of chemicals and reports issued in the program in 1983 and in 1989 is summarized in Table 5.4, based on CPSC's data. The tabulation shows that, according to the CPSC, the components of the Chemical Hazards Program other than STIC were not continued beyond the mid-1980s. The outputs of this program have been used as inputs into various projects conducted by the commission, in areas such as children's chemical hazards, dyes and finishes, school laboratory chemicals, and indoor air quality.

The authority to ban a chemical is one of the most forceful risk management tools that the commission has and is conducted at a fairly high level within the commission. These decisions are usually based on extensive risk assessments. But according to its own tabulations, the CPSC had in force only a few chemical bans by September 1986. These were primarily for asbestos- or lead-containing products and chemicals that had achieved high visibility under other statutes, such as carbon tetrachloride and vinyl chloride.

Table 5.4.   **Progress of the CPSC Chemical Hazards Program**

Program Component	Number of Substances Involved	
	As of June 1983	As of April 1989
Indentification of new chemicals: System for		
Tracking the Inventory of Chemicals (STIC)	951	1304
Preliminary economic reports	549	NA
Recommendations for toxicologic reports	240	NA
Preliminary toxicological reports	40	NA
Recommendations for continued screening	18	NA
Recommendations for in-depth economic and		
toxicologic reports	10	NA

FDA

The National Research Council study (1983) points out that within the FDA, risk assessments are done separately for foods and for drugs. The Bureau of Foods and the Bureau of Drugs perform the majority of the assessments. These bureaus report directly to the Administrator of FDA. Both bureaus are administratively separate from regulatory decisions. The Bureau of Foods conducts risk assessments for the purpose of establishing "action levels" for substances in food. These levels are a type of standard, specifying an allowable concentration of a particular substance in food products or products that come into contact with food. Risk assessments by the Bureau of Foods are conducted in the context of new product reviews by petition from manufacturers, periodic reviews of compounds already approved, and review of "inadvertent contaminants" that become known to the agency through a variety of channels. The Bureau of Drugs performs an analogous role for drugs. Within that bureau, the Office of New Drug Evaluation has the major responsibility for new drug review. This office, in turn, is subdivided into divisions organized by type of drug (National Research Council, 1983, 99).

The exercise of discretion in the use of risk assessment by FDA is made extremely complex by the presence of the Delaney Clause in the FDCA, which requires the agency to regulate to zero risk for certain categories of substances. The Delaney Clause was applied legislatively to three different substances at different times: to food additives in 1958, to color additives in 1960, and to animal drug residues in 1968 (Merrill, 1988, 3). This clause does not have a parallel in any other legislation involving chemical risk. The existence of a zero risk threshold presumably precludes the application of risk assessment, since measuring the level of risk of any concentration of a substance to which the Delaney Clause applies is irrelevant once any level of contamination has been found. Nevertheless, the FDA has exercised its discretionary authority in using risk assessment to demonstrate that certain levels of additives pose only a de minimis or trivial risk, and that to regulate below that level was not the intent of the Delaney Clause. The agency has largely been driven to this position (as have others) because of the increasing sensitivity of detection technologies. FDA's stance with respect to trivial risks has been challenged in court.

OSHA

OSHA is the major regulatory agency in charge of worker risks from toxic substances. Another organization, the National Institute for Occupational Safety and Health (NIOSH), is responsible for conducting the research and risk assessments on which OSHA's standards can be based. NIOSH is located in an entirely different executive department. It is one of the research institutes within the U.S. Public Health Service's Centers for Disease Control in the Department of Health and Human Services. A reason for this organizational arrangement was to separate scientific research from economic and regulatory actions in the development of health standards. The National Research Council (1983) study of the relationship of OSHA and NIOSH revealed that a lack of coordination existed between the two organizations. NIOSH's risk assessment and management activity related to research and criteria development for the standards is not always used for OSHA's regulatory functions because of the organizational separation of the two agencies. OSHA exercises considerable discretion in whether or not it uses NIOSH's studies. For example, in spite of the more than 100 criteria documents produced by NIOSH, few have formed the basis of OSHA standards. OSHA has, in fact, hired its own health scientists to take the place of NIOSH's resources in health effect evaluation studies (McCaffrey, 1982). Greenwood (1984, 116) has underscored this observation, pointing out that over time, OSHA has assumed responsibilities in the area of risk assessment to the exclusion of NIOSH. Greenwood further points out that while OSHA tried to separate the risk assessment and risk management functions organizationally, it never was able to devote the resources to implement that strategy; thus, the two functions became closer to one another over time.

The offices within OSHA that have primary responsibility for managing risk assessments in the decisionmaking process are the Directorate of Health Standards Programs, the Directorate of Safety Standards Programs, and the Directorate of Technical Support. As shown in Figure 5.4, these directorates all report directly to the Assistant Secretary of OSHA.

NRC

The NRC organization plan of 1988 (U.S. Nuclear Regulatory Commission, 1988) was influenced by a reorganization plan under the Carter administration in reaction to the accident at Three Mile Island (Congressional Quarterly, Inc., 1983, 509). The NRC is organized into five offices that report to a Director of Operations, who in turn reports to the chairman of the commission. Prior to this current organization, these offices reported directly to the commission as a whole.

The offices involved in risk management and their functions are as follows (U.S. Nuclear Regulatory Commission, 1988):

- The Office of General Counsel, which reports directly to the commissioners, produces the rulemaking for the commission.
- The Office of Analysis and Evaluation of Operational Data, which reports to the Executive Director for Operations, has a Division of Operational Assessment, which responds to, diagnoses, and conducts training for incidents in nuclear plants. The office also has a Division of Safety Programs, which analyzes operating data for safety considerations.
- The Office of Nuclear Material Safety and Safeguards has as its major focus public health and safety and environmental considerations with respect to nuclear fuel transport, various uses of nuclear materials, and high- and low-level radioactive waste management.
- The Office of Nuclear Reactor Regulation is responsible for policy, program, and procedure implementation and development and implements the regulations of the commission. Included in the procedures are inspection and licensing policies for existing reactors and those under construction.
- The Office of Nuclear Regulatory Research is responsible for the research that is the basis for the development of standards and the development and promulgation of the technical regulations as well.

Thus, risk assessment and related risk management functions are generally scattered throughout the commission, and their location relative to the top is equally variable.

In 1983, Wood (1983, 65) argued that on the whole the commission's responsibility for the management of risks from nuclear power was highly diffused. First, unresolved issues moved from approval stage to approval stage between construction and operation. Second, a communication system between vendors and operators was lacking for safety problems. Third, regulatory responsibility and accountability was weak. Wood gave at that time a poor prognosis for the commission's ability to meet its responsibility primarily because of its organization.

Summary

The existence of separate risk assessment units in government and their placement of the risk management function within the five agencies generally reflect a high regard for the quantitative assessment of risk. They are not, however, located at high levels in the organization where they can directly influence policy. With a few exceptions, many risk assessment functions continue to be decentralized throughout the agency structures, being located where they are used in regulation. Central offices with risk-related functions, however, generally exist in areas responsible for setting standards and conducting research to provide guidance for these individual regulatory functions.

Administrative Discretion in the Use of Advisory Bodies

An important mechanism used in both formal and informal agency decision-making is the development and use of advisory bodies. Cardozo (1981, 3) has defined a governmental advisory committee as "any group of individuals, however selected or constituted, that considers governmental matters and furnishes views and conclusions to government officials or agencies." When representatives of organizations and interest groups outside of government serve on advisory committees, their procedures are governed by the Federal Advisory Committee Act (FACA) (5 U.S.C. App. I (1976)). The purpose of FACA is to prevent the proliferation of committees, to inform the public about the composition of the committees, and to establish procedures for those that exist (Cardozo, 1981, 10).

Scott (1981, 218) has characterized such advisory units as providing lateral connections within organizations (among work groups and departments) for the purpose of information exchange. Such connections, Scott argues, are necessary where information flows are heavy, and where tasks are characterized by a high degree of complexity, uncertainty, and interdependence. The concept of lateral connections is just as applicable between organizations as within them. According to Scott, these connections can take a number of different forms. They can occur as direct contacts among the rank and file or as liaison roles allowing more discrete exchanges. A task force is "a temporary group that is given a specific problem to solve or project to handle" (Scott, 1981, 219), and a project team is a grouping to carry on more regular work assignments. Finally, matrix organizations, which allow formal horizontal linkages among vertical lines of authority in an organization, can formalize lateral connections.

In the area of risk management, advisory bodies are one attempt to deal with the administrative fragmentation that can result from legislation that has developed in a piecemeal and categorical fashion regarding risks. A result of such fragmentation is that the components of a given type of risk (its source, means of transmission and exposure, and effects and target organism) are often managed at several different points within an agency as well as by several different agencies. Advisory bodies can bring together expertise across a wide range of skills, such as statistics, health, environmental science, and engineering, to address all the components of a particular type of environmental risk simultaneously.

Advisory units often appear within and among agencies, and are often mandated directly in legislation. The operations and many of the procedures of advisory groups, especially with respect to the public, are part of formal agency decisionmaking. These advisory units are organized in many different ways, depending on their purpose and the politics of the agencies they serve. They vary considerably in the degree of independence they have. Some advisory committees operate wholly within an agency, and others are external to it. Some are boards that are completely independent of the agencies they oversee.

The Science Advisory Board for EPA is outside of EPA's formal structure, yet its agenda is set by EPA. The National Transportation Safety Board for the DOT, on the other hand, sets its own agenda, deciding what to review within the boundaries of certain criteria. There are a number of interagency committees as well. The Interagency Testing Committee plays a key role in the administration of TSCA. It consists of representatives from the National Institutes of Health, the National Cancer Institute, and the National Science Foundation and agencies with chemical health risk jurisdictions. It makes recommendations for testing under Section 4(e) of TSCA (Stever, 1988). The major advisory committees, both intra- and interagency, for the five agencies studied here and their affiliates are given in Table 5.5. While not discussed here, it is important to point out that all of the agencies use outside reviewers selected on an ad hoc basis for various assignments. In particular, the professional societies play a key role in acting in such an advisory capacity.

The purposes of ad hoc advisory bodies are to centralize and coordinate the management of risks and to expand agency resources for professional expertise. Another important function is to legitimize the agency's decisions by providing independent, expert reviews. One way these advisory groups expand resources is by bringing outside scientists into agency decisionmaking. The major questions raised regarding these entities are:

1. In what kind of decisions should these advisory entities participate?
2. To what degree should these entities make decisions for the agencies they serve?
3. How much control should the agencies exercise over these entities?
4. What initiative should these entities have in defining their own agendas?
5. How should disputes among ad hoc entities or between these entities and the agencies they serve be resolved?

EPA

Risk management within EPA, as well as its coordination between EPA and other agencies, makes use of a series of boards, panels, and other peer group review procedures. Three of the most important entities are the Carcinogen Assessment Group (CAG, now divided into two branches), the Science Advisory Board (SAB), and the Science Advisory Panel (SAP).

Prior to its division into two branches in early 1989, the Carcinogen Assessment Group conducted many of the agency's risk assessments for carcinogens under a variety of statutes. It reported directly to the Assistant Administrator for Research and Development. The National Academy of Sciences report pointed out that in being within the Office of Research and Development, the CAG was administratively separate from the risk managers (i.e., regulators). However, the risk assessments that the CAG conducted were in response to requests by regulators, and the use of these assessments was up to the discretion

Table 5.5.   Summary of Intra- and Interagency Advisory Units

Entity	Relationship to Agency	Agency
Part of formal agency structure		
Directorate of Health Standards Programs		OSHA
Directorate for Health Sciences		CPSC
Office of Health and Environmental Assessment		EPA
Risk Assessment Council		EPA
Separate entities within agencies		
Science Advisory Board		
Clean Air Scientific Advisory Committee		EPA
Scientific Advisory Panel (Pesticides)		EPA
Bureau of Foods, Office of Toxicological Sciences, Cancer		
Assessment Committee		FDA, Bureau of Foods[a]
Toxicological Advisory Board		CPSC
Chronic Advisory Panels (ad hoc)		CPSC
Occupational Safety and Health Review Commission		OSHA
National Advisory Committee on Occupational Safety and		
Health		OSHA
Advisory Committee on Construction and Health		OSHA
External		
NIOSH		OSHA
Drug Evaluation Panels		FDA
National Advisory Food and Drug Committee		FDA
National Toxicology Program, Board of Scientific Advisors		FDA
National Toxicology Program in DHHS		FDA, EPA, OSHA, CPSC, NIH, NIOSH, NCI, NIEHS
Panel on Formaldehyde		EPA, OSHA, FDA, CPSC
Chemical Nominations Group		NIH, NCI, NIEHS, NIP
NAS Committee of Experts		FDA[b]
Federation of American Societies for Experimental Biology		
panels		FDA[c]
Interagency Testing Committee (ITC)[d]		EPA, CEQ, OSHA, NIOSH, NIEHS, NCI, NSF
Advisory Committee on Reactor Safeguards		NRC
Interagency Regulatory Liaison Group (IRLG)		EPA, FDA, OSHA, CPSC

Source: Drawn in part from National Research Council (1983, 88, 103).
[a]Bureau of Foods decides on use of outside panels (National Research Council, 1983, 104).
[b]Mandatory for color additives.
[c]For GRAS substance review, used for cyclamate, saccharin, red dyes No. 2 and 40.
[d]TSCA 4a chemical selection.

of the regulators. The CAG, with the exception of its chairperson, was staffed from within EPA. The CAG did assessments for several regulatory agencies in addition to the EPA, and completed some 150 assessments from its formation in 1976 through 1983, 41 of which were for air quality (National Research Council, 1983, 107). The CAG's responsibilities are now carried out by two assessment branches in OHEA's Human Health Assessment Group. In addition, as pointed out earlier, the Risk Assessment Forum is responsible for agency-wide consensus on risk assessments.

The Science Advisory Board is another entity that actively engages in risk assessment and management. It undertakes a broader range of assessments for air and water quality than the CAG. As Stever (1988) summarizes, the SAB was begun under the CAA. In 1978, its use was expanded to the SDWA by the Environmental Research and Demonstration Act of 1978. Prior to the 1985 SDWA amendments, EPA's revisions to the drinking water standards were first reviewed by the National Academy of Sciences. After the amendments, however, the NAS studies were dropped and only the SAB was consulted on the revisions. EPA exercises considerable discretion in whether it has to use the recommendations of the board (Stever, 1988, 7–34). For example, when EPA proposed regulations for the disposal of sewage sludge under the Clean Water Act, by its own admission it did not use the advice of the SAB in the choice of a groundwater mode, even though it agreed with the SAB's reasoning (U.S. EPA, February 1989, 5772). For review of those regulations, an entirely new committee was formed external to the agency (the "W-170" committee).

Another entity, a separate Science Advisory Panel, undertakes reviews for pesticides. (The separateness of the SAP from the SAB or other entities is mandated under FIFRA.) Both the SAB and the SAP act as peer reviewers for the CAG's assessments. The CAG and the SAB and SAP often differ in the approaches to and outcomes of risk assessments. Resulting disputes often must be resolved at the level of regulators that apply the assessments.

CPSC

The use of advisory committees and boards by the CPSC has changed dramatically, mainly as a result of the Consumer Product Safety Amendments of 1981. The CPSC had operated with three advisory committees during the 1970s. Committee members had to be drawn from other federal agencies, industry, and the general citizenry. These committees were the Product Safety Advisory Council, whose advice the CPSC took at its own discretion; the National Advisory Committee for the Flammable Fabrics Act; and the Poison Prevention Packaging Technical Advisory Committee. In addition, there was a Toxicological Advisory Board. The CPSC was directed to consult with these committees in establishing and prescribing standards (Laporte, Gies, and Baum, 1975, 1093).

In developing carcinogen risk assessments and guidelines, however, the CPSC had not relied on advisory committees the way other agencies had and in fact was not mandated to do so (National Research Council, 1983, 92). The 1981 amendments changed this situation to some extent. The amendments abolished all of the advisory committees, retained the Toxicological Advisory Board, and established a new series of Chronic Hazard Advisory Panels. The Toxicological Advisory Board, set up under the Federal Hazardous Substances

Act, advises the commission on hazardous substance labeling provisions. The Chronic Hazard Advisory Panels are appointed on an ad hoc basis to advise on carcinogens, mutagens, and teratogens. The seven members of each panel are appointed from a list drawn up by the National Academy of Sciences (Congressional Quarterly, Inc., 1983). When the CPSC develops voluntary standards, Consumer Product Safety Standards, or product bans, and the substance being regulated is a suspected carcinogen, mutagen, or teratogen, it must be referred to the Chronic Hazard Advisory Panel prior to the issuance of any rulemaking notices in the *Federal Register*[11] (Stever, 1988, 4–5). The CPSC can also request that the Science Advisory Committees within the EPA Science Advisory Board provide guidance on certain health issues. For example, in 1986 the commission requested guidance on the health effects of nitrogen dioxide in indoor air environments from the SAB's Clean Air Science Advisory Committee (U.S. Consumer Product Safety Commission, 1986, 143).

FDA

Advisory committees and other entities play an important role in the FDA decisionmaking process, primarily by developing and applying risk assessments. Within the FDA alone there are 30 standing advisory committees (Congressional Quarterly, Inc., 1986, 333). These committees advise in the substantive areas under the agencies' jurisdiction (food, drugs, cosmetics, and medical devices), the type of effect (carcinogenic, noncarcinogen toxicity), and safety. One of the advisory groups dealing with carcinogens in foods is the Cancer Assessment Committee of the Bureau of Foods. It consists primarily of medical and scientific personnel from within the agency and decides whether a chemical is a potential carcinogen (National Research Council, 1983, 103). The legislative basis for the determination of carcinogenicity lies in the Delancy Clause and more generally in the safety provisions of the FDCA. The Delaney Clause assumes that no amount of a substance presumed to be a carcinogen is tolerable in food. In a sense it simplifies decisionmaking in that it theoretically allows the agency to make a decision only on the basis of the existence of carcinogenicity, without having to establish the precise levels. The wording of the Delaney Clause gives the FDA considerable discretion through the use of experts in determining what constitutes the induction of cancer when a substance is fed to animals or during other tests (Merrill, 1988, 5). Attention to the Delaney Clause and its effect on agency decisionmaking has increased only recently with the detectability of low levels of substances in foods and the sophistication of techniques of risk assessment (Kessler, 1984, 1037; Merrill, 1988). These two developments have led to considerable debates over acceptable levels of safety.

External review panels are also used by the FDA. The composition of these review panels can be determined by the Cancer Assessment Committee, since it recommends members for such panels. The FDA even goes beyond these ad hoc committees to outside organizations for safety determinations. For example, Kessler (1984) notes that the FDCA does not necessarily require FDA to determine the "generally recognized as safe" (GRAS) status of a substance, which is the foundation of its regulation of food contaminants. FDA can and has relied on other organizations to make these determinations on an advisory basis. The Flavor and Extract Manufacturer's Association reviewed the safety of flavorings, and the Federation of American Societies for Experimental Biology reevaluated the safety of various substances at the request of the FDA (Kessler, 1984, 1040, footnote 13).

OSHA

Besides its relationship with NIOSH, OSHA engages in a number of interactions with internal and external review agencies in the development of its standards. It relies on both permanent and temporary committees. Examples of the temporary advisory committees OSHA established in the past for particular cases include a 1973 committee on 15 carcinogens that established permanent standards, a 1972 pesticides committee (appointed by CEQ) that established emergency temporary standards, and a 1974 committee on coke oven emissions that established an emission standard.

OSHA's two permanent or standing committees are the National Advisory Committee on Occupational Safety and Health and the Advisory Committee on Construction Safety and Health. Both serve in an advisory capacity to the Secretary of Labor. The Occupational Safety and Health committee also advises the Secretary of Health and Human Services (Congressional Quarterly, Inc., 1986, 427).

The Occupational Safety and Health Review Commission (OSHRC) is an independent commission that has had a total of about 90 employees since the mid-1980s. This commission reviews OSHA decisions that are questioned either by employees or employers, and issues orders for penalties that are proposed by OSHA. Examples of decisions that can come before the commission are complaints about the time allowed to correct a violation and complaints about the findings of plant inspections.

NRC

The NRC makes use of a limited number of advisory committees. The NRC's Advisory Committee on Reactor Safeguards is an independent committee of scientists that reviews licensing decisions. Administratively, it reports directly to the commissioners. Its responsibilities are to provide an internal review of safety studies that are the basis for both construction and operating permits. Where a request for assistance is initiated by the U.S. Department of Energy (U.S. DOE), it can also provide similar assistance for the DOE nuclear facilities.

Discretion in Standards-Setting

Chapter 2 examined the extent to which risk statutes prescribe chemicals for regulation, their numerical limits, and conditions of use. These statutory provisions influence agency discretion in several ways:

- *Scheduling*, the degree to which agencies can set their agenda for decision-making against time constraints. Many federal and state agencies are required by law to make a regulatory decision within a fixed amount of time. This can have the effect of reducing agency discretion in the areas that have such time limits.

- *Risk selection*, the extent to which agencies can select the chemicals, activities, and associated risks or health effects they are to regulate. Some legislation will prescribe general categories of substances or activities that should be regulated, leaving considerable discretion to agencies to define the number and type of chemicals within those categories. Other legislation prescribes specific chemicals for regulation.

- *Risk levels*, the extent to which agencies can set levels for pollutants and activities, how they can set those levels, and the degree of certainty of scientific knowledge required before agencies make decisions, given the chemicals and activities that can be regulated.

- *Alternatives*, the latitude agencies have and use to define the risk management alternatives they formulate and implement, particularly in terms of what technologies they can use to implement the laws they oversee.

The five federal agencies differ in the degree of discretion and choice they have and take advantage of with respect to each of these factors.

EPA

Some statutes under which the EPA operates place limits on its discretion, while others give it broad discretionary authority. This is partly a function of the wide variety of statutes the agency administers and the very different legislative history each one has. Some of the limits on agency discretion and allowances for agency discretion in areas related to risk management are highlighted below.

As a general statement on the original limits of EPA's discretion, Marcus (1980, 267) observed that "Congress sought to reduce the risk that the agency would abuse its discretionary authority by placing sharp limits on that authority." On the other hand, MacIntyre (1986, 79) has pointed out that where statutory ambiguity existed, "the administrators of EPA exploited ambiguity to enhance their discretion when seeking amendments to the federal pesticide statute in the early 1970s." Examples of how discretionary authority is constrained or enhanced within EPA are given for the four areas of scheduling, risk selection, risk levels, and alternatives.

Scheduling. Numerous examples exist of statutory limits placed on EPA's authority and discretion. These appear as statutory deadlines for pollution control and impose very stringent limits on discretion in this area. Practically every piece of environmental health–oriented legislation gives dates by which regulations and programs are to be developed. Such deadlines were not typically imposed on older agencies acting under other environmental statutes. Deadlines were usually reserved for individual permits and licenses that prescribed the conditions under which polluters were allowed to operate. Some of the numerous examples follow.

- Originally, CWA set deadlines at 1985 for the achievement of zero discharge of wastewaters into natural waterways, 1976 for the installation by industry of best practicable control technology, and 1981 for best available control technology for water pollution control.
- The CAA of 1970 mandated a 90% reduction in automobile emissions by 1975.
- EPA set a strict schedule for standards-setting under the SDWA Amendments of 1985, discussed in Chapter 2.
- The development of proposed national ambient air quality standards was required within 30 days of the effective date of the CAA of 1970, and final standards were required within 120 days (Lave and Omenn, 1982, 7; Meier, 1985, 151).
- Under TSCA (Section 4), EPA was given one year to react to the chemicals appearing on the Interagency Testing Committee list of chemicals, after which time it had to specify testing rules (Stever, 1988, 2–5).

Risk selection. Chemicals that have statutory limits placed on them with respect to their selection for regulatory action are the exception rather than the rule. While legislation specified when limits should be set, with the exception of RCRA and some sections of TSCA, it rarely spelled out which specific chemicals should be addressed. Nevertheless, the exceptions are quite striking. Examples of some of the few (though prominent) limits placed on the type of pollutants for which standards must be set are given below. Where specific chemicals are not mentioned in statutes targeted for regulatory actions, general categories of chemicals are almost always selected, such as the general categories of toxics, organics, or solvents.

- The requirement that EPA must set water quality standards for 65 toxic substances under the Clean Water Act was the result of a settlement agreement.
- The requirement for the inclusion of an ambient air quality standard for lead under the Clean Air Act resulted from a court case (*Natural Resources Defense Council, Inc., et al.* v. *Train,* 411 F. Supp. 864 (S.D.N.Y., 1976) aff'd 545 F. 2d 320 (2d Cir. 1976.)).
- Cessation of the manufacture of polychlorinated biphenyls by 1978 was mandated by the Toxic Substances Control Act.
- EPA was required to follow the recommendations of an interagency committee

that lists toxic substances for regulation under TSCA in its selection of chemicals to regulate under that statute.

- The requirements under Section 5(a)(2) of TSCA define for EPA what it must take into account when making a determination that a use of a chemical constitutes a "significant new use" and is subject to a "significant new use rule" (Stever, 1988, 2–19).
- The 1984 HSWA to RCRA spelled out specific categories of chemicals, such as organics and metals, for which limits had to be set and applied to various contexts.

Risk levels. EPA has considerable latitude in setting pollutant levels and related conditions under which discharges may occur. It also has considerable latitude in the way it develops these limits.

Under the "prevention of significant deterioration" (PSD) provision of the CAA (which requires that new sources not degrade existing air quality beyond a certain point), the EPA has considerable discretion in what it accepts as the percentage of change in estimated air quality (the PSD increment), as long as ambient air quality standards are not violated. Once the percentage is chosen, however, it cannot easily be modified on a case-by-case basis.

For industries without well-defined wastewater discharge standards, and not discharging any of the toxic pollutants in excess of "reportable quantities," a considerable amount of discretion is allowed EPA permit writers in designing the pollutant levels for wastewater discharge parameters. In addition, the agency can define populations susceptible to risk and can define an "ample margin of safety" as the basis for establishing threshold limits for air emissions by stationary sources and for ambient air quality standards under the CAA (Meier, 1985, 151–2).

To some extent, risk standards also include standards of review for various permits and registrations. One such registration is the Premanufacture Notice (PMN) under TSCA. O'Reilly et al. (1983, 5–21) notes that the legislative history of TSCA allows both EPA and an applicant for a PMN the discretion to interpret the sufficiency of information on the production, processing, and disposal of a chemical (article 5(d)(1) information) and its health and environmental effects on the basis of a criterion of reasonableness (which O'Reilly et al. interpret as incorporating cost considerations).

The courts have allowed EPA considerable discretion in the way it determines risk levels for NESHAPs, though the courts place some general boundaries around this discretion in requiring that EPA use current scientific knowledge. Nevertheless, the EPA is allowed the discretion to define what is meant by adequacy of scientific knowledge and the form that it takes. This was brought out particularly in court cases pertaining to benzene and vinyl chloride.

According to Stever (1988, 8–21), EPA can, under Section 408(b) of FDCA, exercise the discretion to conduct a risk assessment for pesticide residues on raw agricultural products. However, the FDA, which regulates pesticide residues in processed foods, does not have the discretionary authority to conduct risk assessments for these residues.

EPA has considerable discretion in its use of risk assessment to determine risk levels. EPA's highly discretionary use of assumptions underlying risk assessment has been illustrated and criticized by Latin (1988, 129–130) and others in the formaldehyde risk assessment. According to Latin, John Todhunter used assumptions that minimized risk levels as the basis for his ruling that regulation of formaldehyde under TSCA was unwarranted. Latin (1988, 130) underscores the fact that even with EPA's current risk assessment guidelines there is nothing to preclude an individual from making judgments the way Todhunter did. These judgments apply to the weight-of-evidence criteria and the way studies are selected and results are used in a risk assessment. Latin recommends that the application of a weight-of-evidence criterion should be multidimensional and should as a minimum resolve or take a consistent stance on the following issues: (1) whether one good study or numerous weak studies are weighed more heavily, (2) how one weighs the applicability of animal experimentation to humans, (3) quality of the exposure data (range of doses and knowledge of duration), and (4) which extrapolation models are used for low-dose extrapolation.

The EPA guidelines, while attempting to define agency procedures for the use of risk analyses to set chemical risk levels, still left open several critical areas to discretionary choice, such as how risk assessment is to be used for different programs, cost of risk assessments, the sufficiency of evidence, and the form of the numbers. According to Latin, this discretion may not always be beneficial. First, Latin has pointed out that the guidelines are devoid of any consideration of how expensive or time-consuming it might be to conduct a risk analysis (1988, 138). When does an agency decide it cannot undertake an analysis or refine data because it is too expensive? Second, EPA has developed a single set of guidelines for risk assessment to be applied to standards-setting, cleanup levels, and other activities that fall under different statutes with different risk levels implied. (For example, do you apply the same risk assessment techniques to ensure ample and adequate margins of safety?) Third, what constitutes sufficient evidence to undertake a risk assessment that is sufficient for regulatory action? Latin notes that regulation often does not proceed unless sufficient information is in hand, and the delay may cause further harm to human health (1988, 141). This criticism was launched against EPA's initial selection of organic chemicals for regulation under the Safe Drinking Water Act, which limited chemicals for regulation to those for which there were sufficient scientific studies. Fourth, in what numerical form should the risk characterization or calculation be made? Latin (1988, 144) points out that while the 1986 guidelines encourage the use of upper and lower bounds (U.S. EPA, September 24, 1986, 33992, 33998), the benzene regulations of 1984 used point estimates (U.S. EPA, June 6, 1984, 23478, 23493).

Alternatives. Under the SDWA Amendments, EPA must require best available technology to achieve national primary drinking water standards, which is defined as the use of granular activated carbon. That is, no discretion is allowed in choosing and formulating technological alternatives. This follows a tradition that began during the 1970s of prescribing technologies for air and water pollution control directly and limitations on the choice of certain technologies for hazardous waste disposal under RCRA. Under these other statutes, however, alternatives are prescribed more generally. For example, the Clean Water Act and Clean Air Act refer to best available technology. Selection or approval of specific technologies within that general category is up to the agency.

Other. Limits are placed on EPA's authority to tailor standards to particular cases. For example, effluent standards set under Section 301 of the Clean Water Act are applied on a case-by-case basis via permits issued under the National Pollutant Discharge Elimination System (Section 402 of the act). However, standards are well defined for 22 industries and for some 65 toxic pollutants regardless of industry. Where the standards are well defined, little discretion is allowed in their application to particular wastewater dischargers that fall within those categories.

CPSC

The CPSC operates in three areas: the setting of mandatory safety standards and bans, authority to act on imminent hazards, and management of chronic hazards.

Risk selection and risk level. The CPSC can set its own standards and act as a clearinghouse for standards that are offered by outsiders (1105.3(a)(5)), yet the extent to which it has exercised this discretion has been limited.

In setting mandatory standards under the CPSA, the CPSC must demonstrate a standard's necessity in terms of its ability to reduce or eliminate unreasonable risks of injury associated with a given product (16 CFR 1105.3(a)(3)(1975)). Within that constraint, the CPSC can exercise considerable discretion in determining the form of the standard it sets, i.e., packaging, labeling, construction, etc. It can also exercise discretion in terms of the number of standards it sets. The extent to which the CPSC has exercised its regulatory authority in the form of standards and bans under the CPSA and other statutes was criticized initially as not going far enough: after five years of operation, only 11 standards and bans were in force. Those that did exist were not in high-priority hazard areas, and many did not withstand judicial review (Bick and Kasperson, 1978, 33). Over the next four years, however, seven more rules were promulgated (Kasperson and Bick, 1985).

Under the CPSA 1981 amendments, the CPSC is required to work with voluntary standards before implementing mandatory ones. In the 10-year period since it was first started, the commission has been involved in the development of 83 such standards (Kasperson and Bick, 1985). Nevertheless, the CPSC has generally been criticized for exercising discretion in allowing the mandatory standards to lag while relying on voluntary standards, possibly to a greater extent than the law may have intended (Fise, 1987).

The CPSC exercised considerable initiative and discretion in setting a standard for formaldehyde. It was the first agency to take action against formaldehyde once scientific data surfaced on its health effects. EPA and OSHA did not take action under TSCA and the OSH Act, respectively. CPSC issued a product ban on urea-formaldehyde foam insulation in spite of the fact that the commission had to abide by more stringent requirements than the other agencies to take such action (Ashford, Ryan, and Caldart, 1983b, 899). The ban was later overturned.

The CPSC can also exercise discretion with regard to imminent hazard actions under Section 12 of the CPSA. In the first five years of its operation, the commission used this authority only four times. The CPSC has not taken advantage of the degree of authority it could have in the area of chronic hazards either; in the first five years of its operation it only regulated six substances under that provision (Bick and Kasperson, 1978, 34, 37).

The CPSC has the authority to ban products, including chemicals, but has banned only a handful of chemicals (U.S. Consumer Product Safety Commission, 1986), many of which received attention under other statutes.

FDA

Scheduling. In the area of drug review, the FDA can exercise discretion in modifying the allowable time period for review. For example, it can avoid the 180-day requirement for the review of a new drug application by asking companies to submit a request for a voluntary extension or by certifying the application as being incomplete. (The review time starts when applications are complete.) (Quirk, 1980, 208)

In contrast, the FDA's degree of discretion is considerably hampered by the time constraints under which it must render decisions, especially with respect to environmental contaminants. Examples of decisions that were made abruptly (often within a few days) in the area of foods are the cyclamate ban in 1969 and the ban of Red Dye No. 2 in 1976 (Merrill, 1977b, 1005).

Risk selection. The FDA regulates some nine different categories of foods under the Food, Drug, and Cosmetic Act alone (Kessler, 1984, 1035), as well as consumer products covered under other legislation. In carrying out these functions, the FDA exercises considerable discretion in deciding which program to emphasize most. This discretion has been exercised in a systematic fashion by applying three criteria to the program areas: "the risk of harm to the consumer, the extent to which additional FDA resources can increase the protection provided, and the 'public interest' represented by the attention given to the problem by press, consumer, and congressional spokesmen" (Merrill, 1977b, 1001). This priority-setting activity is carried out by FDA staff and by external review panels.

The FDA exercises discretion in the kinds of review procedures it uses. The review procedures for evaluating food contaminants within the nine different food categories vary depending on the type of contaminant. Additives are examined on receipt of a petition and are analyzed by bureau staff. The FDA also exercises discretion in deciding whether an additive is considered approvable on the basis of scientific analysis. If the approval is given, the application is forwarded to the Division of Food and Color Additives. Then it goes to the Associate Commissioner for Regulatory Affairs, where scientific and regulatory judgments are combined. In the case of environmental contaminants in food, the sequence is similar but other divisions are involved as well (National Research Council, 1983, 102).

FDA exercises a considerable amount of discretion in designating substances as "generally recognized as safe" almost by default. No administrative procedures are specified for the agency to identify these substances (Stever, 1988, 8–18). Stever attributes this to the age of the FDCA. After 1977, he points out, regulations were issued governing how a substance could be removed from the GRAS list (1988, 8–20).

The FDA acknowledges that it can exercise discretion in deciding which food and color additives to regulate and to what levels. On August 7, 1986, FDA used its discretionary authority in its decision not to ban two color additives, Orange No. 17 and Red No. 19, and determined that they should be considered safe under the Delaney Clause. Summarizing from Merrill's analysis of FDA's arguments (1988, 7), FDA first argued that the dyes were not food additives since they were only used topically. Second, they were present in low doses, and thus posed an insignificant risk. In order to determine the risk level, FDA used the multistage model to estimate a maximum risk for the two colors at one in 19 billion and one in 9 billion, respectively, and (most importantly from the perspective of the Delaney provision) the risk range could also include zero (Merrill, 1988, 7). Third, FDA, according to Merrill's analysis (1988, 8), argued further that the statute allowed it to make a judgment about what was considered applicable analytical techniques and used a court case as a precedent. Finally, FDA argued that exemptions could be granted on the basis of de minimis risk levels. The District of Columbia (D.C.) Circuit overturned the agency's ruling, arguing that if any risk of cancer was found it was meant to be regulated under the Delaney Clause (Merrill, 1988, 9; *Public Citizen* v. *Young*, 831 F.2d 1108, 1111–12 (D.C. Cir. 1987)). In other words, the presence of the dyes in any concentration was not allowed.

Risk levels. There are a number of examples of FDA's discretionary authority in setting risk levels and modifying them through the use of other criteria, such as cost-benefit analysis.

The determination of carcinogenicity by the FDA's Carcinogen Assessment Committee is done with considerable discretion, i.e., in the absence of formal guidelines, though previous cases are used as guidance (National Research Council, 1983, 103). FDA can exercise the discretion not to set a standard for a substance on the grounds that it poses a de minimis risk, but it has to publish that determination in the *Federal Register* (Stever, 1988, 8–14). In the area of drugs, Quirk has observed that "the FDA has broad discretion concerning the tests it will require and the standards of safety and efficacy it will apply" (Quirk, 1980, 200). This authority is similar to that allowed for food additives (Merrill, 1988, 11).

The FDA theoretically has considerable authority to determine on the basis of risk levels that a substance should be withdrawn from the market, but it has not exercised this authority very much (Merrill, 1988, 9). The FDA also can decide when a drug should be withdrawn from the market, though Quirk has observed that the agency tends to concentrate on initial approvals rather than removals, making it easier for an approved drug to stay on the market. In spite of these broad discretionary powers with respect to the introduction of drugs, Quirk points out that the FDA has no authority over the misuse of drugs. The agency also has no authority in requiring monitoring of a drug once it has been approved, but it can act reactively by withdrawing a drug from the market if adverse effects are discovered.

FDA exercises discretion in the decision to use cost-benefit analysis in establishing risk levels. This particular power and the way it has been used by the agency has been subject to considerable debate. Neither the Occupational Safety and Health Act nor the Food, Drug, and Cosmetic Act explicitly mandates the use of cost-benefit analysis in arriving at decisions about levels for standards or the application of standards to particular cases. FDA has typically shied away from using economic analysis. Its staffing, primarily from the medical profession rather than from economics, has reflected this bias (Quirk, 1980, 211). The courts, however, have taken a different view and have often required a cost-benefit analysis to be performed on a particular decision.

OSHA

Risk selection. OSHA exercises considerable discretion in the development of its three kinds of standards: interim, emergency temporary, and permanent standards. OSHA can modify or supplement the list of standards that NIOSH proposes in its criteria documents. By 1981, however, OSHA had proposed such requirements as standards for only 17 of a total of 250 substances for which NIOSH had prepared documents (McCaffrey, 1982, 73).

The agency has considerable discretion in interpreting and applying the broad criteria for passage of emergency and permanent standards (existence of grave danger and the need for worker protection). An example of the exercise of this discretion was the agency's resistance to establishing emergency standards for the ACGIH list of substances "unless it was sure that employers could comply within six months" (McCaffrey, 1982, 97). In the case of permanent standards, OSHA exercises discretion by deciding when such a standard is needed and what the review procedures for establishing it should be. OSHA has been said to exercise its discretionary authority by choosing to rely on a highly complex review process for permanent standards. This process can involve as many as 19 steps, which has been regarded as going far beyond the requirements for standards development under Section 6(b) of the OSH Act (McCaffrey, 1982, 74). Another example of OSHA's discretionary authority in issuing permanent standards has been identified by the General Accounting Office (GAO). The GAO felt that OSHA had delayed issuing permanent standards on a number of substances by exercising its right to require more information and certainty.

Risk levels. Like FDA, OSHA can exercise considerable discretion in the use of risk assessment to establish risk levels and in choice of methods of analysis, as long as this choice reflects expert judgment. This has been demonstrated in judicial reviews of OSHA's decisions (Oleinick, Disney, and East, 1986, 392). Likewise, as reviewed by Oleinick, Disney, and East (1986, 393), OSHA can also exercise considerable discretion regarding whether to apply a standard as an emergency temporary standard or a permanent standard. Each type of standard often implies a different risk level.

In developing risk levels, OSHA's discretion is limited in a number of ways. First, a number of factors have limited OSHA's ability to use cost-benefit analysis as an input into the establishment of risk levels. The original OSH Act does not directly and unequivocally require cost-benefit analysis. The agency's employment pattern reflected this lack of emphasis: "as of the late 1970s, OSHA employed no more than five professionally trained economists" (Thompson, 1982, 208). Agency actions have been mixed with respect to the inclusion of costs but have tended to favor worker protection over costs. Kelman, for example, explains this in terms of the philosophy of agency officials:

> . . . the most important factor explaining OSHA decisions on the content of regulations has been the pro-protection values of agency officials, derived from the ideology of the safety and health professional and the organizational mission of OSHA. (Kelman, 1980, 250)

In the early 1970s, OSHA tried to exercise its discretion in the area of cost-benefit analysis by making industry responsible for conducting economic analyses (McCaffrey, 1982, 83–85). After 1974, various federal entities tried to reduce OSHA's discretion in choosing whether to use cost-benefit analysis. The Council on Wage and Price Stability, the Council of Economic Advisors, and later the Office of Management and Budget tried to require OSHA to conduct economic analyses for its development and enforcement of standards. This was motivated by rising inflation (McCaffrey, 1982, 63) and was reinforced by the passage of Executive Order 11821 in 1974, which required an analysis of the impact of all rules and procedures on inflation. Economic impact was added to the inflation impact analysis under Executive Order 11949 in 1977 and 12044 in 1978 (McCaffrey, 1982, 65). Prior to these orders, OSHA exercised considerable discretion in determining which cost-benefit analyses to do, how to do them, and when to do them. It exercised this discretion by not quantifying costs and benefits and dealing very little with benefits at all (McCaffrey, 1982, 91–94). In 1981 Executive Order 12291 was passed, requiring all federal agencies to include cost-benefit analysis in their decisions. The promulgation and application of OSHA's cotton dust standards were delayed by this order, however. The courts ruled in OSHA's favor, stating that cost-benefit analysis was not required under the OSH Act (Congressional Quarterly, Inc., 1983, 401; *American Textile Manufacturers Institute, Inc.* v. *Donovan,* 452 U.S. 490 (1981); and other cases).

Other. OSHA's discretionary authority is limited in the area of invoking penalties. This authority is restricted by law, which requires an independent Occupational Safety and Health Review Commission, rather than OSHA, to act on penalties. Criteria for invoking penalties and establishing the size of a penalty are further prescribed by the OSH Act (Congressional Quarterly, Inc., 1983, 407).

NRC

With respect to regulation and standards-setting, the NRC's discretionary authority is one of the broadest of any agency. The legislation that created the NRC did not specify regulatory procedures or safety levels at all or whether and how they should be developed by the commission (Wood, 1983, 6). The NRC relies on professional standards proposed by the American Nuclear Society. These standards, when approved by the American National Standards Institute, are adopted as American National Standards. They cover not only numerical standards for radioactive substances but also various procedural and operating conditions. The fact that a presidential commission had to be appointed to investigate the accident at Three Mile Island is a reflection of NRC's independence, as is the fact that mistakes could only be rectified at very high levels in the executive branch. Much of this independence stems from the historical role of the federal government in the field of nuclear power. Three Mile Island changed this relatively insulated existence abruptly. The extent to which the NRC could oversee existing nuclear plants, not to mention many of those proposed for future development, came seriously into question (Schnaibert, 1982, 111).

In the use of cost-benefit analysis, the NRC exercises considerable discretion over the authority given to it in either the Atomic Energy Act (AEA), as amended, or the Energy Reorganization Act of 1974. Under these acts, the NRC does not have to use cost-benefit analysis in its standards-setting or licensing activities but in fact has done so extensively (Baram, 1980, 403). Use of cost-benefit analysis dates back to the regulations of the AEA recommended by a presidential council. These regulations refer to exposure levels being "as low as is reasonably achievable (ALARA)," which is to take into account economic and social considerations (Baram, 1980, 404). NEPA and the executive orders pertaining to regulatory impact also directed the agency toward the continued use of cost-benefit analysis.

Summary

If one were to go through practically all of the standards listed in Table 2.4, one would find that most of the agencies had little choice as far as (1) having to set standards for the general category of chemicals and (2) when to set them. On the other hand, the agencies had considerable choice regarding which chemicals to regulate (within more general categories of substances that were prescribed), the form of regulation, and the analytical techniques to determine risk levels. The exercise of discretion by the five federal agencies was not always predictable with respect to a protectionist philosophy. It will be shown in the next chapter, however, that judicial review often altered the nature and direction of this discretion.

Discretion in Rulemaking Under the Administrative Procedure Act

The government implements laws through informal and formal administrative processes. Davis (1972, 118) describes informal action as consisting of "conversations, letters, speeches, news releases, advice, advisory opinions, and rulings." It involves agency discretion in deciding what to investigate and in what order, and how to interpret the law in a particular case (Davis, 1972, 89–91). Formal rulemaking, on the other hand, consists of two processes. The first is rulemaking or the formulation of rules, and the second is adjudication or the issuance of orders.[12] This section focuses on the rulemaking function as an illustration of the exercise of discretionary authority by executive agencies.[13]

Rulemaking can take a number of different forms. It can define a process, such as the process of issuing permits, licenses, certifications, and approvals, or the process of risk assessment. It can define criteria and standards against which the performance of activities can be measured. Both of these forms of rulemaking are applicable to risk management and will be described below.

The Administrative Procedure Act of 1946 (as amended) sets forth the norm for formal rulemaking for many federal statutes, not only environmental statutes. It sets minimum requirements for the procedures by which agencies conduct rulemaking.[14] One purpose of the APA is to allow public scrutiny of agency procedures. Breyer discusses the origins of this particular purpose of the APA in terms of public distrust of informal agency procedures:

> The "informal" formulation of important governmental policy is sufficiently common and sufficiently distrusted that it has led to increased demand for more open proceedings and greater participation by "public interest" lawyers who seek to represent consumers or consumer organizations. This effort is but the latest in a long series of procedural changes and requirements, now embodied in the Administrative Procedure Act, designed to make the regulatory process a fair one. (Breyer, 1982, 3)

McCaffrey (1982, 71) has also emphasized one of the act's purposes, that of enhancing the public interest: "Its [the act's] purpose is to give people affected by decisions an opportunity to make their views known so that their interests will not be overlooked or arbitrarily dismissed."

The minimum requirements under the Administrative Procedure Act for rulemaking are to give public notice of proposed and final rules in the *Federal Register*. A period for public comment on the procedures is given when the rules are published in this way. Prior to giving notice of proposed rules, some agencies issue an Advance Notice of Proposed Rulemaking (ANPRM), which is a statement of their intent to propose a rule but does not necessarily solicit public comment the way proposed rules do. As a general rule, agencies must make available to the public through the *Federal Register* a means of public access to the agency's decisionmaking process pertaining to the rule. The *Federal Register*, which is published on weekdays, gives the key personnel to be contacted for information and to whom comments are submitted.

When the regulations are finalized or promulgated, they are codified and published annually in the *Code of Federal Regulations*. Both the *Federal Register* and the *Code of Federal Regulations* are published by the Office of the Federal Register of the General Services Administration and are available through the U.S. Government Printing Office.

The general process of formal rulemaking can be summarized as follows:

1. ANPRM
2. publication of proposed rulemaking and associated requirements, such as a Regulatory Impact Analysis, Environmental Impact Assessment, etc.
3. opportunity to comment, either through submissions directly to the proposing agency or via a hearing
4. agency revisions, where appropriate, based on public comments
5. publication of final rule with statements about how each of the public comments were addressed and the justification for the rule, assuming that the agency decides to go forward with the regulation

Concurrent with the publication of a rule in the *Federal Register* are reviews by other agencies. Review and approval of rules by the Office of Management and Budget is required.

Variations exist in the implementation, interpretation, or application of particular steps in the procedures laid out in the Administrative Procedure Act, primarily as the result of the exercise of discretion by agencies. As shown below, more detailed procedures have emerged under the Clean Air Act as a result of court cases (Melnick, 1983, 11). OSHA has also opted for more complex rulemaking procedures (McCaffrey, 1982, 74). Under the Safe Drinking Water Act of 1974, rulemaking was more complex than specified under APA, but the 1985 amendments simplified the previous procedures in order to expedite standards-setting. These three cases are examined below as they exemplify discretionary action in the direction of developing more complex yet more rigorous rulemaking procedures.

EPA Rulemaking for Clean Air Act
National Ambient Air Quality Standards

Melnick observed that rulemaking procedures under the Clean Air Act had become far more complex during the decade of the 1970s than they were in 1971 when standards for the original six ambient air quality standards were passed (Melnick, 1983, 255–256). The complexity arises primarily in the number and kinds of reviews that are required within EPA before publication in the *Federal Register*. These reviews are often conducted by review boards consisting of members from organizations outside the agency. The steps (compiled from Melnick, 1983 and Berry, 1984) are as follows:

1. An Air Quality Criteria Document is prepared under the direction of the EPA. The criteria document is a review of available literature on the effects of a given substance on human health and the environment. A standard can be recommended in the document. Its purpose is to set forth relationships between varying concentrations of and exposures to the substances and their effects. This document is often of multiple volumes and several hundred pages in length.
2. The draft of the criteria document is then circulated to the Science Advisory Board and the Clean Air Scientific Advisory Committee (CASAC). The purpose of this is to obtain peer review of the document. This particular review is mandated by the 1977 Clean Air Act Amendments and is therefore not discretionary.
3. The SAB and CASAC recommend changes where they deem appropriate, and the criteria document is revised accordingly. The Environmental Criteria and Assessment Office (ECAO) in EPA's Office of Research and Development was established in 1978 to coordinate this effort.
4. The Office of Air Quality Planning Standards (OAQPS), one of the regulatory arms of the agency, together with ECAO, prepares a summary document or staff paper. This undergoes further internal reviews within EPA, including reviews by the Administrator of EPA. Ultimately, the Administrator of EPA decides to publish or not publish a standard. The role of internal reviews

cannot be understated. Many disputes occur within the agency regarding the scientific basis for the standard, the ease of implementation by enforcement officials, and practicality in terms of costs and benefits. Risk analysis has become an important part of this process.

5. If the agency decides to publish a standard, a notice is placed in the *Federal Register*, along with a schedule for a public hearing and comment period. Based on the public comments, an action memo is prepared as backup for a final standard, which is then published in the *Federal Register* as a final rule.

Melnick points out that these variations in the procedures mandated under the APA were in part the result of judicial action. In particular, these complexities were added as the result of two court cases that occurred soon after the passage of the CAA in 1970 (Melnick, 1983, 2): *Kennecott Copper Corp.* v. *U.S. Environmental Protection Agency,* 462 F.2d 846 (D.C. Cir. 1972) and *Portland Cement Assoc.* v. *Ruckelshaus,* 486 F.2d 375 (D.C. Cir. 1973).

The historical development of the NAAQS demonstrates the growing complexity of the NAAQS standards-setting process within the EPA over time (Berry, 1987; Berry, 1984; U.S. General Accounting Office, December 1986a). By way of summary, in 1971 (after the passage of the 1970 act) only 11 major steps were identified in the process. This grew to 25 steps in 1973, 32 steps after the 1977 act was passed, and 44 steps by 1983. Figures 5.14, 5.15, 5.16, and 5.17 show the changes in procedures over time. Berry has attributed much of the increase to the need for the EPA to resolve questions of environmental health risk. In fact, at each of these points in time, he notes, there were major legal developments in the form of judicial review of standards that altered the course of agency rulemaking for NAAQS. He identifies these developments as follows:

- 1973: In *Kennecott Copper Corp.* v. *U.S Environmental Protection Agency,* Kennecott sued EPA over the sulfur oxide standards. This resulted in the addition of a review of the criteria document (which precedes the development of a standard) by an advisory committee and the preparation of a briefing paper, which outlines risk assessment procedures, dose-response information, etc.
- 1977: A lawsuit on the lead standard, initiated by NRDC, Inc., resulted in a much more formalized procedure. The Science Advisory Board became much more active in advising EPA. The first formal dose-response analysis occurred, accompanied by a staff paper assessing exposure.
- 1983: These changes are due to the 1977 amendments and the subsequent development of risk assessment guidelines by the EPA. While the original standards-setting process back in the early 1970s took 45 days, the process now, according to Berry, takes eight years. A key example of the more lengthy process is the particulate matter (PM_{10}) standard.

The growth in the number of steps in the process also accompanies major promulgations of or revisions to NAAQS. The most recent dates for the formulation of or revisions to the standards are as follows:

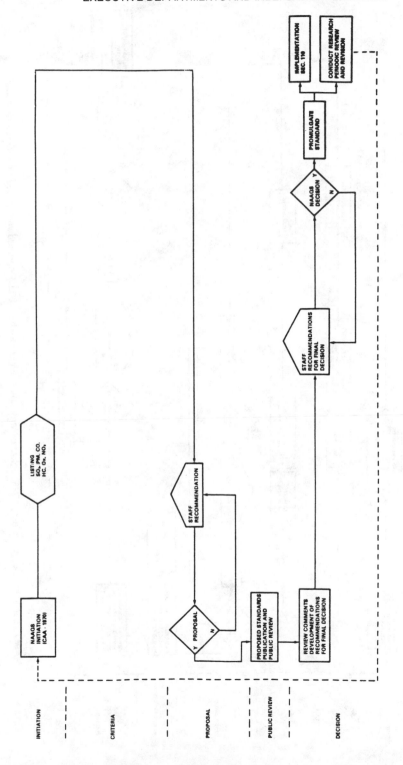

Figure 5.14. NAAQS decisionmaking process, 1971. *Source: Berry,* 1984, Figure 6-3a, p. 201. Reprinted with the permission of the author.

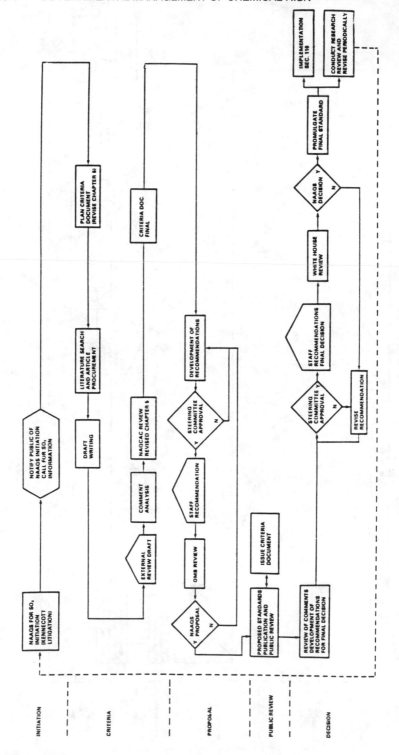

Figure 5.15. NAAQS decisionmaking process, 1973. *Source:* Berry, 1984, Figure 6-3b, p. 202. Reprinted with the permission of the author.

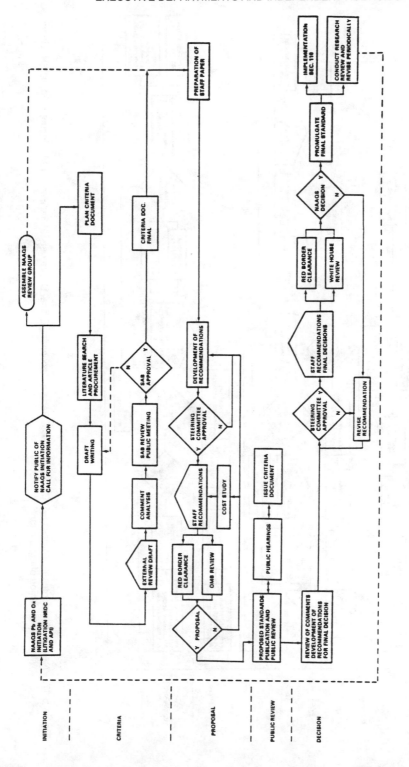

Figure 5.16. NAAQS decisionmaking process, 1977. *Source:* Berry, 1984, Figure 6-3c, p. 203. Reprinted with the permission of the author.

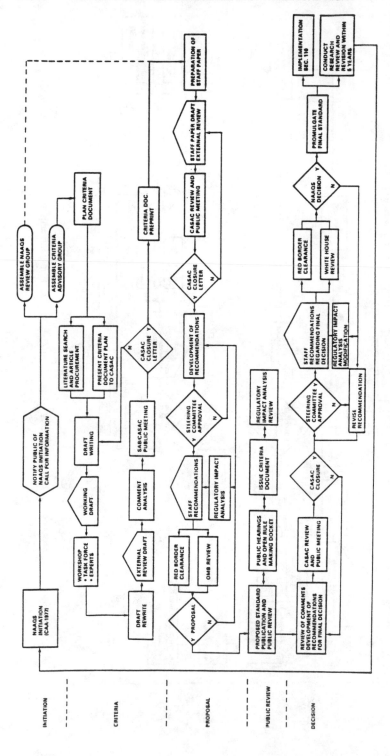

Figure 5.17. NAAQS decisionmaking process, 1983. *Source:* U.S. General Accounting Office, December 1986a; Berry, 1984, Figure 6-3d, p. 204. Reprinted with the permission of the author.

1971: sulfur oxides (SO$_x$)
1978: lead (Pb)
1979: ozone (O$_3$)
1985: carbon monoxide (CO)
1985: nitrogen oxides (NO$_x$)
1987: particulate matter

OSHA's Rulemaking Procedures for Workplace Standards

Kelman (1980, 244) notes the following set of procedures for OSHA's rule-making process.

OSHA can initiate the promulgation of a regulation in several ways. First, it can decide to develop a regulation on its own initiative using its authority under the OSH Act. Second, NIOSH (which is a research organization within HEW) can submit a "criteria document" to OSHA on a particular substance. This document gives scientific information on the effects of various concentrations of the substance and exposure conditions on the health of workers. It also gives recommendations for regulating the substance. A third route by which the development of a regulation can be requested is via a petition from a private party. This applies to the development of an emergency temporary standard.

Once a regulation is proposed, an advisory committee is formed by OSHA, consisting of representatives from various sectors. Such a committee is not required in the legislation. Meetings are open to the public. The committee presents recommendations on the standard to OSHA.

A proposed regulation is published in the *Federal Register,* consisting of all of the information presented by NIOSH and the advisory committee. In addition, other required documents are prepared as necessary, such as a regulatory impact analysis and environmental impact statement. Public hearings are held on the regulation. Oral testimony is recorded and is compiled along with any written testimony submitted during a comment period. The final rule is then published in the *Federal Register* along with a statement of justification.

Thus, by virtue of NIOSH and advisory committee structures, OSHA has developed more complex rulemaking procedures. There are a number of instances in which the APA has influenced regulatory procedure in spite of agency discretion in the application of the act. For example, in spite of the fact that 4,4'-methylene(bis)-2-chloroaniline (MOCA) was considered a carcinogen, OSHA's standard was struck down because of the failure of the agency to properly publish the standard, give the public adequate time to comment, and other reasons (McCaffrey, 1982, 85).

EPA's Rulemaking Under the Safe Drinking Water Act

The SDWA Amendments of 1986 are an example of the simplification of rulemaking to expedite standards-setting. As recounted by Stever (1988, 7–31 to 7–34), prior to the amendments, National Interim Primary Drinking Water Regulations were promulgated as temporary standards by the EPA. These could only be finalized after a National Academy of Sciences study. The NAS did not want to back thresholds for carcinogens, so in the case of certain organics, this created a logjam in the system. As a way of circumventing this, EPA issued informal health advisories called "Suggested No Adverse Effect Levels" (SNARLs). Stever points out that the SNARLs bypassed APA procedures but had a major influence over state rulemaking. In 1978, legislation introduced review by the SAB into the standards-setting process under SDWA. The 1985 amendments allowed EPA to revise the standards in the form of national primary drinking water standards, bypassing the use of NAS, but EPA still used the recommendations of its SAB at its own discretion.

Summary

The bottom line in agency rulemaking is that many of the APA procedures were modified dramatically in the direction of introducing many more steps. This arose from the complexity and uncertainty of information, mandatory and voluntary use of advisory committees, court cases, and public pressure for participation in the review procedures. As a result, standards were often delayed in emerging from the administrative agencies, but were probably more defensible. This was not only true of the three programs illustrated above, but occurred in other programs as well, such as PMN procedures under TSCA and the Special Reviews conducted for pesticides suspected of posing an environmental health risk (U.S. General Accounting Office, April 1986).

AGENCY ADJUSTMENTS TO INTERNAL PERFORMANCE

In the introductory chapter, several guidelines were presented as criteria for viewing the output of a program or organization. These were the timeliness of actions, their match against objectives, and the effectiveness of their design in light of unanticipated consequences, even though objectives may have been met. The critical issue is not the normative evaluation of agency output but rather the extent to which the agency acknowledges or recognizes its own internal performance or outputs and reacts to them to make adjustments in the risk management process.

Timeliness

CPSC decisions have been considered slow (U.S. General Accounting Office, April 1987). This has been indirectly attributed to a high turnover rate among the commissioners, i.e., a lack of continuity in leadership, which has been identified as a shortcoming of the commission system in general. In reaction to this, the GAO undertook a study of the commission and reviewed and evaluated several alternatives to the commission system, such as reduction of the number of commissioners from five to three, a single administration, and integration with the FDA. A change in the structure of the commission would require a change in Section 4 of the CPSA (U.S. General Accounting Office, April 1987, 10). Another factor contributing to the rate of standards-setting as well as to its stringency was a 1981 amendment to CPSA. The amendment required the CPSC to consider the use of voluntary standards submitted by industry prior to setting mandatory standards. The commission's lack of action or delayed action because of this policy has been under attack since the amendment was passed (Fise, 1987, iii). Finally, the statutory requirement that the CPSC issue an Advance Notice of Proposed Rulemaking as part of its rulemaking process (which is usually a discretionary action for other agencies) will tend to slow rulemaking (Fise, 1987; Christoffel and Christoffel, 1989, 336). In fact, Fise actually tracked a substantial decline in rulemaking activity by the commission since the passage of the amendments. CPSC has not yet reacted to these regulatory obstacles.

The EPA Rebuttable Presumptions Against Registration (which have now been replaced by a Special Review) for a pesticide under FIFRA take many years to complete. In addition, the revocation of approval for a pesticide can be very time-consuming (except in the case of an emergency suspension). EPA first approached the problem, through the 1978 amendments to FIFRA, by allowing the review of chemicals (active ingredients) in lieu of individual pesticide preparations, which was a generic approach to pesticide review. This speeded up the process. As of 1986, about 124 registration standards were developed by EPA for the active ingredients, though these standards have not been based on quantitative risk assessments (U.S. General Accounting Office, April 1986, 26–27).

The standards-setting process as discussed extensively in this chapter has, in many program areas, taken years to promulgate final, defensible standards. From an administrative point of view, this is attributable to the various layers of review required to set a standard. Changing scientific knowledge has introduced considerable uncertainty into the process. Congress has addressed the problem legislatively, but on a program-by-program basis. For example, the virtual stalemate in setting drinking water standards was broken by strict scheduling for standards-setting under the SDWA Amendments of 1986. NESHAPs and NAAQS have been slow in developing under the Clean Air Act because of lengthy reviews. While regulatory programs requiring permits for activities to occur have shifted the burden of proof to the applicants, the burden of proof is still with government when it comes to setting standards.

Because of FDA's procedures for new drug approval, the process had been lasting an average of eight to nine years. At one time, 86% of the drugs that underwent the process were rejected (Meier, 1985, 84). The application for approval of six artificial color additives that have a coal tar or petroleum base has been postponed 26 times over the past 25 years (Burros, 1985). During the

1970s, the FDA had already begun to modify some of its procedures to expedite the review process. These measures were primarily aimed at reducing redundancy in test data (Quirk, 1980, 220). As a solution to the problem, by the early 1980s, the FDA had introduced fast-tracking procedures that resulted in a doubling of the number of new drugs approved between 1980 and 1981 (Demkovich, 1982, 1250). Unfortunately, this fast-tracking system and the approval of drugs within that system (e.g., Oraflex, an arthritis drug, and Zomax, a pain killer) were subsequently linked to a number of deaths, which led to allegations of industry bias in the agency (Meier, 1985, 85).

Program Design Relative to Objectives

The EPA's program for regulating the operation of hazardous waste management facilities is designed so that a number of exemptions are allowed under the program, and the exemptions are often regulated under other programs. This can potentially result in a program that regulates hazardous wastes under a highly variable set of standards. The EPA has not reacted to this problem as yet.

FDA has interpreted its responsibilities under the Delaney Clause as pursuing its regulatory functions through the review and approval of new products, not as evaluating existing products. Thus, the design of the program (at least as the agency has interpreted it) precludes going after what could be the major portion of the problem, which could not have been the intent of its objectives. The FDA has not modified its program to direct it more toward existing problems as yet.

In a number of instances, regulation of risk levels for a given chemical is divided up among several different agencies because very fine distinctions are made regarding where the chemical is found. Compounding the situation even further is the fact that the different statutes under which different agencies operate lead to variations in analyses and risk levels. Two examples of this problem were given earlier in the book. First, nitrosamines, a carcinogen, have been found in baby pacifiers, which are regulated under the CPSC, and in nipples, regulated by the FDA since food flows through them (Fise, 1987, 25). As Fise has pointed out, this divided jurisdiction resulted in the CPSC establishing a 60-ppb limit for nitrosamines in pacifiers in December 1983, while the FDA established a 10-ppb limit for nipples in January 1985. Another example is the case of pesticide residues in foods. As Stever (1988, 8–21) points out, the EPA is responsible for regulating pesticide residues in raw agricultural products and can use risk assessment to set limits for such residues. The FDA, on the other hand, is responsible for pesticides when they appear in processed foods. In that context, they are considered food additives and are subject to the Delaney Clause (zero tolerance). It is unlikely in either one of these examples that it was the objective of the statutes to produce such differences. Such anomalies are usually handled on a case-by-case basis when they are discovered.

Design of Programs to Avoid Contingencies

The provisions of the Delaney Clause that are the basis of FDA regulation of food contaminants did not anticipate changes in detection limits such that infinitesimal amounts, previously assumed to be zero, are now subject to regulation. While the FDA has reacted to this constraint through discretionary action (described above, in the application of de minimis risk levels), it has not succeeded in gaining judicial support of this position in the courts.

The permit programs designed to restrict discharge of hazardous pollutants into the air and water were not directly designed to anticipate changes in knowledge regarding the toxicity of chemicals. Theoretically, agencies and legislators have anticipated this by providing for the renewal of permits on a periodic basis to allow such knowledge to be integrated into permit discharge conditions. However, the process of developing discharge standards has not always kept up with the state of knowledge, and revision of permits has not coincided with revisions in standards.

CONCLUSIONS

Tradition

The rate of agency formation during the peak period for federal risk management reflected a growth rate that was traditional within government. The rate of growth of the risk bureaucracy reflects in a general way one aspect of the intensity of the administrative process. This growth reached its peak in the last decade—during the 1970s—when the major laws were first enacted. Only one major agency, however, appeared in the early 1980s, even though legislation continued to be passed. To some extent the slowdown in the growth of new agencies may mask a proliferation of administrative entities at the programmatic and bureau levels within agencies. To answer the question of whether the growth of new agencies was in fact supplanted by the proliferation of bureaus within these agencies requires empirical study. Certainly the size of the staff did not grow substantially during the 1980s and in a number of instances declined.

Several dozen statutes set the parameters under which the five major agencies involved in risk management operate. In the area of setting standards, discretionary action occurred largely in the development and measurement or expression of risk levels and the techniques by which these levels are derived, while the broad categories of risk that agencies were to examine (by selecting substances to regulate and scheduling those regulations) were relatively more prescriptive. This deference to agency judgment in how to regulate within the constraint of what and when to regulate is not an unusual approach to regulation. The next chapter explores the way in which the judiciary has shaped agency discretion, largely in reaction to the outcomes of the administrative processes and parameters set by legislation.

Agency reactions to their own performances are mixed; however, a recognition of these outcomes by government appears to have motivated some changes. First, bureaucratization of procedures has taken its toll in slowing down processes such as standards-setting in order to accommodate more peer reviews. Some fast-tracking has begun and has required legislative changes. Second, program design targets are often missed, but administrative interpretations and legislative changes have aimed at refining targets to be more feasible. Third, contingency planning in terms of responses to unexpected adverse effects of risk policies has not been explicitly built into administrative processes. However, built-in controls at the beginning of program development, such as mandatory regulatory impact analysis and environmental impact analysis, have continued to avoid some of the more prominent impacts to the extent that they are identified. While these are clear signs of agency reaction to feedback, whether these reactions are part of the normal course of evolution of agency process is difficult to say.

Initiative

Agencies have exercised initiatives in several areas: in the use of risk assessment, the design of the risk management function, and the use of advisory groups to provide input into the design of risk levels.

Within the tradition of allowing agencies the discretion of deciding on the method of regulation, the agencies have exercised considerable initiative in the extent to which they have used risk assessment to set standards, as the basis for the design of control measures, and for other risk policy decisions, such as setting priorities among risk problems. These initiatives have not always met with success, however. In many cases, the way agencies have applied risk assessment and chosen assumptions has been challenged in the courts and in public commentary on rulemaking.

Agencies have also shown considerable initiative in designing into their structure risk management units to carry out and review these assessments. Yet the level of commitment to these functions is somewhat obscured by the fact that risk management has not been separated from other regulatory functions in budgets and in the allocation of personnel.

NOTES

1. In addition to the specific works cited, much of the information on agency characteristics and organization is drawn from the Office of the Federal Register (1988) and Congressional Quarterly, Inc. (1983, 1986).
2. The description of the formal procedures of CSPC in this section was drawn from

Laporte, Gies, and Baum (1975), Congressional Quarterly, Inc. (1983), and other references (cited in the text).

3. The federal government is required to set up its own workplace standards and submit reports to OSHA for review annually. States similarly are given the right to pass their own laws and set up their own program in this regard. OSHA enforces the federal law until the states develop their own programs (Congressional Quarterly, Inc., 1986, 417).

4. The example given by Brickman, Jasanoff, and Ilgen (1985) of how new agencies are created to assume new functions is the creation of three agencies—administrative entities within the Department of Labor, the Occupational Safety and Health Review Commission, and the National Institute for Occupational Safety and Health.

5. Interestingly, Kaufman notes that statutes are not the most common way in which organizations have been created; in fact, he states that the role of statutes in agency formation has declined over time. Departmental orders, executive orders, reorganization plans, and other administrative actions, rather than legislation, played key roles in agency formation (Kaufman, 1976, 42).

6. Budgetary figures were adjusted for inflation using the GNP deflator (Executive Office of the President, 1989).

7. The origin of discretion is, in fact, delegation, which has a long history. According to Greenwood (1984, 3–4), the history of administrative discretion is a history of the gradual delegation of powers from Congress to the agencies to intervene in the affairs of private individuals.

8. Vig (1984, 63), for example, has pointed out how the judiciary's new role in reducing agency discretion emerged from th eenvironmental laws of the 1970s.

9. Interpreting organizational location as an indication of the importance of the function must be done cautiously. An intermediate layer of administration or "middle level of command" may be more important than top-level functions. Some intermediaries, if well placed and given sufficient resources, can act as powerful spokespersons for the functions that report to them. For an early but extensive development of this theme in the context of the military and the relationship between communication and command, see Deutsch (1966, 154).

10. Greenwood's study of OSHA and EPA (1984) discusses the location of the risk assessment function within these two agencies.

11. The *Federal Register* is a compilation of rules and regulations from all federal agencies, along with histories, commentaries, and other background information. It is published Monday through Friday.

12. Adjudication is a quasi-judicial process performed by administrative agencies that often involves issuing orders to enforce provisions of statutes under agency jurisdiction. Adjudication is described in Section 551(7) of the Administrative Procedure Act as an "agency process for the formulation of an order." In the APA (Article 551(6) and (7)) an order is defined as "the whole or a part of a final disposition . . . of an agency in a matter other than rule making but including licensing." Adjudication is defined in the act as the "agency process for the formulation of an order." Thus, an order is the outcome or purpose of adjudication. Rules are a result of rulemaking and orders result from adjudication.

Adjudication can also be used in the rulemaking process under two conditions: when a statute requires that a hearing be held as part of the process or when facts are in dispute (Davis, 1972, 194). A number of statutes also directly allow an administrator to issue administrative orders. The Administrative Procedure Act imposes minimum requirements on the adjudication process, and other procedures are developed by agencies through formal rulemaking or by the courts. Davis points out that one of the purposes of adjudication is to reserve for the courts only those situations that are clearly illegal or point to the severe problems with agency discretion (Davis, 1972, 212–213).

13. Davis points out that most agency action is informal rather than formal; in fact, about 80–90% of administrative processes are informal. In spite of this fact, he states, most of the attention is paid to the formal processes (Davis, 1972, 88).

14. Amendments to the APA include the Freedom of Information Act, the Sunshine Act, and the Privacy Act. Major changes in the APA occurred in 1982.

CHAPTER 6

The Judiciary

INTRODUCTION

It is appropriate to conclude this analysis of governmental processes with respect to risk policy with the judiciary. It is here—in the course of judicial review—that the process begins all over again. The judiciary plays a major role in formulating policy through its review of the implementation process (Mazmanian and Sabatier, 1983, 3). Judicial review has been defined as "the power of a court to determine the legality and constitutionality of an action of a government official, agency, or legislative body" (Heffron and McFeeley, 1983, 293). Judicial reviews occur where disputes arise between government and the citizenry or among private citizens (Brickman, Jasanoff, and Ilgen, 1985, 100). In the area of environmental health risks the courts formulate policy by interpreting laws and administrative procedures for risk management. This role has varied considerably over time in response to changes in the legislative and political context for risk management. Many have argued that in the risk area, inconsistencies in judicial review have made agency procedures more complex, while others have maintained that judicial review keeps the administrative process in balance. Agencies, often the target of the review process, are sometimes mixed in their reactions to the review process.

The trends in the intensity and direction of judicial activity and the factors that influence the role of the courts provide an important context for understanding judicial decisions in the area of environmental health risk. The trends that are examined here include the intensity of judicial activity in the risk area in terms of caseloads; the prominence of agencies, particularly the EPA, as parties in lawsuits on environmental risk; the success of government agencies in case outcomes; and the existence of specialization in the courts, particularly the prominence of the United States Court of Appeals for the District of Columbia Circuit in handling risk cases. Aside from these structural and procedural trends, the effect that trends in judicial philosophy regarding substantive reviews and positions on the use of risk assessment have on case outcomes is also examined.

In line with the overall theme of the book, the conditions influencing the judicial review of risk cases are traced in terms of those that are traditional to government and those that represent new initiatives. Conditions that are traditions in government pertain to the total caseload in the federal courts since the early 1970s as a factor in the rate of growth of risk litigation, changes in judicial philosophy, the overarching structure of the court system, and citizen influence. Characteristics of the risk issue itself that reflect new initiatives as determinants of the caseload include the nature of legislation and rulemaking, the resource base of plaintiffs, and the role of dispute resolution.

ORGANIZATION OF THE JUDICIAL BRANCH[1]

At the federal level, the judicial branch functions in adjudication, along with agencies, and manages the court system. In its adjudicatory role, the court system expedites the resolution of disputes between government entities, between private individuals and groups, or between government and the private sector. The judicial branch "is the final arbiter of the balance of power between the states and the federal government, and between the executive and legislative branches of government" (Wenner, 1984, 197). This function is exercised in the context of four different kinds of law: constitutional, civil, criminal, and administrative (Neely, 1981, 4). Most cases involving health, safety, and environmental risks fall under the latter three types of laws.

Organizationally, the judicial branch consists of the Supreme Court, the Lower and Special Courts, the Administrative Office for the Courts, and the Federal Judicial Center. The Lower Courts within the judicial branch consist of courts of appeals and district courts. These are the U.S. Courts of Appeals (one within each judicial circuit); the U.S. Court of Appeals for the Federal Circuit, established in 1982 with nationwide jurisdiction; and the District Courts or trial courts, with at least one per state. In addition to the Lower Courts are Special Courts established by Congress to act on particular kinds of cases. The Courts of Appeals are divided geographically into 12 districts. These 12 districts include the District of Columbia (D.C.). States and territories are assigned to particular districts. Wenner (1982, 100–102) shows how the districts vary in size and degree of socioeconomic and political homogeneity.

Congress plays a major role in shaping judicial activity. While the Constitution lays out the broad powers of the courts, it also confers on Congress the power to exercise considerable discretion in determining the structure and authority of the court system. Congress can create inferior courts to exercise certain judicial powers in addition to those exercised by the Supreme Court. In addition, Congress can define by statute certain appellate jurisdictions for the Supreme Court (Office of the Federal Register, 1988, 66). Congress also has the authority to set up the National Courts to hear cases on a particular matter (Office of the Federal Register, 1988, 73). Finally, Congress can reduce court jurisdiction under particular statutes by passing special legislation to circumvent existing statutory restrictions (Wenner, 1982, 26).

TRENDS AND PATTERNS IN JUDICIAL REVIEW

The courts play a more extensive part in risk management now than they have in previous decades. In 1979, Bazelon underscored this level of activity, noting that "hardly a sitting in our court goes by without a case from EPA, FDA, the OSHA, or the NRC" (Bazelon, 1979, 278). A cross-national comparison of chemical regulation noted the uniqueness of the U.S. federal courts with respect to their very active role in the regulation of carcinogens: a very large number of carcinogens, such as asbestos, vinyl chloride, DES, aldrin, dieldrin, and 2,4,5-T, have been brought before the judicial system (Brickman, Jasanoff, and Ilgen, 1985, 119–120).

The courts' activity in risk management can be defined in several ways: in terms of (1) trends in the number and types of cases, (2) the distribution of cases among different federal courts and the differences in agencies that are parties in the cases, (3) the breadth or scope of risk issues covered in judicial reviews, and (4) judicial philosophy toward risk. The first two concerns pertain to structural or organizational attributes of the system and the latter two emphasize the more substantive nature of the reviews.

Caseload Trends

Trends in the number and types of cases processed provide a guide to the direction of judicial review of risk cases. The overall numerical caseload trends can be further analyzed by type of case, outcome (who wins and who loses), and agency involved in the case.

Overall Trends in Number of Cases

The courts' workload can be measured in several different ways: in terms of number of cases filed, number of cases disposed, or number of written opinions. Of the three, the number of cases filed is more commonly used as an indicator of caseload, primarily because of the accessibility of the information.

Since the beginning of the era of environmental and health-related federal statutes in the late 1960s, the number of cases has risen dramatically, almost paralleling the growth in environmental legislation. A search of federal cases that specifically used "risk assessment" or "risk analysis" as key words was undertaken.[2] Only 39 cases specifically addressing risk assessment were found,[3] reflecting both the relative newness of the use of the technique and the fact that its use by federal agencies is discretionary. There was a steady increase in the number of risk assessment cases per year over the period from the mid-1970s (when cases first began appearing) through 1988. The tabulation of these cases by year is shown in Table 6.1. The disproportionately larger number of cases appearing in 1981 is reflective of the passage of the Superfund legisla-

Table 6.1.   Cases Mentioning Risk Assessment or Risk Analysis

Year	Number of Cases
1976	3
1977	0
1978	1
1979	1
1980	0
1981	6
1982	3
1983	3
1984	3
1985	3
1986	6
1987	7
1988	3 (through May only)

tion, CERCLA, which explicitly referred to the use of risk assessment as a way of characterizing potential health risks at hazardous waste disposal sites.

Historically, the number of cases appearing annually in the environmental area has been constant. An extensive study by Wenner (1982) of 1900 environmental cases brought in the context of over 60 laws between 1970 and 1979 revealed that since 1972, the number of all cases brought under these laws per year remained constant at about 200 (Wenner, 1982, 21). The scope of that study is not entirely analogous to environmental health risk cases considered here. That study excluded a number of statutes that are risk-related, such as the Occupational Safety and Health Act and the Food, Drug, and Cosmetic Act (Wenner, 1982, 8). In addition, about 765 of the 1900 cases were brought under only one law—the National Environmental Policy Act (Wenner, 1982, 6)— which had not been particularly focused on environmental health during the time period covered by the study (though in subsequent years NEPA cases have addressed health). Just considering air and water pollution cases tabulated by Wenner (1982, 21) (i.e., subtracting the NEPA cases), there is a clear increasing trend over the period (rather than a constant number of cases), broken only by a slight decline during 1977 and 1978.

Thus, from the evidence available on court case trends in the general environmental area and in the area of risk assessment, it appears that the number of cases per year involving human health risks from environmental chemical contaminants has been growing. The key point is that the litigation is continuing at a pace paralleling the rise in new laws and regulations.

Trends by Type of Legislation

The analysis of cases conducted here of all laws since the 1940s involved a search of cases under the key word "risk" (as distinct from "risk assessment").

First, the number of cases that appear using the key word "risk" is far greater than the number of cases that appear using "risk assessment." This is expected from the fact that risk is explicitly mentioned in the statutes and is often a basis for bringing cases or at least is referenced in cases. Risk assessment, on the other hand, is not usually mentioned in the statutes, and its use is discretionary as mentioned above. Second, the findings pertaining to the distribution of cases by statute more or less confirm the pattern found in other studies of the emergence of cases by environmental statute. The distribution of such cases by statute that emerged from the search through 1988 is shown in Table 6.2. As can be seen from the table, by far the largest proportion of total cases as well as the subset of risk-related cases was brought under the Clean Water Act. While it has not been the target of recent risk-related court cases, the Clean Water Act was one of the first statutes to set the stage for risk issues in legislation. The 1980 water quality criteria, for example, were among the first environmental criteria to be based on detailed risk analyses for each substance. The Clean Air Act and CERCLA (Superfund) followed in that order in terms of the proportion of cases brought. Looking at these three statutes (CWA, CAA, and CERCLA), it is clear that the large number of risk-related cases roughly parallels the large volume of cases brought under these statutes in general. It is instructive to note that about two-thirds of the cases brought under the three statutes were risk-related. This proportion was matched only by cases brought under pesticide legislation and atomic energy legislation.

The Oleinick, Disney, and East review of cases under statutes administered by CPSC, OSHA, and EPA (1986) shows roughly similar findings. The results derived from that study are summarized by statute in Table 6.3. The largest number of cases are for a variety of statutes including the Clean Water Act, the Clean Air Act, and solid and hazardous waste statutes. Unfortunately, the lack of disaggregation of their data by law precludes breaking down the findings further for individual statutes.

According to the Wenner study of all environmental laws (1982, 5-7), the leading laws under which cases were brought over the decade were the National Environmental Policy Act (40.3% of the cases), the Clean Water Act (16.5%), and the Clean Air Act (10.9%). Thus, the dominance of the Clean Water and Clean Air Acts (when NEPA cases are excluded) as tabulated by Wenner agree with the findings above. (CERCLA was passed just at the time the Wenner study was conducted and was not included in it.) A partial explanation of the pattern of case development by statute offered by Wenner is that there is a relationship between when cases are brought and when a piece of legislation has been passed. Typically controversy (and hence, the emergence of cases) occurs just after an act has been passed, in order to refine and shape the administration of the law (Wenner, 1982, 22). She notes that NEPA is the one exception to this, having been characterized by continuing litigation.

Table 6.2.   Risk-Related Cases by Statute

Name of Legislation (Date of Enactment[a])	Total Number of Cases Under Act[b]	Number of Cases on Risk-Related Issues	Risk Cases as Percentage of All Cases
Clean Water Act (1972)	1119	796	71%
Clean Air Act (1970)	756	460	61
Comprehensive Environmental Response, Compensation, and Liability Act (1980)	308	191	62
Resource Conservation and Recovery Act (1976)	207	88	43
Federal Insecticide, Fungicide, and Rodenticide Act (1972)	160	119	74
Occupational Safety and Health Act (1970)	126	54	43
Atomic Energy Act (1974)	110	76	69
Toxic Substances Control Act (1976)	82	32	39
Food, Drug, and Cosmetic Act (1938)	63	25	40
Safe Drinking Water Act (1974)	57	17	30
Hazardous Materials Transportation Act (1975)	12	6	50

[a]While the date of enactment given reflects the most significant recent change, the tabulation of cases goes back to initiation of that area of law. For example, Clean Water Act cases would include those brought under water pollution control legislation dating back to 1948.
[b]Tabulated through May 1988.

Table 6.3.   Summary of Findings of Risk Management Cases by Statute, 1970–1982

Legislation	Number of Cases	
	Court of Appeals	Supreme Court
OSH Act	31	3
CPS Act and related statutes	16	0
FIFRA—Registrations, cancellations, suspensions, emergency orders	14	0
TSCA—standards	3	0
Other[a]	57–302	5
TOTAL	121–366	8

Source: Summarized from Oleinick, Disney, and East (1986), Table 3.
[a]The category "other" includes CAA, FWPCAA, NCA, RCRA, SDWA, other TSCA, and other FIFRA cases.

Trends by Agency

The trends in numbers of court cases by federal agency are consistent with the way litigation and statutory responsibilities are divided up by statute among agencies.

O'Brien's review of court statistics shows that the EPA was involved in the largest number of cases. A total of "approximately 22 percent of all administrative appeals in 1980 involved regulatory action by four agencies involved in environmental litigation. These were the Environmental Protection Agency (10 percent), the Federal Energy Regulatory Commission (8.3 percent), the Occupational Safety and Health Administration (2.9 percent), and the Nuclear Regulatory Commission (0.6 percent)" (O'Brien, 1986a, 42, citing Administrative Office of the U.S. Courts, 1981). Since EPA has jurisdiction over the three statutes accounting for the largest caseload (see Table 6.2), as well as having jurisdiction over a total of at least nine major environmental statutes, it is not surprising that overall it dominates other environmental agencies with respect to caseload.

Studies by Wenner (1984), Davies (1984), and others also indicate that cases involving administrative agencies have been dominated by EPA. However, these studies further point out that in the early 1980s, the number of referrals from the EPA was strongly influenced by the political changes going on within the agency. Gorsuch altered the enforcement function within the EPA by drastically reducing the number of staff and reorganizing it. The effect on referrals to the Justice Department was described as follows: "Not surprisingly the effectiveness of the office was reduced considerably: between June 1981 and June 1982, EPA cases referred to the Justice Department declined by 84 percent, compared with the previous 12 months, and civil penalties imposed by EPA dropped 48 percent" (Davies, 1984, 149, referencing Kurtz, 1983). These cuts occurred most prominently for cases regarding stationary sources of air pollution (a decline from about 100 to about 40 between 1979 and 1983) and water pollution control (80 to 40); slight declines occurred in cases under other programs. In addition to case referrals, the number of administrative enforcement orders declined as well (Davies, 1984, 153).

In interpreting these findings, it should be kept in mind that not all cases referred are heard in the courts. It has been pointed out, for example, in the context of referrals from EPA to the Justice Department, that "20 times more cases are referred than are eventually adjudicated in court. The Justice Department functions as a gatekeeper, selecting some cases for litigation but dropping many others because of the department's limited resources or priorities that are different from those of EPA's attorneys" (Wenner, 1984, 186).

Patterns of Winners and Losers

Through the period from 1970 through 1982, the government maintained a strong lead as victor in environmental cases, whether it has been involved as defendant or plaintiff; it won 67% of cases as plaintiff and about 62–63% of cases as defendant (Wenner, 1984, 185).

In the Wenner study, the percentage of cases won each year that were pro-environmental remained constant at about 50% (Wenner, 1982, 29). Several findings of that study address risk issues.

The largest number of cases brought by industry against government pertained to the legitimacy or justification of standards under the clean air and clean water legislation. The cases questioned the consistency of application to different firms within the same industry, to different industries for the same substance, and in different parts of the country. While government won many of the cases, industry gained delays in enforcement (Wenner, 1982, 51).

Melnick points out that a number of agencies adapted to judicial reviews of their procedures. In fact, some agencies, particularly the attorneys within those agencies, actually considered reviews beneficial (Melnick, 1985, citing Pederson, 1975, 59–60 and Rodgers, 1979, 699). The EPA, for example, a major party in many environmental suits, has been considered generally supportive of judicial review of its procedures in spite of the fact that it often lost major cases. Melnick points out:

> One might expect that an agency that has lost as many major court cases as the EPA would take a jaundiced view of judicial intervention. Such is not the case. While complaining about some decisions, EPA officials generally credit the courts with improving the agency's competence and programs. (Melnick, 1983, 379)

This high degree of acceptance is attributed by Melnick to the generation of a number of new programs and offices within the EPA as a result of the litigation; that is, he argues that litigation contributed to the growth of the organization:

EPA officials speak highly of the courts precisely because court decisions have played such a major role in shaping the agency's structure, strategy, and sense of mission. By forcing the EPA to carry out new programs, the courts created new program offices within the agency. . . . In short, a variety of court decisions have produced new bureaucratic units whose task it is to respond to court demands. Since they derive their jobs, their influence, and sense of purpose from court action, it is not surprising that officials in these offices tend to praise judicial review. (Melnick, 1983, 379–380)

This was particularly true for regulatory activities relating to health risks:

Court decisions stressing the need for more evidence on the health effects of regulated pollutants and requiring the EPA to set standards for additional pollutants led to expansion of the agency's research staff, the creation of special task forces (such as one for airborne lead), and the establishment of a new office to write criteria documents. (Melnick, 1983, 380)

Another reason why EPA viewed litigation relatively positively is that occasionally EPA has used judicial review to obtain additional funds for unfunded but mandatory statutes. This occurred, for example, in the case of *Natural Resources Defense Council, Inc.* v. *Train*, 396 F.Supp 1393 (D.C. Cir. 1975), (the toxics consent decree), where EPA had to develop standards for many more chemicals than had been expected under the water pollution control program (Melnick, 1983).

However, the impact and acceptance of judicial review within a given agency like the EPA was not entirely uniform. Melnick points out that while the general counsel's office tended to obtain more work from court cases, enforcement divisions tended to be somewhat more frustrated, as was top management. The top management of these two offices within EPA often took the criticism for losing cases, and technical professionals were also blamed because their work was judged inadequate (Melnick, 1983, 381).

Case Distribution by Court

Congress is one of the major forces determining the distribution of cases among the courts. It has the authority to determine the jurisdiction of the courts, that is, which court hears a case, and whether an action is subject to judicial review (Heffron and McFeeley, 1983, 294). Congress often expresses these preferences directly in the legislation it enacts. Which courts hear different cases is particularly significant to case outcomes, since different courts are said to vary in their philosophies and leanings toward the environment. It is believed that case outcomes can differ depending on the court in which the case is heard. Wenner points out, for example, that the courts in the north-central and northeastern sections of the country show the greatest support for the environment, while those in the west and south show the least (Wenner, 1982, 119). This is supported by data on voting behavior in Congress as well (cited in Chapter 4).

Plaintiffs can also, under some circumstances, determine the distribution of cases among the courts, since there is some latitude on the part of plaintiffs as to where they bring a case. If, for example, a plaintiff is an organization with multiple offices, the case may be brought in any one of the areas in which the offices are located. Similarly, in cases involving the government, a case may be brought where the government resides. The location ultimately depends on the nature of the case.[4]

Appeals Courts

Within the federal court system, the appeals courts hear many risk-related cases. Between 1970 and 1983 alone, Oleinick, Disney, and East (1986, 383) tabulated over 1000 OSHA, CPSC, and EPA cases in the federal courts of appeals system. Within the appellate system, the D.C. Circuit Court of Appeals hears the majority of certain environmental risk cases (Molton and Ricci, 1985, 4). Oleinick, Disney, and East have observed a similar trend in the cases heard involving the three agencies they studied: between 1971 and 1983, the D.C. Circuit accounted for 20% of the OSHA, CPSC, and EPA cases, while its share of total cases heard by the appellate courts was only 4%. One reason for the dominance of the D.C. Court in many of the key federal cases involving risk is statutory requirements that the cases be heard in that court. For example, Section 108 of SARA requires that judicial review of civil penalties relating to financial responsibility be conducted by the D.C. Court of Appeals (U.S. Congress, House of Representatives, 1986b, 207). A second reason is a tradition that the D.C. Court has had in hearing agency rulemaking cases in general; many environmental cases pertain to agency rulemaking.[5] The second most popular circuit (in terms of numbers of risk-related cases) has been the Fifth Circuit. This circuit heard 14% of the cases involving OSHA, CPSC, and EPA. This is not unusual, however, considering that its overall caseload was quite high, accounting for 25% of all cases heard by the appeals courts (calculated from Oleinick, Disney, and East, 1986, 383). In other words, the large percentage of risk-related cases in the Fifth Circuit is probably driven by the generally large caseload that this circuit receives.

Supreme Court

The Supreme Court has heard only a few environmental cases relative to its overall caseload. The caseload of the Supreme Court over its term (usually nine months) has been estimated at over 5000 cases that are decided on; in addition, each justice can act on an average of 1200 applications that are typically filed (Office of the Federal Register, 1988, 66). In 1920, in contrast, the Supreme Court heard only 565 cases. Of these cases, less than 10% are reviewed; written opinions are prepared for only 3% (O'Brien, 1985, 667). Highlights of Wenner's review of Supreme Court decisions are summarized as follows:

The Supreme Court reversed the circuits and districts in more cases (27) than it upheld them in (20). While it overturned many positive decisions for the environment, the Court tended to uphold the lower courts in their negative decisions. . . . In the [Supreme] Court's 31 unanimous and per curiam decisions, more environmental defeats (16) were recorded than were victories (12). (Wenner, 1982, 150–152)

Judicial Philosophy in Decisionmaking

Of greater significance than the trends in the sheer numbers of cases and their distribution among the various courts is the style of judicial review during different administrations, under different judges, for different cases, and during different time periods.

The Scope of Judicial Review: Procedural vs Substantive Reviews

The Administrative Procedure Act defines the formal rules of judicial review of agency decisions for which there are no judicial review standards or procedures. The APA had defined two standards for the judicial review of agency decisionmaking: (1) the arbitrary and capricious standard and (2) the substantial evidence standard. The distinction between the two standards was clouded and perhaps eliminated in *Pacific Legal Foundation* v. *Department of Transportation*, 593 F.2d 1338 (D.C. Cir.), *cert. denied* 444 U.S. 830 (1979). It has been argued that the elimination of the two standards and absence of anything to take their place has had the effect of placing an obligation on agencies to establish a better factual and methodological base for their decisions (O'Brien, 1986a, 44).

Two views of the scope of judicial review characterize the overall philosophy of the courts in deciding on risk issues. The first is that the courts should only focus on the procedures used to arrive at technical decisions. The second argues that courts should undertake, in addition to administrative or procedural reviews, de novo or at least searching reviews of scientific information, that is, the courts should be charged with conducting substantive or factual reviews and thereby arrive at their own decisions about risk.[6]

Both points of view have been argued vociferously by court justices, administrative agencies, and the public. Much of the debate in the judiciary occurred in the D.C. Circuit Court of Appeals (O'Brien, 1986a, 35). This legal debate occurred particularly in the context of judicial review of environmental decisions and often set forth precedents for other areas of social policy.

Technical/substantive reviews: the "hard look" doctrine. In the mid-1970s, Judge H. Leventhal strongly advocated a role for the courts that would emphasize technical reviews by justices. That is, the courts should take a "hard look" at the problems directly. In emphasizing this type of a review, the judiciary was reacting to the pervasiveness of scientific and political uncertainty (regarding judgments about risk acceptability) in risk decisions (O'Brien, 1986a, 38). This role was advocated to the point of even proposing organizational changes in the court system, which would allow for special personnel and specialized courts that would handle such technical analyses (Leventhal, 1976, 432; 1974, 509).[7] The hard look approach was rejected by the Supreme Court in *Vermont Yankee Nuclear Power Corp.* v. *Natural Resources Defense Council, Inc.*, 435 U.S. 519 (1978).

The interest in the issue of substantive technical reviews by the courts is reflected in the data on cases. According to one study, the largest number of cases over the decade of the 1970s was brought on substantive rather than procedural grounds: "In every year from 1970 to 1979, at least 80 percent of all cases adjudicated by the federal courts were resolved on a substantive legal issue, with one exception. In 1971, 21 percent of the cases were decided on procedural grounds alone" (Wenner, 1982, 23).

It has been claimed that, at least in the area of pesticides and pharmaceuticals, the move toward reviews of the substance of agency decisions under the hard look doctrine has induced agencies to collect better information. Rather than being pressured into improving their own information base, one way that agencies have reacted is to impose more stringent requirements on industry to provide the information: "In many instances stronger controls were imposed on data submitted by the regulated industry" (Hadden, 1986, 13).

Procedural reviews. Judge D. L. Bazelon (1979, 279), on the other hand, has been the strongest proponent of the argument that the court should not second-guess administrative agencies on technical matters. According to Bazelon, courts should defer to reviewing agencies with respect to whether they are following their substantive mandates:

> In reviewing regulatory decisions, the court does not reweigh the agency's evidence and reasons. Nor does it decide whether the factual conclusions and policy choices are correct. Courts lack the technical competence to resolve scientific controversies; they lack the popular mandate and accountability to make the critical value choices that this kind of regulation requires. The court's role is rather the monitor of the agency's decisionmaking process—to stand outside both the expert and the political debate and to assure that all the issues are thoroughly ventilated. (Bazelon, 1981, 792)

This sentiment was reiterated in a dissenting opinion by Justice Thurgood Marshall in the OSHA benzene standard case, where he argued that the courts had no right to meddle in areas beyond their technical abilities and in particular should not be second-guessing agency expertise (Vig, 1984, 77).

Wenner observed that the Supreme Court expressed the same philosophy during the Reagan administration:

The Supreme Court in the 1980s gave clear signals to lower federal courts that they should restrain themselves when asked to review decisions made by federal administrative agencies. At the same time it reduced congressional oversight capacity toward the same agencies. (Wenner, 1984, 197)

Vig (1984, 77) also has pointed out that it is the philosophy aired in Marshall's dissenting opinion that characterized the Supreme Court's subsequent approach to the hard look doctrine, reflected in *Baltimore Gas & Electric Co. v. Natural Resources Defense Council, Inc.*, 462 U.S. 87 (1983).

The argument against substantive reviews has often been expressed in terms of the separation of powers in government: that de novo substantive reviews of agency decisions violate executive agency authority (O'Brien, 1986a, 47). Opponents of substantive review have also argued that risk policy decisions either primarily hinge on questions of fact or are politically based. If the decisions are heavily fact-laden, then the courts are often viewed as having insufficient expertise and having an inappropriate adversarial approach to deal with questions of scientific facts. If the decisions are primarily political, then the courts are considered inappropriate once again because the elected representatives are ostensibly supposed to make political judgments (Vig, 1984, 61, 65; Green, 1980; Shapiro, 1979).

Judicial Philosophy and Case Outcomes

Application of the hard look doctrine does not necessarily lead to consistent outcomes in judicial review of the use or extent of risk assessment and other quantitative techniques in agency decisionmaking (Vig, 1984, 70). The application of the hard look doctrine to risk assessment matters varies from court to court. Vig points out that "if there is a common thread, it is that the highest court will only insist that regulatory agencies meet what the judges perceive as statutory minima in establishing risk significance" (Vig, 1984, 73). It has been observed, ironically, that:

> . . . though Judges Bazelon and Leventhal disagreed on the standard of review, their votes on cases on which they served on the same panel have been identical except that their reasons differ, that is, Judge Bazelon finding procedural error when Judge Leventhal found substantive defects, and the reverse. (Oleinick, Disney, and East, 1986, 387, citing McGarity, 1979b)

Several cases demonstrate how the courts can take different positions when they exercise the hard look doctrine in the review of health risks of chemicals. These different opinions often emerged during the same time period. The cases are reviewed in the next section in connection with the overall consistency of judicial decisions, but a few examples highlight the differences in approach taken under the same doctrine applied to the analysis of chemical risks. Cases that tended to favor restricting judicial review to procedural issues were *Vermont Yankee* and one of the Reserve Mining cases (*Reserve Mining Company v. U.S. Environmental Protection Agency*, 514 F.2d 492 (8th Cir. 1975)), where regulation was supported in the absence of clear scientific evidence. Examples of cases where more substantive reviews were favored, particularly regarding risk assessment, are as follows:

- In *Ethyl Corp.* v. *U.S. Environmental Protection Agency*, 541 F.2d 1 (D.C. Cir. 1976), the hard look doctrine supported EPA's regulation of lead additives in gasoline but did so in the absence of a conclusive and detailed scientific basis for the regulations (Vig, 1984, 71).

- In *Natural Resources Defense Council, Inc.* v. *Nuclear Regulatory Commission*, 547 F.2d 633 (D.C. Cir. 1976), *rev'd*, *Vermont Yankee* v. *Natural Resources Defense Council, Inc.*, 435 U.S. 519 (1978), risk assessment procedures and findings for the analysis of risks of spent fuel were found deficient, especially the conclusion of zero risk associated with deep disposal. The case was referred back to the NRC for further fact-finding (Vig, 1984, 74).

- In the case of formaldehyde, a CPSC ban was thrown out because of alleged deficiencies in the risk assessment methodology.

- In *Industrial Union Dept.* v. *American Petroleum Institute*, 581 F.2d 493 (5th Cir. 1980), the Fifth Circuit ruled that OSHA had not demonstrated "significant risk." The court did not allow regulation where it felt insufficient evidence existed. In the Supreme Court's review of the decision (448 U.S. 607 (1980)), the significance test used by the lower court was upheld. OSHA had the responsibility of estimating risk quantitatively. Opinions written on what was meant by enough and conclusive evidence were considered ambiguous by the court (Vig, 1984, 73–75; O'Brien, 1986a).

- Several petitions were filed by NRDC in the U.S. Court of Appeals for the District of Columbia Circuit and then directly to EPA, questioning the basis for the withdrawal by the EPA of certain emission standards for benzene and the setting of other standards (U.S. EPA, July 28, 1988, 28504). EPA denied NRDC's petition. A few years later, the court ruled en banc in *Natural Resources Defense Council, Inc.* v. *U.S. Environmental Protection Agency*, 824 F.2d 1146 (D.C. Cir. 1987) that EPA's revision of a standard for vinyl chloride was improper in that it was governed by cost and technological feasibility without first making a determination of what constituted a safe emission level. While the court left it up to EPA's discretion to determine an acceptable level of risk, EPA was required to set such a limit on the basis of appropriate scientific and other expert opinion.

Consistency of Judicial Reviews

The reaction against factual or substantive reviews by the courts may simply reflect the reaction to the uneven and inconsistent interpretations of fact from court to court even when the same case was at issue. The following examples illustrate this point:

- *Benzene.* Hadden observes in the benzene case (*Industrial Union Dept.* v. *American Petroleum Institute*, 448 U.S. 607 (1980) that "no two Justices used the same reasoning to reach a conclusion" which "suggests that an active role in reviewing the validity of technical data poses threats to the court's own institutional legitimacy" (Hadden, 1984, 13; also see Vig, 1984; Jasanoff, 1985).
- *Lead.* The D.C. Circuit first ruled in favor of industry, and then "after EPA petitioned for a rehearing, the District of Columbia Circuit, sitting en banc with all the judges present, reversed itself and held for regulation of lead in gasoline" (Wenner, 1982, 65).
- *Asbestiform fibers.* In the *United States* v. *Reserve Mining Company* case (514 F.2d 492 (8th Cir. 1975)), Vig (1984, 64) observes that "federal and state courts at four different levels each approached questions of risk assessment differently and arrived at widely different conclusions as to the hazards of water- and air-borne asbestiform fibers (Vig, 1979). Although the Court of Appeals for the District of Columbia has greatest responsibility for reviewing federal administrative decisions, its efforts at formulating new doctrines for evaluating uncertain risks have largely been frustrated by other circuits and the Supreme Court."

The consistency of judicial reviews can vary from one court to another for a variety of reasons. In cases involving EPA, Marcus (1980, 292) notes that the courts "interpreted the law by using different principles at different moments." In the case of OSHA, McCaffrey (1982, 136) points out that the values of a particular court very much influence the nature of standards-setting—the courts required cost/benefit analysis in two cases and not in a third.

Melnick attributes the variation between agency decisionmaking and the outcomes of the judicial review process to general trends in administrative law: first, the Administrative Procedure Act has been interpreted as allowing many more groups to participate in agency decisionmaking; second, there is an increasing burden of proof on agencies to explain their actions under the APA; third, there is an increased willingness on the part of the courts and the public to counter agency interpretations of administrative procedures (Melnick, 1983, 11).

Influence of Judicial Review on Risk Policy and Decisionmaking

Judicial Interpretations of Quantitative Risk Assessment: Hazardous Waste Site Identification and Remediation

The requirement for using risk assessment has most commonly come under review either directly or indirectly by the federal courts in cases under CERCLA and SARA.[8] Risk assessment began to be commonly used to evaluate remedial action alternatives under CERCLA. When SARA was passed, the requirement to use risk assessment was rescinded: only health assessments and not full-blown risk assessments were mandated under SARA. However, in hazard identification, SARA required that in identifying indicator chemicals, the relative risk of chemicals should be taken into account, and a relative risk analysis implicitly requires risk assessment. In practice, full risk assessments, rather than just health assessments, do accompany remedial investigation/feasibility studies under CERCLA/SARA.

A number of cases under CERCLA and SARA have made determinations as to whether clarifications or interpretations of the concepts "imminent," "substantial," and "endangerment" (that are used as qualifiers for the concept of risk) require a quantitative risk assessment. The phrase "imminent and substantial endangerment" is a major criterion used in CERCLA to determine if action under Section 106 of the statute should be taken and also whether actions under Section 7003 of RCRA should be taken as well. Ironically, the phrase is not defined in the statute, and the courts have had to go back to Congressional intent in order to interpret it. The Federal Water Pollution Control Act (FWPCA) (Article 504(a), 33 U.S.C. 1364(a), 1982) apparently loosened the traditional "imminent and substantial harm" to "imminent and substantial endangerment" (Silver, 1986, 90).

Some examples of the cases in which the question of whether risk assessment should be used and under what circumstances it should be used to clarify these terms are described below (from Stever, 1988, 6–36).

United States v. *Conservation Chemical,* 619 F.Supp 162 (W.D.Mo. 1985): Even prior to SARA's apparent substitution of health assessment for risk assessment, this decision in a case brought under CERCLA appeared to lessen the need for quantification of risks via any method, including risk assessment, when applying the "imminent and substantial endangerment" test. In *United States* v. *Conservation Chemical,* the court ruled that the term "imminent and substantial endangerment" does not require a quantification of risks.[9] The court concluded that Congressional intent with respect to defining what risk meant required that where an error in the endangerment standard is made, it should be in the direction of greater protection of health and welfare. The court interpreted this to mean:

Thus, just as the word imminent does not require proof that harm will occur tomorrow, and the word endangerment does not require quantitative proof of actual harm, the word substantial does not require quantification of the endangerment (e.g., proof that a certain number of persons will be exposed, that "excess deaths" will occur, or that a water supply will be contaminated to a specific degree). Instead, the decisional precedent demonstrates that an endangerment is substantial if there is reasonable cause for concern that someone or something may be exposed to a risk of harm. (Stever, 1988, 6–36)

United States v. *Northeastern Pharmaceutical and Chemical Co., Inc.*, 579 F.Supp. 823 (W.D.Mo. 1984): In this case EPA's conclusion that there was an imminent and substantial endangerment, reached without the use of quantitative risk assessment, was evaluated. Drums of dioxin, toluene, and trichlorophenol were removed to a temporary storage site on a farm as a means of reducing the imminent and substantial endangerment from the chemicals. The endangerment existed because of the nature of the bedrock below the farm, which made it possible for chemicals to reach drinking water supplies. The court opinion of *United States* v. *Conservation Chemical* states: "Thus, without considering the concentration of the hazardous substances at the point of exposure, the number of persons or other living organisms who might be exposed, or the extent to which excess deaths or illnesses would occur, the court found that the conditions of the site presented an imminent and substantial endangerment" (Stever, 1988, 6–36).

United States v. *Seymour Recycling Corp.*, 618 F.Supp. 1 (S.D.Ind. 1984): In this case the court concluded that since groundwater flowed from the disposal site in the direction of a subdivision, thus making it possible for contaminants to reach people, it represented an endangerment under Section 106 of CERCLA. "Because hazardous substances are, by definition, capable of causing serious harm, a substantial endangerment may exist whenever the circumstances of a release or threatened release of a hazardous substance are such that the environment or members of the public may become exposed to such substances and are therefore put at risk. For very hazardous substances, such as those which are toxic at low concentrations or known or suspected carcinogens, a substantial endangerment will arise when small amounts are released or threatened to be released" (Stever, 1988, 6–36).

Reserve Mining Company v. *U.S. Environmental Protection Agency*, 514 F.2d 492 (8th Cir. 1975): In this case "the Eighth Circuit interpreted 'endangerment' as used in FWPCA to mean threat or risk of harm rather than the harm itself . . . " (Silver, 1986, 89), that is, the probability of harm and not so much whether that harm had occurred. In addition, the court based the existence of a probability of harm or a health risk on reasoning from theories related to carcinogenesis rather than actual empirical evidence of that fact. Furthermore, the final ruling in the case determined that since the harm had not yet occurred, the risk was a potential one, not imminent. The court used this as the basis for not immediately closing the plant (O'Brien, 1988, 98–99).

United States v. *Midwest Solvent Recovery, Inc.*, 484 F.Supp. 138, 142 (N.D.Ind. 1980): In this case the court interpreted "serious harm" to mean "the presence of 14,000 fifty-five gallon drums containing various chemical wastes 'with dangerously low flashpoints,' evidence of leaching into topsoil, and other evidences of abnormally high levels of chemicals" (Hinds, 1982, 17).

United States v. *Hardage*, No. CIV80-1031-W (W.D.Okla. December 2, 1980), 13 E.R.L. 20189: In this case the fact that dangerous chemicals were allowed to be released and that these releases pose a threat to human health was considered enough to meet the "imminent hazard" requirement under Section 7003 of RCRA.

Environmental Defense Fund, Inc. v. *U.S. Environmental Protection Agency*, 548 F.2d 998 (D.C. Cir. 1976): In this FIFRA case involving "unreasonable risk," "imminent hazard" was defined as existing "where continued use during the time required for the cancellation proceeding would be likely to result in adverse effects on the environment" (Silver, 1986, 91).

While the quantification of risk may not be required for a determination that an imminent and substantial endangerment exists, the components of a risk assessment have often been spelled out in court cases pertaining to the "imminent" provision under Section 7003 of RCRA (a section that is similar to Section 106 of CERCLA in terms of the proof required [Stever, 1988, 6-35]). The following cases illustrate this point.

United States v. *Vertac Chemical Corp.*, 489 F.Supp. 870 (E.D.Ark. 1980): This case involved the imminent hazard provisions of both the Clean Water Act and RCRA. It used the *Reserve Mining* case as an analogy. Under the original RCRA, imminent hazard was interpreted as "a reasonable medical concern over the public health"—no actual harm was demonstrated. But 1980 RCRA amendments sponsors defined imminent hazard as existing when a "risk of serious harm is present" and raised the level of proof required (Hinds, 1982). In a 1985 case, "the court stated the two elements that must be considered [in determining if a hazard is imminent] are the toxicity of small concentrations of the substance involved and the likelihood that there will be human or environmental exposure to it if nothing is done."[10] These two elements—toxicity and likelihood of exposure—are the two components of a risk assessment calculation.

United States v. *Northeastern Pharmaceutical and Chemical Co.*, 579 F.Supp. 823 (W.D.Mo. 1984): This case specifically spells out proof as involving evidence that allows an "assessment of the relationship between the magnitude of risk and harm arising from the presence of the hazardous waste" (Stever, 1988, 6-38). This implies a risk assessment.

Sterling v. *Velsicol Chemical Corp.*, 855 F.2d 1188 (6th Cir. 1988): Rodenhausen (1989, 6) interpreted the stance of the federal courts with respect to the use of risk assessment in connection with this case. According to Rodenhausen, in the *Sterling* case, the Court of Appeals made a distinction between two kinds of information. The first is called "generic" information. It refers to probabilities or the potential of harm. Harm is based on three factors: " '(1) the intrinsic toxicity of the chemical substance . . . ; (2) a pathway from defendant's operation to the plaintiff . . . ; and (3) actual exposure of the plaintiff to significant levels of contamination . . . ' A second level of information is 'specific causation,' where there is a 'reasonable medical certainty' that the contamination caused the individual plaintiff's illness or injury." This is considered the most difficult kind of information to obtain. The damages are usually divided in terms of these two kinds of proof of causation (Rodenhausen, 1989, 6).

The cases cited above are generally consistent with the kind of reasoning that Rodenhausen presented for the *Sterling* case. One could infer from many of the cases above that generic information does not necessarily require quantification of risks where the issue is site remediation. The questions that remain are still (1) at what level damages should be set, (2) how damages should be divided, and most importantly, (3) how individual harm should be proven.

Thus, it appears that in recognition of the uncertainties posed by risk assessment and the general uncertainties as to whether something with potential for harm will in fact result in harm if left unabated, the courts have receded from seeking proof of the certainty of harm to seeking proof of the potential for harm.

Judicial Review of Risk Parameters and Levels in Standards-Setting

Judicial review has had a different impact with respect to the use of risk assessment when it comes to setting standards. Judicial review has altered the number of parameters that describe environmental quality and the risk level specified for a given parameter under standards-setting processes. Courts get directly involved in extensive reviews of risk assessment procedure when it involves decisions about which chemicals to regulate and at what level. An example of judicial review of the number of chemicals regulated in the water pollution control area was a settlement that required EPA to issue water quality standards for 65 toxic substances under the Clean Water Act, in addition to the standards set for conventional water pollutants. In the occupational safety area there are two examples in which the court questioned a risk level. The court questioned the reduction of the benzene standard from 10 ppm to 1 ppm on the grounds that there was an inadequate cost/benefit analysis to justify the reduction (*Industrial Union Dept.* v. *American Petroleum Institute,* 448 U.S. 607 (1980)) and upheld the right of OSHA to upgrade standards in the vinyl chloride case *(Society of Plastics Industry* v. *OSHA,* 509 F.2d 1301 (2nd Cir. 1975)).

Design of Programs/Administrative Oversight

The judiciary performs a major role as final arbiter of governmental disputes. There has been a shift over the past century in the proportion of cases pertaining to dispute resolution among private parties vs questions of statutory and constitutional law. The shift has occurred in the direction of a greater judicial role in dispute resolution (O'Brien, 1985, 669) and oversight with respect to government administration.[11]

The scope of its authority over the design of implementation programs is reflected in the diversity of decisions that have come within the scope of judicial review. A representative example of this is the judicial review of Nuclear Regulatory Commission decisions. The following kinds of NRC administrative and technical decisions have come under judicial review, as summarized by Cook (1980, 38–9): emergency core cooling system regulations, siting policies, radiation emission standards, and the relationship of the licensing procedure to other federal statutes. In *Sierra Club* v. *Ruckelshaus,* 344 F.Supp. 253 (D.C. Cir. 1972), a new implementation program for the "prevention of significant deterioration" of air quality was designed by the courts. In another case, EPA was required to develop pollution control programs for five pollutants (Melnick, 1983, 2; *Natural Resources Defense Council, Inc.* v. *Train,* 545 F.2d 320 (2nd Cir. 1976)).

Within the context of the judicial review of administrative procedures, a variety of opinions have been expressed regarding the appropriateness and stringency of such reviews. As a consequence of the debate over this issue, the impact of these reviews on administrative process has not been consistent. Cooper highlights a number of instances of the judicial review of agency administrative procedures, and gives numerous examples of how court decisions have enhanced administrative discretion rather than restricting it. He argues that "law is a discretion-reinforcing agent, a fact that argues for improved judicial-administrative relations and against continued hostility" (Cooper, 1985, 643). The case of *Vermont Yankee Nuclear Power Corp.* v. *Nuclear Regulatory Commission,* 435 U.S. 519 (1978), is particularly cited as having discouraged the courts from going beyond explicitly statutory mandates in defining administrative procedure (Cooper, 1985, 646) and by inference is a case that has upheld agency discretion (O'Brien, 1986a). Melnick, in contrast, argues that rather than a partnership existing between the courts and administrative agencies, there has been a "judicial usurpation of administrative authority" (Melnick, 1985, 653). In connection with the EPA he has argued that the courts have not increased its autonomy significantly (Melnick, 1983, 379).

Increased Influence of the Judiciary in Risk Decisions

It can be argued that the judiciary exercises by far the greatest discretion in policies that appear on its agenda, because, as has been pointed out earlier, it is the ultimate arbiter of disputes. The judiciary's influence on risk management decisions was pervasive, not so much in terms of the number of cases that were reviewed but rather in terms of the reach of its influence over time. It underscored the responsibility of government to define and warn people of levels of harm (*Irene Allen* v. *United States*, 588 F.Supp. 247, p. 338, as described by Cooper (1985, 650)). It required that the agencies use the latest scientific evidence and expert judgment, yet left it up to the agency to determine precise levels of harm. This pattern of judicial dominance in decisionmaking about risk is by no means a consistent one, and can be described as cyclical at best, varying over time and among courts, cases, and administrations.

Some have argued that the reach of the courts into risk policy is inappropriate (O'Brien, 1982, 101). Many have argued the need for retraction on the basis of the inconsistent interpretations the courts have given in risk-related cases (Jasanoff, 1985, 165). Studies of the technical backgrounds of judges support a more limited role. Others argue that the role of the courts is an important one, but should aim at reinforcing the appropriate use of risk evaluation practices (Goldsmith and Banks, 1983).

CONDITIONS AFFECTING THE ROLE OF THE COURTS IN RISK MANAGEMENT

A number of conditions, some originating within the context of government in general and some unique to the risk issue and representing new initiatives, have influenced the role that the courts play in risk management.

Factors associated with judicial review that reflect a traditional governmental context include growing caseloads in general, judicial procedure, structure of the court system, and citizen access to the courts. The general rise in the caseload of the courts is partly a function of changes in judicial procedures within the court system and increased citizen access to the courts.

In contrast, a number of conditions affecting caseload are unique to the risk issue, often generate new initiatives, but are by no means unusual in the course of evolution of a policy and its implementation. These conditions include the large volume of laws passed during the 1970s and amended during the 1980s and the large volume of agency rulemaking accompanying these laws, the vagueness of those laws with respect to specific underlying scientific validity and how this motivates the judicial reach of the courts, increased organizational capacity or resources of the plaintiffs, and the use of dispute resolution to supplement and substitute for court action.

Conditions Originating in the Traditional Governmental Setting for Judicial Action

A number of factors traditional within government can be linked to the general growth in cases and to their outcomes. These are the general volume of cases and trends in litigation in general, judicial procedures that determine how cases are brought, the allocation of cases within the court system, and citizen access to the courts.

Volume of Cases in General

Court cases pertaining to risk policy have appeared annually at a steady rate since 1970. Putting this in context, however, environmental and risk-related cases comprised a small percentage of the total caseload of the federal courts in any given year. A liberal estimate of the percentage that environmental cases were of total cases in the federal court system was about 1% in 1979, according to Wenner, for all environmental cases. Wenner (1982, 21, 24) pointed out that during the 1970s the number of cases did not vary with changes in legislation. Another study of only risk-related cases showed a similar trend. The total number of cases involving EPA, OSHA, and the CPSC only averaged 1.3% over all of the federal district appeals court circuits between 1971 and 1983 (Oleinick, Disney, and East, 1986, 383). Given the small percentage that risk cases are of the total caseload of federal courts, it is difficult to determine whether that percentage is growing relative to total caseload. One observation that can be made from the small, constant percentage is that risk cases are merely an outcome of the actions of a society very prone to litigation in general. Furthermore, the tendency toward litigation in general may be growing, as O'Brien (1988, 3) observes:

> Our society is litigious and our culture is adversarial. In 1980–1981, for instance, there were 180,576 civil and 31,287 criminal cases filed in federal courts, more than twice the number of a decade earlier.

Thus, the caseload in general not only is large relative to risk cases but appears to have been growing independent of the risk issue.

Judicial Procedures

New procedures in the way cases are heard often arise that can affect the overall caseload. These procedures emerge out of the general governmental context and are available to risk cases as well as any other type of case. For example, the procedures that allowed consolidation of court cases enabled over 600 suits (representing the interests of over 2.4 million people) filed with regard to Agent Orange exposure in Vietnam to actually be consolidated. This was apparently a novel approach in toxic tort law, since toxic tort law cases typically focused on individual injury. The Agent Orange case became a mass toxic tort (Schuck, 1986). While the Agent Orange case was unusual in terms of the number of plaintiffs under a single suit, if a trend like that continues it could influence the nature and form of the caseload. The willingness of plaintiffs to organize themselves for a mass toxic tort case may ultimately depend on the magnitude of the awards received per plaintiff in that setting as compared with being an individual plaintiff. Requirements that plaintiffs give up their future right to bring suit as a condition for settlement also affect caseload in the direction of reducing it.

However, in contrast to the previous examples, most of the new procedures tend to increase court caseload. The major ones primarily have to do with rights of appeal and who has access to the courts. Cases considered "frivolous," such as the so-called "pauper" cases, can drastically increase the number of cases filed (O'Brien, 1986b, 211). Changes in citizen access to the courts can also influence caseload and are discussed in the next section.

Allocation of Cases Within the Court System

The underlying structure of the judicial system can promote inconsistencies in the way cases are heard. O'Brien (1988, xii) points out:

> The decentralized structure of the federal judiciary promotes litigation strategies for competing interest groups that work against the development of coherent and uniform regulatory policies [which is] the inescapable price of resolving disputes over the regulation of risk in a pluralistic society in which the cultural values of science, law, and politics collide in determining the direction of policies for managing health-safety and environmental risks.

Which court prevails in a given case depends on several factors. For example, political preferences can determine which courts prevail. Melnick has observed that while environmental groups and the EPA tend to favor the D.C. Circuit, businesses tend to favor local courts (Melnick, 1983, 56). Industry, however, often files in both district and circuit courts simultaneously in order to have the most flexible options (Wenner, 1982, 54). In general, the district judges are closer to political constituencies, depend on them for advancement more than circuit court judges, and have narrower constituencies (Wenner, 1982, 120). Appellate courts appear to temper these differences (Wenner, 1982, 144).

The limited capacity of certain courts to hear cases can result in a redistribution of caseloads. For example, in discussing the role of the Supreme Court in environmental cases, Wenner points out:

> The Supreme Court, although the final arbiter, is not the only decisionmaker in the federal judiciary. The Court can review only a minority of decisions made by even one circuit in any given year. Hence the federal circuits, which tend to be more favorably disposed toward environmental values, continue to wield some power. (Wenner, 1984, 196)

In spite of the decentralization of the court system, certain courts can by statute and practice dominate in hearing certain cases. As discussed earlier, cases pertaining to the risk issue appeared to concentrate in certain courts, such as the D.C. Court of Appeals. On the one hand, this distribution was partly a function of factors specific to the risk issue, such as statutory design. On the other hand, it was also related to the tradition of the D.C. Court in hearing agency rulemaking cases in general, regardless of the type of issue.

Statutory specifications have reinforced the role of the D.C. Court in the environmental area. Environmental laws can directly go to a court of appeals, and this is spelled out explicitly in many federal laws dealing with environmental health risks. One review of chemical litigation pointed out that in some instances, cases must be heard only by the D.C. Circuit (for example, cases brought under the Clean Air Act, which accounted for a large number of the environmental cases brought in the 1970s):

> The CAA required anyone who wished to object to national primary or secondary ambient air quality standards, new stationary sources' performance standards, or motor vehicle emissions standards created by EPA to initiate a case in the U.S. Court of Appeals for the District of Columbia Circuit. Accordingly, over one quarter (59) of all clean air cases in the 1970s were adjudicated in the District of Columbia Circuit. (Wenner, 1982, 65)

Cases pertaining to regulations under SDWA must be brought in the D.C. Court (SDWA, Section 1448 a(i)). The Clean Water Act, in contrast, does not specify the D.C. Circuit as having exclusive jurisdiction, and cases have tended to be scattered over a variety of circuits (Wenner, 1982, 68).

Yet, it is usually argued that the basis for this concentration of environmental cases in the D.C. Circuit is its specialization as a court that deals with agency procedure in general. The D.C. Circuit has "been given special powers and functions to oversee administrative agencies of the federal government and to determine national policy in some important policy areas, including many issues relevant to the environment. Its workload includes cases involving issues that arise in every geographic area of the United States" (Wenner, 1982, 103). Because of this specialized role, the D.C. Circuit has acquired a large body of administrative law and considerable expertise in this area (Brickman, Jasanoff, and Ilgen, 1985, 103; Melnick, 1983, 387).

Thus, there are several factors traditional to the structure of the judicial system that can influence outcomes of risk cases. First, the court system is decentralized, and which court prevails at any given time can be politically determined as well as prescribed by judicial procedure and workload of individual courts. Second, using primarily the D.C. Circuit to hear environmental cases has arisen not only because of statutory prescriptions but also because of the tradition of its use for agency rulemaking cases. While the use of a single court to hear similar cases ostensibly would lend consistency to decisions, the changing nature of information, influence, and the judges that hear the cases still adds a certain level of inconsistency to the process.

Citizen Access to the Courts

To the extent that risk cases are brought by citizens, one factor said to influence their number and outcome is these groups' increased access to the courts. Several scholars in the area of judicial process have observed the link between caseload and increased citizen access to the courts (O'Brien, 1986a, 38). This access has been allowed to citizens largely by Congress and through court policy. It is referred to as "standing to sue" and is one of several tests that determine whether a case is suitable to be heard in court.[12] As Morgan and Rohr point out, the citizen access phenomenon has many different dimensions, all of which have been in the direction of increasing access. This trend began well before the chemical risk cases entered the courts: "During the 1950s and 1960s, the thick wall of obstacles that blocked judicial access to citizens seeking remedies for government transgressions eroded in the face of a series of landmark U.S. Supreme Court decisions" (Morgan and Rohr, 1986, 221).

First, standards regarding the type of proof needed to show injury and the kinds of injuries that can be litigated have relaxed. Injuries had primarily emphasized financial damages, but then were expanded to aesthetics, conservation, and recreation, as well as to other areas of the public interest (Morgan and Rohr, 1986, 223). O'Brien has made this same point (O'Brien, 1986a, 34). This expansion in the kinds of injuries allowed to be heard in the courts was accompanied by a rise in the number of individuals and groups of citizens allowed access to the courts.

Second, citizens not economically or personally affected by particular actions were granted the right to sue. In the *Stringfellow* case, the Court of Appeals upheld the right of a citizens' group to supplement immediately injured parties in bringing suit against the owners of a landfill who were accused of releasing hazardous wastes (*United States* v. *Stringfellow*, 783 F.2d 821 (9th Cir. 1986)).

Third, the basis for bringing tort liability claims was considerably expanded to include the right of citizens to sue government for a wrong action. The basis for claims expanded (1) from personal liability to government liability and (2) from being restricted to infringements of constitutional law to including infringements of statutory law as well (Morgan and Rohr, 1986, 225–226). According to Morgan and Rohr, this resulted in greatly restricted discretion for administrative agencies and displayed the court's basic distrust of administrative discretion.

Fourth, Congress has incorporated clauses expanding the right of citizens to sue in much of the legislation pertaining to chemical risks, though the laws have not changed substantially in this regard for at least a decade.

The standing to sue has by no means been a stable situation over time or for different courts. Attempts to prevent environmental groups from gaining access to the judiciary were discouraged by Justice Earl Warren; citizen access to the courts became more formal as citizen right to sue was written directly into legislation in the 1970s (Wenner, 1984, 181). Differences in the willingness of district courts to hear cases brought by national environmental organizations without personal injury claims, for example, were resolved in favor of NRDC by the Supreme Court in *Train* v. *Natural Resources Defense Council, Inc.*, 421 U.S. 60 (1975) (Wenner, 1982, 114). The Burger court, however, acted to restrict such general citizen access to the courts in a number of cases between 1975 and 1984 and restricted private groups from suing in other ways (Cooper, 1985, 648).

Accompanying increased court access by citizens was the right of citizens winning cases to obtain payment of attorney's fees and related costs for bringing suits under statutes involving environmental health risks. For example, under the Clean Air Act Section 304(d) courts "may award costs of litigation (including reasonable attorney and expert witness fees) to any party, whenever the court determines such award is appropriate." Many court cases have interpreted this in favor of citizens' groups.[13]

All of these developments with respect to citizen access were not unlike developments in citizen access in other ares of government.

Conditions Originating in Attributes of the Risk Issue

Effect of the Volume and Clarity of Legislation

As discussed in Chapter 2, the number of risk-related laws has grown at an almost exponential rate, accompanied by a rise in the regulations associated with these laws. This produced an increase in the activity of the judiciary in scrutinizing bureaucratic behavior to prevent abuses of new powers (Melnick, 1983, 3). The trends in numbers of laws are discussed in some detail in Chapter 2.

In addition to the number of laws passed, certain attributes of these new laws have contributed to increased judicial activity. For example, the growing scientific information required in the laws, together with the lack of agreement in the scientific community on this information, elevated the level of scientific uncertainty, and the uncertainty has increased judicial intervention into regulatory decisionmaking (O'Brien, 1986a, 38). Furthermore, statutory ambiguity, which is inevitable when a large number of laws is passed over a short period of time, induces greater judicial activity (MacIntyre, 1986, 76; MacIntyre, 1985; Vig, 1984, 62). Laws that tend to be the most litigable are (1) those whose statutory language is vague and subject to considerable interpretation and (2) those mandating agencies and private parties to meet strict deadlines that go unmet and result in a flurry of cases on procedural grounds. Sabatier and Mazmanian (1979) have developed a typology based on statutory vagueness, described in Chapter 4. MacIntyre has commented that laws passed in this century "have typically been vague and ambiguous" (MacIntyre, 1986, 67). The influence of these factors on the increase in caseloads is by no means clear-cut, and many arguments have been offered as alternatives to the judiciary as passive recipient of ambiguous and uncertain laws. It has been argued, for example, that by guarding consumer interests and other public interests they invited litigation (Vig, 1984, 62; Shapiro, 1979). Another argument is that it was "the inadequacy of the courts as a mechanism for protecting society against future risks that produced the new legislation of the 1970s" (Vig, 1984, 63).

The scientific uncertainty underlying much risk legislation has also been considered one of the contributing factors to the increased judicialization or reach of the courts. O'Brien (1988, 149) attributes the "judicialization of regulatory politics" and "judicialization of the administrative process" in the risk policy area (where it involves scientific controversy) directly to issues of scientific uncertainty inherent in the risk issue. First, while consensus in the scientific community occurred over certain issues relating to the ostensible relationship between chemicals in the environment and cancer, there was considerable disagreement on the causes and method of demonstrating the relationship accurately using animal or human data. Certainly the existence of this uncertainty and the extent to which it was a pervasive part of risk policy has been underscored by many others (Greenwood, 1984; Marcus, 1983). This uncertainty was left unresolved by the administrative agencies and thus found its way into the courts.

Intensity of Agency Rulemaking

Increased administrative rulemaking accompanying the rise in the number of laws generated cases. In an overview of cases brought against some of the leading federal agencies involved in risk management, the following trends were observed (O'Brien, 1986a, 41–42):

> The increase in agencies' health, safety, and environmental rule making in turn led to increasing litigation in federal courts. Even though the exact contribution of such regulatory challenges remains difficult to discern, litigation of science-policy regulations undoubtedly figures prominently in the rise of federal court case loads. The number of civil suits challenging regulations and enforcement proceedings encountered by the Environmental Protection Agency, for instance, rose from less than 20 in 1973 to almost 500 in 1978. Challenges to all administra-

tive actions, moreover, rose 28.8 percent in 1980 from the previous year and constituted 14.4 percent of the total federal appellate caseload.

Organizational Capacity of Plaintiffs

Increased organizational capacity of the plaintiffs can reinforce citizen access to the courts. Many of the early cases were brought by environmentalists against government or industry (Wenner, 1982, 40). Considerable growth in nationally organized environmental groups occurred over the decade of the 1970s and continued through the 1980s, as discussed in Chapter 2. Wenner (1982, 59) notes that the degree of organization or organizational strength of an environmentalist plaintiff dramatically affects the extent to which cases are won— national organizations have a significantly greater win rate than private environmentalists or ad hoc organizations have had.

Dispute Resolution as an Alternative to Judicial Review

The growth in costs and time delays associated with resolution of disputes through the court system for both sides involved in litigation has contributed to the popularity of out-of-court settlements via various negotiation techniques. When negotiation is chosen after court cases are filed, it obviously does not affect the number of filings but substantially affects the workload of the courts. When it is chosen prior to or instead of a court filing, it affects both the filings and the workload of the courts.

Negotiated procedures have grown considerably in popularity and are often included in laws or rulemaking. These statutory references formalized a practice that had already been occurring, where government acted as arbitrator. For example, Negotiated Testing Agreements were proposed by EPA in 1982 and 1983 under TSCA to replace the issuance of testing rules, but this was struck down in court (*Natural Resources Defense Council, Inc. v. U.S. Environmental Protection Agency*, 595 F.Supp. 1255 (S.D.N.Y. 1984), cited in Stever, 1988, 2–7 and 2–8). Under TSCA, disputes between manufacturers and between parties who are in dispute about who should pay for data submitted under TSCA for chemical testing (when one party uses another party's data for an exemption) are required to be submitted to the American Arbitration Association. Negotiated orders can be issued under TSCA Section 5(e) for violations of Significant New Use Rules (Stever, 1988, 2–11). Many of these requirements originated in court cases. CERCLA requires disputes among Potentially Responsible Parties (PRPs) to be submitted to the American Arbitration Association also. A similar requirement exists under FIFRA (FIFRA, Section 3(c)(1)(D)).

A number of conditions are associated with the effective use of bargaining and negotiation, either within the context of or in lieu of the court system.

Breyer (1982, 178–179) points out that bargaining works under the following conditions:

• There is a strong incentive to reach an agreement, such as a work stoppage.

• The parties are of equal strength.

• Decisions are amenable to decentralization (i.e., do not require centralization and uniformity), since the outcomes of bargaining may vary by situation.

O'Hare, Bacow, and Sanderson (1983, 154–157) have developed a set of eight conditions in the context of negotiation for selection of sites for controversial facilities. The issues in site selection may or may not involve health risks; in fact, many of the conditions they set forth are not met by these kinds of risks. The conditions they identify and their relevance to issues pertaining to environmental health risks from chemicals can be summarized as follows:

• There are few parties involved in the dispute. This condition is often not met in the case of toxic torts, though the organization of injured parties under mass toxic torts and the involvement of spokespersons can reduce the number of parties in the dispute.

• Opponents are well organized. There has been a growing trend in the area of environmental health risks to organize plaintiffs.

• The opponents are geographically defined. The reasoning here is probably that it is easier to identify parties applicable to a dispute when geographical boundaries delimit the areas of controversy. In the case of long-term, latent health impacts, geographical definition may be impossible if people have migrated from the original source of the harm.

• Impacts are clearly traceable to the project. The move to strict liability doctrines in common law is based on the fact that chemical risks are difficult to trace to a particular source.

• Outcomes exist that are mutually acceptable to each party. The widely varying nature of awards and the uncertainty in cause-and-effect relationships in chemical risk cases makes acceptability difficult to obtain.

• Recreation of the status quo is possible, that is, impacts can be prevented altogether or reversed. This is rarely possible when the impact of a chemical risk is cancer from after-the-fact exposures.

• The parties involved in the dispute are capable of offering a binding commitment.

• There is an absence of initial hostility. In some communities, chemical risk issues are often preceded by a long-term animosity toward the activity that produced the hazard, especially where the community does not benefit directly from that activity. On the other hand, where an activity is critical to the economic well-being of a community, such hostility may be less.

Nelkin and Pollak (1980, 73) have recommended: "Mediation works best when two major protagonists share a minimum common interest that will lead to a mutually satisfactory compromise." They further point out that antagonistic groups are not appropriate to reaching compromise, since they are usually not well organized and do not share common values.

Other conditions for effective negotiation can be added as well:

• The parties involved in the negotiation each have something of value to the other one.

• There is a reason or incentive for each party to enter into a negotiation, that is, each party has something to lose by not entering into the negotiation process.

• Attitudes toward the risk being negotiated are fairly stable over time.

• Those that are bargaining have the support and commitment of their constituencies.

CONCLUSIONS

This chapter first explored the intensity and direction of judicial review in the area of chemical risk management. Based on a small set of air and water cases, the number of court cases in the risk area appears to be increasing but remains only a small, constant percentage of the total caseload in the United States. In contrast, the number of cases brought per year that explicitly reference risk assessment, while small in total, has grown substantially between 1980 and 1988. The total federal caseload has been on the rise as well, which implies that one factor in the rise in risk litigation is a growing tendency toward litigation in society in general. Other factors, such as access of citizens to the courts and legislative uncertainties, also appear to be operating as determinants of change in the risk caseload. These factors have been traditional influences on government.

Environmental cases have been known to concentrate in certain courts, though this is more likely a function of other aspects of the cases than of their environmental orientation per se (except where a statute specifically mandates that a particular court hear cases under the statute). That is, the dominance of certain courts appears to be more a function of a judicial procedure that exists independently of the risk issue rather than because of it.

The role and influence of the judiciary in the risk area has been notable in terms of the extent of its review, especially of how agencies interpret risk and apply risk assessment. However, there has been a lack of consistency in many of the court decisions regarding methodology, risk terminology, and risk levels. Nevertheless, judicial review has resulted in changes in legislation in the direction of greater clarity (or at least specification of risk parameters), increased organizational capacity of citizen intervenors, modifications of agency procedures, and attempts to reach consensus in the use of risk assessment through negotiation. These actions represent relatively new initiatives.

NOTES

1. Except where indicated otherwise, much of the description of the organization and functions of the judiciary is drawn from Office of the Federal Register (1988, 63–80; 359–396).
2. The search of these cases and others cited in this chapter was conducted using WESTLAW (West Publishing Co., St. Paul, Minnesota) during the spring of 1988. "Risk assessment" and "risk analysis" were used as the key words under this particular search. This part of the search excluded cases that only referenced the word "risk."
3. This represents the total number of cases explicitly using the term "risk assessment." The total number of cases using the term "risk," however, was many times larger, as shown in Table 6.2.
4. D. W. Stever, Jr., May 8, 1989, personal communication.
5. D. W. Stever, Jr., May 8, 1989, personal communication; O'Brien, 1986a, 35.
6. Vig (1984, 66–70) has added another dimension to this two-fold, traditional dichotomy in court philosophy. He adds judicial review of facts (risk measurement) and values (risk acceptability) as a further breakdown of the substantive/procedural dichotomy.
7. Other writings that have discussed the "hard look" approach to judicial review are Stever (1988) and those reviewed in Melnick (1983, 388) and O'Brien (1986a, 36). See also Whitney (1973, 473), Yellin (1981, 489), and Rodgers (1979, 705–706).
8. In applying the de minimis concept and the related concept of significant risk, risk assessment is implied and encouraged, but not mandated under OSHA, TSCA, and the CAA. (See Ricci and Cox, 1987, 87, 89.)
9. The question of whether to use risk assessment would probably not come up in the context of substantial endangerment. Stever (1988, 6–36) has argued that the substantial endangerment provision rarely is an issue. Most of the issues that arise under the statute are assumed to involve substantial endangerment, since they deal with the most dangerous chemicals. Thus, substantial endangerment is presumed to exist without need for a risk assessment to prove it.
10. Quotation is from the court opinion in *United States* v. *Conservation Chemical*, 619 F.Supp. 162, 164–165 (W.D.Mo. 1985).
11. The limits of the oversight function have been subject to considerable debate. The authority of the judiciary to exercise oversight functions was proclaimed in *Marbury* v. *Madison*, 5 U.S. 137, 168–173 (1803), but was limited as well in that case (O'Brien, 1986a, 30).
12. Morgan and Rohr (1986, 222) summarize these "threshold tests" as including "ripeness, exhaustion, primary jurisdiction, and political question" in addition to standing.
13. See, for example, *Pennsylvania* v. *Delaware Valley Citizens' Council for Clean Air,* 478 U.S. 546 (1986). This case supported the citizens' group's appeal for the coverage of attorney's fees in connection with a suit filed to compel the State of Pennsylvania to implement a vehicle inspection and maintenance program under the Clean Air Act. The U.S. District Court for the Eastern District of Pennsylvania (581 F.Supp. 1412 (1984)) awarded the fees, including a multiplier for certain phases of the work based on the quality of the work. (Summary is drawn from *Environmental Reporter Cases,* Vol. 24, pp. 1577–1580.)

CONCLUSION

CHAPTER 7

Traditional Approaches and New Initiatives

The federal government has used traditional mechanisms and new initiatives to manage chemical risks for at least two decades. A major theme of this book has been the evaluation of the balance between these two approaches. While this does not address the larger, normative question of what balance is the most appropriate one for governmental action in managing risk, it provides a knowledge of how government has been operating in the past, which is an important foundation for making new adjustments.

The balance that has been achieved in government between traditional approaches and new initiatives has been evaluated from three perspectives that characterize governmental action. These are the intensity and direction of action, the degree of bureaucratization of the management process (viewed in terms of degree of discretion and openness), and government's use of the outcomes of its own actions in risk management as feedback for readjustments in its institutions. These perspectives have been applied to the major components of federal risk management—lawmaking and incentive-based systems of management undertaken by the three branches of government. Before presenting the culmination of these endeavors, the findings of the previous chapters with respect to application of these perspectives to governmental behavior that is traditional or represents new initiatives are briefly highlighted below.

The pace of governmental activity gives a rough indication of the degree and rate of changes that are being pursued in government. Governmental activity in the risk area has been relatively intense, particularly in lawmaking, agency formation, and judicial review, but not necessarily any more intense than the rate at which these activities appear to be traditionally pursued in government. Governmental regulatory activity in general has been driven by factors such as increased executive and staff involvement in policy development, increased requirements for substantiation and justification of administrative rulemaking for the courts and the Office of Management and Budget, and increased sophistication of analytical and evaluative techniques used for making decisions (West, 1988, 774–777). The direction of governmental risk management activity has displayed some distinct patterns, such as a shift in emphasis in lawmaking from purely environmental protection concerns to health risks. This has been followed by the incorporation of economic concerns into risk management. New directions have been taken in areas such as financial incentives for risk management, applications of the common law doctrine of strict liability and federal common law to risk problems, and the use of the "hard look" doctrine in the judicial review of agency decisionmaking. While many of these directions have been new in their application to risk problems, they have been drawn from strategies that were customary in government.

There is a large theoretical foundation relating degree of bureaucratization to various aspects of organizational behavior, which can be linked to the behavior of organizations engaged in risk management. The institutions engaged in risk management have displayed both bureaucratic and discretionary behavior. Laws contained both heavily prescriptive elements (for example, timetables for agency action) and elements that allowed a high degree of discretion (such as the choice of measures and methods to evaluate risk). Agency actions have also displayed a mix of highly bureaucratic regulatory programs and discretionary actions in the use of procedures for defining and measuring risk. Whether bureaucratic or discretionary, the behavior of the institutions that arose around the risk issue has reflected a mix of traditional government processes and new initiatives, but has tended more toward the traditional. New regulatory programs in the form of standards-setting processes and programs for the issuance of permits, licenses, registrations, certifications, and other approvals were largely patterned after earlier programs developed for environmental protection. (However, some new provisions in part have reflected lessons learned from previous failures.)[1]

Finally, the legislature, the line agencies, and the judiciary often demonstrated a sensitivity to the outcomes of their previous actions by making readjustments in the design of their operations. Congress made readjustments by passing new laws, agencies made readjustments through alterations in their programs and administrative organization, and the judiciary was able to modify its decisions through changes in judicial philosophy. While new initiatives were taken in the process of readjustment, the mechanisms used to accomplish these changes were largely traditional to government.

These three perspectives have up until now treated attributes of governmental risk management as separate dimensions. These governmental attributes actually are interrelated and are highly interactive as governmental decisions move from initial lawmaking to legislative readjustments. In order to understand these interdependencies, governmental action is portrayed next in terms of stages in decisionmaking on risk issues involving the three branches of government. Future directions of risk management are then drawn from the evaluation of the way government uses traditional mechanisms and new initiatives and the choices that are open given some of the challenges to policy development and implementation that have been identified.

THE STAGES IN RISK MANAGEMENT

Risk management processes viewed from the perspectives adopted here are highly interactive. In order to capture the dynamics of these interactions, risk management processes are portrayed in a series of stages in governmental decisionmaking as shown in Figure 7.1. The dynamics of these interactions over time further clarify the nature of the balance between custom and initiatives. A major factor that distinguishes each of the stages from one another is the degree to which government draws on traditional mechanisms vs undertaking new initiatives.

The series of stages in Figure 7.1 begins with a natural progression of laws being passed by Congress under pressure from the appearance of crises and interest group reaction to crises.[2] New agencies are formed in the process of lawmaking or old ones are adapted to the needs of new requirements. These laws define responsibilities and authorities and assign them to agencies. Some of these requirements allow considerable discretion, and some are heavily prescribed. Agencies then act on these authorities and responsibilities. Several developments can occur in the course of implementation. A mismatch between agency resources, capabilities, and the original legislative prescriptions often results in what government interprets as poor performance. Also, a mismatch between the exercise of discretion and legislative intent can similarly affect performance. One result of these mismatches is that some of the initiatives, particularly those pertaining to financial incentives and mechanisms for risk management, do not gain legislative support or support from the institutions of society. Another effect is that both discretionary and nondiscretionary actions can be challenged through judicial action. Agencies also make adjustments through changes in programs and rulemaking. They do so either by voluntarily using the feedback from the outcomes of their actions in order to avoid what they perceive as poor performance or under pressure from Congressional oversight and judicial review. Finally, the process of lawmaking repeats itself so that adjustments can be made in response to these unintended or undesirable directions. This is how normal governmental processes operate in the risk area.

Defining the Stages and Their Implications for Risk Management

Stage 1

Many laws were generated in a short time by traditional governmental responses to interest group pressure. In a traditional manner, a process of lawmaking was initiated in response to public pressures and environmental crises that continually reinforced one another. This has been characteristic of the agenda-setting function that touches off the policy process or what Mitnick

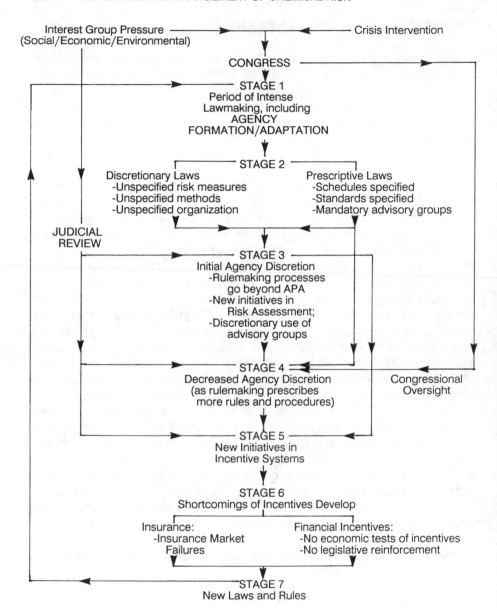

Figure 7.1. Stages in risk management in government.

(1980, 169) refers to as "issue creation." Congress, in the course of agenda-setting, has often been motivated to alter agency behavior when there is a public outcry, protest by strong interest groups, a real danger to public health and safety, or a real disaster (Foreman, 1984, 213). Legislators have transferred these concerns into law in a manner traditional to government; as Mitnick (1980, 169) puts it, legislators engage in issue expansion by passing many laws over a short period of time. This is how the branches of government respond to risks when they manage them with rules.

It was pointed out in earlier chapters that these laws were passed at a steady pace but not necessarily at a greater rate than laws were passed in general. Laws also created agencies, which continued to proliferate, a phenomenon observed throughout the federal bureaucracy and especially among regulatory agencies. In the risk area, however, this leveled off by the end of the 1970s. Court caseloads challenging laws or agency implementation of them grew steadily, but appeared to stay at a constant percentage of the total caseload of the federal courts, implying that total caseloads were growing at a rate similar to caseloads pertaining to risk issues. Thus, the patterns associated with intensity of risk management activity in lawmaking, agency formation, judicial review, and incentives appeared typical of general practices in government with respect to these activities. In fact, the pace of general governmental activity in these areas might even have been adding momentum to the level of intensity of similar activities within the risk area.

However, many of the public pressures initially placed on government that influenced the pace and direction of lawmaking originated and operated outside of its rule system for chemical risks. For example, after the early 1970s, when legislation had a strong health orientation, economic conditions in the country and political reactions to them gradually changed. As a result, regulatory reform movements of the late 1970s continuing into the 1980s (constituting another interest group) perpetuated a trend that actually had its origins in the 1960s and earlier, requiring the balancing of health benefits against economic costs. Yet through the 1980s an emphasis on health was still very pervasive in the language of the statutes, along with economic objectives. All of this activity, originating largely outside of the lawmaking system itself, not only increased the intensity of lawmaking and judicial review but affected its direction as well in terms of relative emphasis on health risks associated with chemicals. These new directions, however, were largely reflective of the way government normally reacts to similar external forces. While new directions emerged, they were not unlike the paths taken in other areas, particularly in the environmental protection movement that immediately preceded the era of risk management.

Stage 2

Congress reacted to scientific uncertainty with highly prescriptive laws. An overview of environmental health risk legislation reveals that the scientific bases for causal relationships between proposed regulatory strategies and human health were highly uncertain. The uncertainties largely pertained to knowledge of the effects that alternative chemical exposure levels have on human health and the lack of consensus in the scientific and political community on how to deal with uncertainties that could not be resolved.

It was pointed out in earlier chapters that effects of this uncertainty included the vague statutory definitions of the risk concept and inconsistencies in its definition from statute to statute. A second outcome was highly variable acceptable risk levels defining desirable human health for the same chemical in different settings.

In spite of the inconsistencies and lack of clarity, Congressional response was largely prescriptive. This reaction is consistent with classical theories of organization and management (Thompson, 1967; Simon, 1969): in its lawmaking function, Congress attempted to reduce uncertainty by constricting the latitude of concerns in the form of developing highly prescriptive requirements regarding what should be regulated, when, and how much. This behavior was also exhibited in the way Congress managed lawmaking. It divided up the risk issue among its long-standing, traditional, highly decentralized system of committees. As pointed out in Chapter 4, this is a normal response by Congress when it is faced with new issues (Weiss, 1989, 412–413). The generation of many individual pieces of highly prescriptive legislation was, in fact, an outcome of that mode of managing the legislative process. Thus, as in the case of the pace of governmental activity in Stage 1, Congressional action, in terms of both the generation and design of prescriptive laws and the management system that produced them, also could be characterized as a traditional response by government.

Stage 3

Initial agency discretion was promoted by Congress. In contrast to the prescriptiveness in some legislation, Congress in its customary fashion also delegated the responsibility of administering the statutes to agencies. In that process, new agencies were created or old ones were adapted to new requirements. Delegation of responsibilities to the executive agencies enhanced discretionary action by the agencies. These responsibilities were delegated to agencies either by prescribing such discretion explicitly in the laws or by not specifying responsibilities and authorities within the laws, i.e., by omission. Both scarcity of resources (forcing priority-setting) and abundance of resources (creating organizational slack resources) also enabled agencies to exercise discretion.

Agencies were able to exercise considerable discretion in the selection of particular chemicals to be regulated within the more general areas specified in legislation. They could select the form in which chemical risks were to be expressed and the method of measuring risk, such as risk assessment, as long as it was scientifically and socially defensible. The development and application of the science of risk assessment to chemical risks was a major new initiative that emerged from discretionary authority. While initiatives in developing standards and in rulemaking were generally the result of highly discretionary actions, they occurred within the boundaries set by the Administrative Procedure Act.

As was the case with Congress, when executive agencies were faced with scientific uncertainties and laws that were vague with regard to their management, they displayed a classic organizational response. Like the legislature, they tried to reduce uncertainty in their environments by circumscribing it, through some new initiatives in the use of analytical procedures (such as risk assessment), and through the discretionary use of administrative entities, such as advisory committees, to stretch their resource and knowledge base.

As early as the mid-1970s, agencies began to use their discretionary authority to develop the science of risk assessment to reduce scientific uncertainty in the area of chemical risks.[3] They legitimized the process and the regulations based on this scientific information through the use of advisory bodies. By using risk assessment in developing environmental standards, they elaborated and extended the requirements for rulemaking specified in the Administrative Procedure Act, acting on their own initiative.

During the 1980s, the use of risk assessment was formalized directly and indirectly by various legislative and administrative actions and became commonly used. According to Kraft (1986, 416), President Reagan's Executive Order 12291 indirectly encouraged risk assessment (as well as cost/benefit analysis) by requiring the balancing of risks and costs of governmental actions. In spite of this emphasis, risk assessment was never formally mandated by the order. Similarly, CERCLA (until the Superfund Amendments and Reauthorization Act), as well as other legislation, continually referred to the analysis of risk but did not mandate the particular techniques embodied in risk assessment. This left the use of risk assessment within the realm of agency discretion.

Stage 4

Administrative discretion was ultimately reduced by prescriptive requirements in legislation, judicial review, and the proliferation of agency rulemaking. In spite of the greater amount of discretion obtained through delegation and the allocation of resources, administrative discretion gradually decreased for a variety of reasons.

First, the development of highly prescriptive requirements in legislation contributed to a decline in administrative discretion. Over the past two decades, legislation has specified deadlines for rulemaking and compliance, the general areas of chemicals to be regulated, and numerical limits for those chemicals to an extent rarely seen before.[4] Such legal prescriptiveness has reduced agency discretion in many areas of risk regulation. The number of schedules, categories for regulation, and specification of control technologies in many of the risk laws probably has few parallels in governmental regulation.

Second, the continual codification of requirements by the agencies themselves in the course of rulemaking also constricted discretion. While rulemaking eased implementation by reducing procedural uncertainty, the agencies now had to live up to and in many cases were constrained by their own requirements, which had become formalized.

Third, both prescriptive laws and increased codification in rulemaking reinforced the decline in administrative discretion by invoking Congressional oversight and judicial review of the new procedures. Judicial intervention increased as new requirements were specified and especially as agencies took initiatives in areas such as risk assessment. The science of risk assessment became more refined; new detection technologies and the science of human health slowly tried to narrow the range of uncertainty in standards. However, the judicial demands for proof at times increased at an even faster rate than the improvements in many areas, and refinements in detection technology made greater demands on scientific judgments about cause and effect. When judicial admonition of the use or misuse of risk assessment outstripped agency developments in that science, agency discretion was constrained still further. Agency action was particularly constrained when judicial requirements were placed on when and how risk assessment was to be used.

In another area—economic balancing provisions in legislation—agency discretion at first appeared to be broadened, since the legislation did not specify how and when such balancing should occur, and the economic tests were uncertain. The judicial review of these provisions, however, tended to curb agency discretion in interpreting economic balancing, as it did in the area of risk assessment.

One outcome of the use of risk assessment, advisory committee reviews, and judicial reviews was a delay in the process of standards-setting. While judicial action ultimately slowed those parts of the system down that relied on risk assessment, it redirected agencies toward refinements in the method of risk assessment and its underlying scientific bases. The judicial effect on agency discretionary authority was interpreted (especially by the oversight functions of Congress) as poor performance in the form of missed prescribed deadlines, failure to set standards, and low compliance with existing regulations. Table 7.1 (discussed in more detail below) shows that many agencies, in fact, perceived that this poor performance was the outcome of their decisionmaking processes.

Stage 5

Stringent requirements encouraged agencies to move into the area of incentive-based systems. As a reaction to (1) stringent rules agencies felt had been placed on them by regulation by directive and (2) their growing inability to meet these requirements (often reflected in episodes perceived as poor performance), agencies encouraged the use of incentive-based systems in an attempt to reduce risks from the outset or at least avert their consequences. Agencies also developed these incentive systems on their own initiative, i.e., through the exercise of agency discretion. Government's role in the development and implementation of these incentive systems was substantial. While incentive systems for risk management had been largely implemented by the private sector, government's role was to set the rules under which this activity occurred.

Many initiatives were taken to develop new incentive systems or the rules for their use, but these occurred in areas that were customary to government operation. For example, a wide variety of financial mechanisms appeared and were applied to the risk issue in familiar areas of insurance, taxation, marketable rights, and grants and loans, but the extent and type of applications to risk were often unique and innovative. Some incentive systems, such as grants and loans, were provided for in legislation, while others emerged from agency discretion.

While these developments may have been initially linked to agency discretionary action, often starting out as optional initiatives by governmental agencies, they too soon became rulebound and were often implemented with highly prescribed conditions and formulas. More likely, they would become subject to traditional conditions that were often inappropriate to their nature.

Stage 6

Incentive-based systems of risk management proved inadequate. Uncertainties in the legislative base and scientific basis for assessing risk, along with the use of the strict liability doctrine to circumvent these limitations, drove more cases to the courts. Plaintiffs tended to demand increasingly higher awards, and the courts generally reacted in favor of claimants. The insurance market, operating largely outside of government but under its rules, in many cases limited insurance coverage and increased its cost, thus failing to support the risk takers. While the courts were partly reacting to toxic tort claims, they were probably reacting more to developments or traditions that had been set in other areas of insurance, such as medical malpractice and automobile insurance. The costs of insurance were driven up, insurance availability declined, and conditions of coverage became more stringent. As a result, incentive systems in the form of insurance became just as highly bureaucratized and prescriptive as regulation by directive.

Uncertainties in the social, economic, and environmental impacts of incentive-based alternatives to direct regulation other than insurance, along with uncertainties in the underlying political motives behind them, contributed to a lack of legislative reinforcement. As a result, incentives largely went unsupported. Some legislative reinforcement did occur in the area of air emissions trading, however, where benefits were considered important and well-defined enough to test and implement the mechanisms. The absence of insurance could have driven risky activities in the direction of risk reduction to rejuvenate the insurance market.

Stage 7

A new round of regulation through lawmaking emerged. Amendments to major pieces of legislation occurred every few years between the mid-1970s and the end of the 1980s. This pattern of lawmaking emerged in part because of the limitations of mandatory regulations and voluntary efforts at implementing incentive-based systems. More importantly, though, it reflected the way government adapts by using feedbacks from the outcomes of its prior activities. Such actions correspond to Mazmanian and Sabatier's stage of statutory revisions at which point the policy implementation process is terminated and the whole cycle presumably could start over again (1983, 38).

The Role of Feedback

Systems that comprehensively monitor the quality of the ambient physical or natural environment to detect changes in its condition and violations of legislative goals have been largely absent, unavailable, or ineffective. This situation is due in part to a mismatch between the level of resources needed to sustain these functions and the demands created by scientific and engineering uncertainties in monitoring technologies. In light of the problems constraining the use of direct performance measures, governmental institutions have relied on more easily definable, controllable, or manageable program objectives such as a means of gauging compliance with the prescriptive deadlines and controls mentioned above. By circumscribing performance measures in this way, outcomes would be evaluated more modestly, in terms of which goals get accomplished, how long it takes to accomplish them, and the extent to which unintended impacts occur that defeat the purpose or intent of the initial action.[5] While these may be more attainable performance targets, they prevent a linkage between governmental action and environmental quality (U.S. General Accounting Office, November 1988b, 4).

The way that the system responds to the outcomes of risk management processes is a major driving force of the stages shown in Figure 7.1. Table 7.1 highlights a number of outcomes of Congressional, administrative, and judicial risk management processes as they interact with one another within the federal government. These outcomes fall within the performance objectives presented in the introductory chapter—feasibility of goals, goal attainment, timeliness of attainment, and unintended impacts that impede performance, though more emphasis is placed on timeliness in the examples in Table 7.1. The responses of government shown in the table primarily reflect a reliance on traditional processes rather than new initiatives; furthermore, these responses vary for different branches of government. For Congress, the response is typically the passage of new laws and the exercise of various authorities, such as oversight and review of leadership, to direct government action in certain ways. For the executive agencies, it often involves administrative actions under existing law

Table 7.1.   Traditional Governmental Responses to Program Outcomes

Outcome	Conditions Under Which Outcome Occurred	Government Reaction
Highly Prescriptive Behavior		
Slow EPA response under FIFRA in withdrawing pesticides; slow registration procedures under FIFRA's RPAR process[a]	Highly bureaucratic rulemaking process under the registration and cancellation or special review process (formerly RPAR).	Congress amends FIFRA law; EPA uses (and, according to GAO, overuses) conditional registrations to speed up the process: EPA uses generic approach emphasizing active ingredients only.
Passage of relatively few standards under NAAQS and NESHAPs of the CAA by EPA.[b,c,d]	Highly bureaucratic rulemaking (many reviews, many stages) as a prerequisite for decisions.	State delegation of rulemaking partially circumvents the problem; GAO oversight exposes the problem.
Finalization and promulgation of few drinking water standards under SDWA by EPA.	Complex mandatory rulemaking procedures with extensive reviews and approval preclude rapid standards-setting.	Congress passes SDWA amendments, which impose schedules on standards-setting.
EPA delay in issuing PCB labeling and disposal regulations under TSCA by 7–18 months.[a]	Ostensible lack of resources for program development.	GAO oversight publicizes the problem.
Cleanup not occurring fast enough nor completely enough at Superfund sites.[e,f]	Agency leadership deficient; bureaucratic constraints prevent a rapid response to the problem.	Congress uses its traditional authority of reviewing agency leaders, resulting in dismissal of Gorsuch; oversight.
CPSC standards set too slowly;[a,g] CPSC rulemaking power underutilized; between 1984 and 1988, not a single rule is promulgated under CPSA.[h]	Statutory constraints imposed by complex rules.[h,i]	GAO initiates study of CPSC decisionmaking, and legislative changes are recommended to CPSC structure.
Standards used by CPSC have to be voluntary; mandatory rules only used where voluntary ones do not exist.[h]	Statutory constraints posed by 1981 amendments.	Congressional action suggested.

Table 7.1,   continued

Outcome	Conditions Under Which Outcome Occurred	Government Reaction
Slow FDA action in regulating chemical risks in food once they are identified as potential risks.	Highly bureaucratic rulemaking process slows standards-setting; agency reliance on action levels rather than standards.	Congressional action often considered.
OSHA has not taken advantage of criteria documents prepared by NIOSH to develop standards.	Highly bureaucratic and prescriptive rulemaking processes.	OSHA recently takes initiative in expanding rulemaking.
The pace of OSHA standards-setting is very slow—as of May 1988, GAO noted that 52% of unissued standards had been in the process for over four years.[j]	Cumbersome rulemaking procedure.	OSHA embarks on a mediated rulemaking process, which GAO still finds ineffective.

Highly Discretionary Behavior

Outcome	Conditions Under Which Outcome Occurred	Government Reaction
Laxity in asbestos enforcement under TSCA by EPA.	Agency discretion directs resources elsewhere under claims of limited resources.	Congress passes special legislation—Asbestos Hazard Emergency Response Act.
Lack of consistency between EPA criteria for the pesticide registration process and registration cancellation process.[k]	Agency discretion in rulemaking.	Congress amends FIFRA law.
Inadequate implementation of the CAA SIP, e.g., using elevated stacks to reduce estimated risk levels.	State agency discretion allowed by EPA for the design of SIP strategies.	Judicial review questions the SIPs; EPA rescinds SIPs and requires corrections.
Poorly backed or unjustified standards in terms of defensible risk levels by OSHA and EPA for benzene and formaldehyde.	Discretionary use (or non-use) of risk assessment (e.g., OSHA relies on its generic carcinogen policy to justify not using risk assessment for a benzene standard).	Judicial review turns risks estimates back to OSHA for more rigorous analysis—judicial substantive review with deference to agencies.
EPA criticized for not using risk assessment procedures and recent health studies in its benzene analysis.[a,c]	Uncertainty in the science of risk assessment, particularly in the database for benzene; agency reliance on risk assessment and inherent weaknesses of it.	New scientific data and methods developed; EPA rulemaking for benzene substantially revised in 1988.

Table 7.1,   continued

Outcome	Conditions Under Which Outcome Occurred	Government Reaction
Risk levels used as cleanup targets at Superfund sites are not consistent from site to site.[f]	Uncertainties in the data and methods of risk assessment.	SARA amendments charge ATSDR with doing site-specific health assessments.
Few PMN applications exist under TSCA; those that exist are often incomplete.[a]	Discretionary testing requirements under TSCA often negotiated.[i]	Congressional action may be considered.
Delay of 17 months by the DOE in issuing siting regulations under NWPA.[a,m]	Uncertainties in the scientific basis of siting decisions and discretionary attempts at defining more precise criteria.	GAO oversight publicizes the problem.
CPSC priorities for standards not in chemical hazard areas.[a]	Agency discretion in the prioritization of activities.	Public pressures result in reorientation to chemical issues.
CPSC product bans rarely used.[a]	Agency discretion in the prioritization of enforcement actions.	No action yet.
Slowness in FDA color additive decisions on standards.	Requirements of scientific proof imposed by a combination of administrative, judicial, and legislative requirements.	FDA uses administrative discretion to introduce fast-tracking.

[a]Zimmerman (1987b, 262–265).
[b]Berry (1984).
[c]U.S. General Accounting Office (December 1986a).
[d]U.S. General Accounting Office (1983).
[e]U.S. General Accounting Office (November 1988b).
[f]U.S. Congress, Office of Technology Assessment (1988).
[g]Bick and Kasperson (1978).
[h]Christoffel and Christoffel (1989, 337).
[i]Fise (1987, iii).
[j]U.S. General Accounting Office (1988a, 5).
[k]U.S. General Accounting Office (1980).
[l]U.S. General Accounting Office (1982).
[m]U.S. General Accounting Office (January 1985).

(such as permit denials or imposition of fines). For the judiciary, it involves interpretations that can steer governmental action in directions different from those the agencies have chosen. Thus, while government's actions ultimately result in the passage of new laws, there are many intermediate readjustments, such as administrative actions, judicial actions, and the like.

In Table 7.1, the conditions that contribute to these outcomes are expressed in terms of one of the major concepts used to characterize government throughout the book, namely, degree of bureaucratization, discretion, and openness. Table 7.1 shows that while actions are largely traditional, a relationship is apparent between the degree of bureaucratization and whether actions draw primarily on traditional government or new initiatives. When outcomes considered undesirable by any given branch of government are organized by degree of bureaucratization, certain patterns of response by government are apparent. While it is difficult to draw strong generalizations from these examples, they do suggest some general patterns. First, when outcomes that are considered adverse are associated with highly bureaucratic conditions, the responses within government that are traditional ones commonly occur in the form of legislative changes initiated by Congress and administrative actions aimed at carrying out the mandates of existing law. Second, when adverse outcomes are associated with conditions that reflect the use of agency discretion, government tends to respond with a less predictable and less systematic set of actions, whether bureaucratic or discretionary. Initiatives may emerge here, but traditional lawmaking is just as likely. Third, as time goes on, judicial, Congressional, and administrative responses ultimately tend to impose more constraints on the system, e.g., by placing boundaries around or circumscribing the discretion. In other words, they tend to drive the system toward a more highly bureaucratic, less discretionary condition. Traditional mechanisms are more consistent with this type of bureaucratic behavior than discretionary behavior.

Summary

The degree of bureaucratization of governmental organization, management, and decisionmaking is a major way of characterizing government operations as they design and operate institutions for risk management. This activity has been further characterized in terms of its degree of discretion and openness, defined as the range of highly bureaucratic vs flexible organizational behavior. By way of generalization, governmental processes that are characterized by a high degree of bureaucratization are those that tend to be traditional ones rather than new initiatives, older programs rather than newly established ones, programs that are or have to be relatively narrow in their focus (given the complexity of management), and programs that are naturally subdivided or subdividable into discrete and highly certain pieces.[6] In contrast, governmental processes are more open, flexible, and discretionary when there is a high degree of uncertainty and no particular pressure to resolve the issue, that is, under circumstances where no one is doing any better.

Examination of governmental risk management in terms of stages in decisionmaking sheds light on the role of bureaucratization in decisionmaking and its relationship to traditional behavior vs new initiatives. The discussion of governmental decisionmaking in terms of the stages in risk management (above) reveals that when new issues emerge, governmental entities usually exercise a high degree of discretion at the outset. This is a traditional governmental reaction to the high degree of uncertainty accompanying any new issue. Congress makes use of its discretionary lawmaking and oversight capacities. Agencies can take advantage of the discretionary powers given to them directly by Congress in the laws, through the resources allotted to them, and through the vagueness and loopholes in the laws by choosing their regulatory targets and extending the time frames for action. Innovations in the area of incentive-based systems for risk management emerge as guidelines; a few make their way into some regulations (blurring the distinction somewhat between regulation by directive and incentive-based systems). The judiciary exercises considerable discretion in trying or settling a large number of claims as well as challenging agency decisions.

As they move through the stages outlined above, the branches of the federal government, as is customary, initially use their discretionary authority to deal with uncertainty. Discretionary action encourages initiatives under conditions where traditional mechanisms are absent or not applicable. Congress exercises its discretion by generating heavily detailed requirements for substances and activities whose connection with human health is often vague and not fully justifiable (or by limiting agency discretion in certain areas). The agencies exercise their discretionary authority to deal with uncertainty through the development and application of risk assessment to identify and measure risk, the encouragement of technological advances in chemical detection systems, the use of nonmandatory advisory bodies, and the formalization of rulemaking processes and peer review procedures beyond what the Administrative Procedure Act requires. Discretionary action can stretch agency resources and fill in where managerial mechanisms and policies are as yet unspecified. But as the early years of the Reagan administration demonstrated, discretion can only adapt management strategies to available resources and cannot easily overcome basic resource limitations and lack of political support.

In the course of trying to reduce uncertainties and exercising discretion, however, the system becomes more bureaucratized as it becomes more rulebound. The structure of information, the way it is used, and the way rules are set are formalized. Even the discretionary use of incentive systems ultimately becomes highly bureaucratized and rulebound. While some have argued that rulemaking legitimizes agency discretion (West, 1988, 775), it can constrain or at least put boundaries around discretion as well. The result is ultimately a new wave of lawmaking, which can either reopen the system or continue to tighten the rules under which it operates.

Thus, the stages in risk management outlined above and the behavior of the system analyzed throughout the book show that risk management becomes highly bureaucratic as a way of making risk problems more manageable in the face of complexity and uncertainty and when identifiable pieces can be isolated. In other words, a system that starts out with a considerable amount of discretion becomes highly bureaucratized over time, with a mode of operation that is highly prescribed. This is consistent with how many organizations evolve, and is largely the result of government drawing on its traditional processes as an approach to risk management.

A high level of bureaucratization in and of itself is not bad; highly bureaucratic systems are useful for well-defined tasks, tasks that are amenable to definition by rules, and tasks that have relatively predictable and acceptable outcomes and clear means to achieve them. Such conditions do occur at some points in the management of risk policy. Under some circumstances, however, a high degree of bureaucratization can lead to performance that is inconsistent with management under uncertainty. The chemical risk issue imposes a high degree of uncertainty that often precludes an overreliance on bureaucratic systems of management. As a result of recurring and persistent uncertainty, a system has to break out of its highly bureaucratic mode and often relies on another cycle of lawmaking and rulemaking, since discretionary authority is absent. Bureaucratic systems may also be precluded where the demands of the tasks, however circumscribed or narrowed in function, outstrip the actual or perceived capacity of the regulatory system. At this point, even the flexibility that incentive systems ostensibly offer may not be powerful enough to encourage voluntary compliance to compensate for a lack of regulatory capacity.

Thus, the management of health risks from chemicals embarks on a course between highly bureaucratized, rulebound management systems and open, flexible, and highly discretionary systems. The course is often a treacherous one, because it breeds intense controversy among the participants. The evolutionary framework that has been presented for risk management, as a system that moves between traditional actions and new initiatives, closure and openness, and discretion and prescription, is consistent with frameworks that explain the behavior of organizations, management, and decisionmaking in other areas of government and in contexts other than government.

This exploration of the management of environmental health risk policies has revealed that regardless of which perspective is taken (i.e., looking at intensity of actions, bureaucratic behavior, or the use of feedback), traditional approaches of "normal" government have tended to prevail over initiatives in government. Where new initiatives have been taken, they have emerged primarily within the context of traditional approaches. This system of management works only insofar as risk problems can be identified and managed in relatively discrete ways.

RAMIFICATIONS FOR FUTURE RISK MANAGEMENT

The management of chemical risks has not fit easily into the governmental regulatory structure. Regulation is designed for fairly well-defined objectives or targets and easily accessible means or procedures to carry them out, whereas risk management is often characterized by uncertainties in its scientific base, in judgments about appropriate risk levels or regulatory targets, and in the effect that regulatory strategies will have on overall risk levels. The risk bureaucracy within government has gone in many different directions to find solutions, but has initially drawn on traditional mechanisms.[7]

The traditional approach may work where the problems are clearly identifiable and subdividable into individual pieces and for which analogies in other areas of government can be established. To some extent the risk problem has been subdividable into discrete components (e.g., setting separate requirements for different chemicals in different contexts). In the evolution of risk management this may have been a necessary way to proceed—identifying and learning about the pieces first, before attacking the whole. A lot had to be learned about the behavior of chemicals in the environment as distinct from laboratory settings and the nature of human exposure and tolerances.

Governmental action has paralleled the way the problem has been subdivided. During the 1980s, Congress has tended to strike out in a lot of small ways in the area of lawmaking. It has done so by passing piecemeal, separately titled amendments to laws (for example, the Asbestos Hazard Emergency Response Act, the Lead Contamination Control Act, the Ocean Dumping Ban Act) to cover what it has seen as deficiencies in existing legislation, rather than attempting a wholesale overhaul of existing laws. Government agencies have also designed regulatory programs and incentive systems on a program-by-program basis. Finally, the judiciary has been criticized for its often piecemeal approach to cases involving risk and risk assessment.

The stages of risk management show the result of all of this: government largely is caught up in a loop as it continues to repeat the stages portrayed in Figure 7.1. This phenomenon is traditional to the way that the management of policy or policy implementation proceeds—while initiatives do occur, they occur within the larger context of normal government. This is the legacy of government. It is here to stay.

Different theories have arisen to explain reasons why this fragmentation in policy implementation has occurred. Some scholars of governmental process, for example, have viewed government as reactive to processes originating in a larger social matrix (Cobb and Elder, 1983, 12). Taking this further, governmental behavior is consistent with the theories of management and organizational behavior outlined earlier. First, the response within government to bureaucratize the decision process has largely paralleled the organization of the set of interests to which it reacts, which is highly specialized. The degree of specialization of these interests, and their number, have increased in response to the complexity of environmental health risks and due to the resources they have been able to command. Fragmentation of issues has encouraged governmental entities to respond in a fragmented manner. Second, government is largely dependent on scientific information and its translation by experts. The technical demands of managing chemical risks have been formidable, and it has taken a long time for government to learn about the nature of the problem and to propose some solutions. Furthermore, a high degree of uncertainty has characterized the scientific underpinnings of the risk issue, and debates exist among experts who often present opposing viewpoints that are not easily amenable to resolution. Even conventional disciplines, such as chemistry, toxicology, and epidemiology, have had to develop new subdisciplines (with the prefix "environmental") to address many of the scientific problems presented by chemical contamination. The reliance on information and the approach taken by government to attempt to resolve uncertainties in the context of risk management has also led management in the direction of subdividing issues.[8]

Yet there has been a price in proceeding this way. Critical problems have constantly been pushed aside and continually emerge as major obstacles to the design and implementation of risk management strategies. As the science of risk assessment and the management system for risks become more refined, they will have to grapple with some of these important unanswered questions that were recognized very early in the quest for risk management as in other policy areas. Some examples of these questions are:

1. How do we measure and apply levels of acceptable risk, confounded by the fact that attitudes and beliefs regarding acceptable risk are often not stable over time and do not correspond to behavior?
2. What decision rules should be invoked in the attainment of acceptable risk levels? Do we regulate the chemicals, the activities producing them, or the context or exposure conditions? Or do we make people more resistant to effects after exposure?
3. How much are we willing to spend to achieve these goals, and how should the costs be distributed?

Risk preferences, choice of regulatory decision rules, and cost quantification and distribution depend on how problems and solutions are organized. These questions have been difficult ones for government to address within its traditional approaches to risk management. Moreover, their relationship to governmental fragmentation is unclear.

A management perspective has helped to identify fragmentation and understand its implications. The fragmented approach to government's management of the risk issue may work reasonably well at the beginning of a problem when the components are not well understood, but new directions are now needed to orchestrate the pieces and to cope with issues, such as risk acceptability, that have posed major problems for risk management. A chemical-by-chemical approach to risk management (subdivided still further by a programmatic emphasis) has led to some inconsistencies and at the very least to regulatory burdens involving surveillance and analysis of uncertain impacts that are beyond government's capacity to address. Incentive systems are no panacea either, since they can generate their own system of rules, loopholes, and regulatory demands in terms of monitoring, surveillance, and proof. Furthermore, innovations that were originally put forth as new technological or managerial approaches to environmental protection have not been followed up to any great extent; some have actually had unanticipated negative side effects of their own. Overall, our reliance on chemicals has not decreased, it has just shifted to new chemicals.

Now that government has had a chance to understand and react to chemical risk problems on a piecemeal basis, it has to address the larger picture. This may encourage government to call for new initiatives in the context of a more macroscopic approach—one that emphasizes, for example, more program integration. Program integration typically strives to design management strategies to meet several objectives simultaneously. Once again, a management perspective can provide direction for an integrated approach.

The call for integration is by no means novel.[9] Integrated risk management has been attempted in many forms before. Consolidated permit programs were tried and then quickly abandoned because of their complexity. Integration was tried as an organizational approach to state and local government management, and superagencies became a popular way of achieving this. Many of these agencies, however, soon broke apart because they were unwieldy. Finally, cross-media studies and environmental modeling efforts, which were commonly pushed in agency research and development programs, were often too complex to handle at one time (Zimmerman, 1987b, 262); focus shifted toward smaller, more specialized models that were easier to work with. Government's management of chemical risks has yet to draw lessons from the theoretical work in public policy implementation and the management of public policy.

New approaches to integration, however, are heading in the direction of these management perspectives. For example, the interactions of chemicals in the environment are being explicitly recognized in the conceptualization of risk problems; attempts are being made at regulatory consistency, scientific consensus, and the building of institutions designed to deal with complexity. Each of these approaches poses its own set of problems, but as a group they are noteworthy beginnings.

Recognition of the interaction of chemicals in the environment is a major driving force for an integrated chemical risk management policy. Interactions can occur synergistically or antagonistically and should be a consideration in any risk management policy. The regulations passed by EPA in 1986 on chemical mixtures were a beginning. Some of the specific methodologies cited have been the target of considerable criticism, however. Furthermore, they do not have simple solutions, such as product substitution, since the number of combinations of chemicals that define mixtures is formidable.

A policy in support of regulatory consistency aims at systematizing the methods and assumptions used in setting environmental and human health standards from program to program. Attempts at regulatory consistency during the 1980s have appeared in the form of requiring a search of applicable regulations before new regulations are developed. In line with this approach, some laws passed during the 1980s frequently cross-reference standards set in earlier legislation as taking precedence over newer ones. Applicable or relevant and appropriate requirements (ARARs) under Superfund and a similar approach taken in regulations proposed in 1989 for sewage sludge management under Section 404(b)(3) of the Clean Water Act (U.S. EPA, February 6, 1989; 40 Code of Federal Regulations Part 503) are examples of attempts at regulatory integration by means of searches for other regulatory precedents. There are, in fact, formal guidelines that have been issued under CERCLA/SARA for the format of such regulatory searches (U.S. EPA, October 1988, Appendix E).

Agencies and the courts are also attempting to coordinate procedures for setting standards. For example, in the mid-1980s the courts examined EPA's method for assessing risks of vinyl chloride and required the agency to redo the calculations. In anticipation of similar criticism of its then-impending regulations on benzene emissions, the EPA took upon itself to rescind and review the risk assessment for benzene as well.[10] Advocates of the regulatory consistency approach argue that it should be made an explicit governmental policy (Travis and Hattemer-Frey, 1988).

Efforts at scientific consensus or forums to minimize and manage dissent or at least promote compromise are another example of an integrated risk management policy. (See, for example, Nelkin and Pollak, 1980.) Chapter 5 identified the use of such forums both within and outside of federal executive agencies. Such efforts usually are accompanied by or have counterparts in management strategies that rely on bargaining and negotiation (O'Hare, Bacow, and Sanderson, 1983). However, even if consensus on a given scientific issue could be achieved to a point where policies could be managed, some dangers of taking such consensus too literally have been recognized. There are many examples in the history of science of where certain ideas and hypotheses prevailed for a time and then were refuted as new evidence emerged. While consensus among experts narrows the range of measurements for policymakers, the consensus can be wrong (and can actually act as a barrier to developments that prove the prevailing consensus wrong). For example, Freudenberg (1988) has pointed out that the consensus in the late 19th century on measurements of the speed of light turned out not to include the value that was later discovered with a more refined technology. In spite of the limitations of any attempt at estimating scientific phenomena and predicting the future state of any set of facts pertinent to risk management, these attempts are nevertheless important undertakings. As Boulding says, in connection with models of biological evolution, "evolutionary models are not very good at predicting the future," but this is a result of "the uncertainty of the real world, not a defect in the model" (Boulding, 1982, 17). According to Boulding, models are useful in providing valuable clues to the future. In other words, they are the best we can do.

In order to be more successful, integrated and coordinated management should incorporate several management strategies that are currently not practiced to any great extent. One such strategy is a shift from a top-down structure of management to one that legitimizes coordination among professionals dealing with similar technical questions but located in different lines within a particular organization or in different organizations. These lateral linkages (which currently only occur on a voluntary and ad hoc basis through professional organizations) promote more of a forum for scientific debate and could reduce or narrow the range of scientific uncertainty.

Another management principle is the simplification of work routines by enabling a manager to reduce the number of pieces being worked with at any given time. This reduction is necessitated by the large number of chemicals, activities, and financial complexities one works with at a given time in the chemical risk area. A management principle that is applicable to such a strategy is Simon's concept of the use of "subassemblies" or stable intermediaries, which potentially promote stability and save time in work routines within organizations (1969, 93). Mintzberg (1979, 108–129) has carried this concept further by identifying types of groupings and criteria for using them in an organization. Such groupings could be organized on the basis of similar chemicals appearing in different contexts, similar activities, similar risks, or activities occurring in the same geographic area. Many of these groupings have been tried by government at different times, but mostly on an ad hoc, discretionary basis, and there is little evaluation of the relative performance of these alternative ways of organizing work. These subassemblies, however, could conflict with the promotion of lateral or horizontal linkages discussed above, since subassemblies may be difficult to design into systems that are not hierarchical, i.e., that allow lateral exchanges (outside the line of authority) within and between organizations. In fact, subassemblies may not work simultaneously for lateral (staff) and vertical (line) authority structures, because different groupings are often required to support staff and line functions. That is, the same units required to perpetuate fairly circumscribed regulatory functions that lend themselves to bureaucratic and hierarchical organization are not necessarily the same units that promote a constant flow of information within the scientific community regardless of organizational affiliation.

Risk management strategies have so far primarily emphasized middle management responsibilities in government and industry, such as regulation, technology development and design, and similar functions. Risk responsibilities at the level of operations and maintenance have been organized differently and originate from a different professional tradition. The concerns of risk management among middle managers must reach the level of operations to ensure that system operators and maintenance personnel are working with the same assumptions as those who are planning, designing, and siting systems with potential health risks. This may involve the design of new communication channels, new modes of communication, and new personnel to accomplish these missions akin to the "knowledge brokers" that have provided an interface between government and scientists.

Finally, the risk management system has to be kept flexible enough to accommodate new technologies. It has been claimed by scholars of the technology process that the timing of regulatory decisions and the scheduling of development of new technologies are not in step with one another, and in particular that relatively little time goes into assuring a smooth transition between bench-scale and full-scale applications (Berg, 1989). These factors often result in the failure of new technologies to be considered as regulatory alternatives because they are either ignored altogether for lack of complete information, considered suspect because of errors potentially resulting from underinvestment in testing, or not developed sufficiently to be competitive with well-established technologies.

Whether these changes will involve small steps that are once again within the reach of traditional governmental processes or will require major changes in the paradigm under which government operates is the next challenge for risk management.

NOTES

1. For example, programs prompting quicker action, such as emergency remedial measures for the cleanup of hazardous waste sites, were examples of attempts to at least formally remediate some of the failings of earlier regulatory programs.
2. An alternative view is that interest groups generate and use crises in order to direct lawmaking into certain areas.
3. The science of risk assessment actually started much earlier in other areas, such as engineering and nuclear power.
4. This trend in legislative philosophy is consistent with the Sabatier and Mazmanian (1979) category of "quantitative tilt," which implies a high degree of clarity in legislation.
5. Some recent performance measures have been even more modest than this, restricting the evaluation of performance to merely a demonstration of influence over policies and programs that is sustained over some reasonable time period (Pugliaresi and Berliner, 1989, 379). Such a definition circumvents the issue of effectiveness as a component of performance. The GAO (November 1988c, 9) notes that there has been a pervasive decline in the ability of government to conduct program evaluations. They attribute this to a decline of 22% in agency evaluation staff between 1980 and 1984, which is a faster rate of decline than that for overall agency staff declines in government. Furthermore, funds allocated to the program evaluation function declined as well.
6. Another more normative view associates the concept of bureaucratization with complexity and time-consuming procedures or mechanisms that in reality are too complex to be effective (Christoffel and Christoffel, 1989, 338).
7. Mazmanian and Morell (1988) have made a similar observation in toxic waste management—that for many years, as one might expect, government acted in predictable ways in shaping and implementing policy.
8. Others emphasize the self-perpetuating rather than the reactive nature of the governmental system and view government largely as a set of processes imposed upon problems as they arise.
9. Mazmanian and Morell (1988), for example, emphasize the importance of integra-

tion in the area of toxic waste management, though their emphasis is on source reduction as a means of integration.

10. However, contrary to this appearance of an attempt at integrating risk assessment methodologies for two different chemicals, the agency clearly stated in its proposed regulations for benzene in 1988 that the approach taken in the risk assessment should not be construed to apply to any other legislation.

Bibliography

Abraham, K. 1982. "Cost Internalization, Insurance, and Toxic Tort Compensation Funds," *Virginia Journal of Natural Resources Law* 2:123–148.

Abraham, K. 1986. *Distributing Risk* (New Haven, CT: Yale University Press).

Ackerman, B. A., and W. T. Hassler. 1981. *Clean Coal/Dirty Air* (New Haven, CT: Yale University Press).

Administrative Office of the U.S. Courts. 1981. *Management Statistics for United States Courts* (Washington, DC: AOUSC).

Aharoni, Y. 1981. *The No-Risk Society* (Chatham, NJ: Chatham House).

Aidala, J. 1985. "The Accident in Bhopal, India: Implications for U.S. Hazardous Chemical Policies" (Washington, DC: U.S. Government Printing Office, October 3).

Alexander and Alexander, Inc. 1987. "A Guide to the Liability Risk Retention Act of 1986," *Insurance Advocate* (February 14).

Allison, G. T. 1971. *Essence of Decision* (Boston: Little, Brown and Co.).

American Chemical Society. 1988. *Detection in Analytical Chemistry: Implementing Theory and Practice* (Washington, DC: ACS).

Anderson, F. R., et al. 1977. *Environmental Improvement Through Economic Incentives* (Baltimore: Johns Hopkins University Press).

Ashford, N. A., C. W. Ryan, and C. C. Caldart. 1983a. "A Hard Look at Federal Regulation of Formaldehyde: A Departure from Reasoned Decisionmaking," *Harvard Environmental Law Review* 7:297–370.

Ashford, N. A., C. W. Ryan, and C. C. Caldart. 1983b. "Law and Science Policy in Federal Regulation of Formaldehyde," *Science* 222:894–900.

Atiyah, P. S. 1982. "A Legal Perspective on Recent Contributions to the Valuations of a Life," in *The Value of Life and Safety*, M. W. Jones-Lee, Ed. (Amsterdam: North-Holland Publishing Co.), pp. 185–200.

Badaracco, J. L., Jr. 1985. *Loading the Dice* (Boston: Harvard Business School Press).

Baram, M. S. 1980. "Cost-Benefit Analysis: An Inadequate Basis for Health, Safety and Environmental Regulatory Decisionmaking," *Ecology Law Quarterly* 8:377–435.

Baram, M. S. 1982. *Alternatives to Regulation* (Lexington, MA: Lexington Books).

Baram, M. S. 1986. "Chemical Industry Accidents, Liability, and Community Right to Know," *American Journal of Public Health* 76:568–572.

Baram, M. S. 1987. "Chemical Industry Hazards: Liability, Insurance and the Role of Risk Analysis," in *Insuring and Managing Hazardous Risks: From Seveso to Bhopal and Beyond*, P. R. Kleindorfer and H. C. Kunreuther, Eds. (New York: Springer-Verlag New York, Inc.), pp. 415–441.

Baram, M. S. 1988. "Corporate Development of an Integrated Hazardous Waste Management Strategy: Some Preliminary Legal Considerations" (Boston: Boston University School of Law).

Bardach, E. 1987. "Persuasion and Tax Code Enforcement," paper presented at the Association for Public Policy Analysis and Management conference, Washington, DC, October 29–30.

Bardach, E., and R. A. Kagan. 1982a. *Going by the Book: The Problem of Regulatory Unreasonableness* (Philadelphia: Temple University Press).

Bardach, E., and R. A. Kagan. 1982b. "Introduction," in *Social Regulation*, E. Bardach and R. Kagan, Eds. (San Francisco: Institute for Contemporary Studies), pp. 3–19.

Bardach, E., and R. A. Kagan. 1982c. "Liability Law and Social Regulation," in *Social Regulation*, E. Bardach and R. Kagan, Eds. (San Francisco: Institute for Contemporary Studies), pp. 237–266.

Barke, R. 1986. *Science, Technology and Public Policy* (Washington, DC: Congressional Quarterly, Inc.).

Barnard, C. I. 1938. *The Functions of the Executive* (Cambridge, MA: Harvard University Press).

Barron, J. 1986. "40 Legislatures Act to Readjust Liability Rules," *New York Times* (July 14).

Bazelon, D. L. 1979. "Risk and Responsibility," *Science* 205:277–280.

Bazelon, D. L. 1981. "The Judiciary: What Role in Health Improvement?" *Science* 211:792–793.

Berg, D. 1989. Panel presentation on regulation and new technologies at the 82nd Annual Meeting of the Air & Waste Management Association, Anaheim, CA, June 25–30.

Berliner, B. 1982. *Limits of Insurability of Risks* (Englewood Cliffs, NJ: Prentice-Hall, Inc.).

Berliner, B. 1988. "Components Impeding Insurability of Hazardous Waste Facilities and Approaches to Combat Them," paper presented at the Conference on Risk Assessment and Risk Management Strategies for Hazardous Waste Storage and Disposal Problems, University of Pennsylvania, Wharton School, May 18–19.

Bernstein, M. H. 1955. *Regulating Business by Independent Commission* (Princeton, NJ: Princeton University Press).

Berry, M. A. 1984. "A Method for Examining Policy Implementation: A Study of Decisionmaking for the National Ambient Air Quality Standards, 1964–1984," PhD Thesis, University of North Carolina, Department of Environmental Sciences and Engineering, Chapel Hill, NC.

Berry, M. A. 1987. "Use of Risk Assessment in the Development of Ambient Air Quality Standards," paper presented at the 87th Meeting of the Air Pollution Control Association, Session 43, New York, June 21–26.

Bick, T., and R. E. Kasperson. 1978. "The CPSC Experiment: Pitfalls of Hazard Management," *Environment* (October), pp. 30–42.

Black, B. 1988. "Evolving Legal Standards for the Admissibility of Scientific Evidence," *Science* 239:1508–1512.

Bolton, J. R. 1977. *The Legislative Veto: Unseparating the Powers* (Washington, DC: American Enterprise Institute).

Bosselman, F., D. A. Feurer, and C. L. Siemon. 1976. *The Permit Explosion* (Washington, DC: The Urban Land Institute).

Bosworth, B. P. 1981. "The Economic Environment for Regulation in the 1980s," in *Environmental Regulation and the U.S. Economy*, H. M. Peskin, P. R. Portney, and A. V. Kneese, Eds. (Baltimore: Johns Hopkins University Press), pp. 7–24.

Boulding, K. E. 1982. "Irreducible Uncertainties," *Society* 20(1) (November/December).

Break, G. F. 1980. *Financing Government in a Federal System* (Washington, DC: The Brookings Institution).

Breyer, S. 1982. *Regulation and its Reform* (Cambridge, MA: Harvard University Press).

Brickman, R., S. Jasanoff, and T. Ilgen. 1985. *Controlling Chemicals: The Politics of Regulation in Europe and the United States* (Ithaca, NY: Cornell University Press).

Broder, I. E., and J. F. Morrall, III. 1983. "The Economic Basis for OSHA's and EPA's Generic Carcinogen Regulations," in *What Role for Government?* R. J. Zeckhauser and D. Leebaert, Eds. (Durham, NC: Duke University Press), pp. 242–254.

Brown, M. S. 1987. "Communicating Information about Workplace Hazards: Effects on Worker Attitudes Toward Risks," in *The Social and Cultural Construction of Risk*, B. B. Johnson and V. T. Covello, Eds. (Dordrecht, Holland: D. Reidel Publishing Co.).

Bruff, H., and E. Gellhorn. 1977. "Congressional Control of Administrative Regulation: A Study of Legislative Vetoes," *Harvard Law Review* 90:1369–1440.

Bryner, G. 1986. "Administrative Law and Administrative Discretion," in *Administrative Discretion and Public Policy Implementation*, D. H. Shumavon and H. K. Hibbeln, Eds. (New York: Praeger Publishers), pp. 50–64.

Burros, M. 1985. "The Saga of a Food Regulation: After 25 Years, Still No Decision," *New York Times* (February 14), p. C1.

Business Week. 1984. "Union Carbide Fights for its Life," December 24.

Butterfield, F. 1988. "Trouble at Atomic Bomb Plants: How Lawmakers Missed the Signs," *New York Times* (November 28), pp. A1, B10.

Cannon, M., and D. M. O'Brien. 1985. *Views from the Bench: The Judiciary and Constitutional Politics* (Chatham, NJ: Chatham House).

Cantor, R. A., and R. Bishop, Jr. 1987. "Using Information from Toxic-Tort Litigation to Value the Health and Safety Consequences of Regulatory Decisions," draft manuscript (Oak Ridge, TN: Oak Ridge National Laboratories, September 30).

Cardozo, M. H. 1981. "The Federal Advisory Committee Act in Operation," *Administrative Law Review* 33.

Cheek, L., III. 1988. "Insurance Issues Associated with Cleaning up Inactive Hazardous Waste Sites," paper presented at the Geneva Association and Risk and Decision Processes Center conference, Philadelphia, University of Pennsylvania, May 18–19.

Christoffel, T., and K. K. Christoffel. 1989. "The Consumer Product Safety Commission's Opposition to Consumer Product Safety: Lessons for Public Health Advocates," *American Journal of Public Health* 79(3):336–339.

Cobb, R. W., and C. D. Elder. 1983. *Participation in American Politics: The Dynamics of Agenda-Building* (Baltimore: Johns Hopkins University Press)

Cobler, J. G., and F. D. Hoerger. 1985. "Analysis of Agency Estimates of Risk for Carcinogenic Agents," in *Risk Analysis in the Private Sector* (New York: Plenum Publishing Corporation).

Cohen, M., J. G. March, and J. Olsen. 1972. "A Garbage Can Model of Organizational Choice," *Administrative Science Quarterly* 17(1):1–25.

Cohen, R. E. 1983. "The Gorsuch Affair," *The National Journal* (January 8).

Congressional Quarterly, Inc. 1982. *Guide to Congress* (Washington, DC: Congressional Quarterly, Inc.).

Congressional Quarterly, Inc. 1983. *Federal Regulatory Directory 1983–84* (Washington, DC: Congressional Quarterly, Inc.).

Congressional Quarterly, Inc. 1986. *Federal Regulatory Directory*, 5th ed. (Washington, DC: Congressional Quarterly, Inc.).

Cook, C. E. 1980. *Nuclear Power and Legal Advocacy* (Lexington, MA: Lexington Books).

Cooper, J., and P. A. Hurley. 1983. "The Legislative Veto: A Policy Analysis," *Congress and the Presidency* 10 (Summer).

Cooper, P. J. 1985. "Conflict or Constructive Tension: The Changing Relationship of Judges and Administrators," *Public Administration Review* 45:643–652.

Costonis, J. J., and R. S. DeVoy. 1975. "The Puerto Rico Plan: Environmental Protection Through Development Rights Transfer" (Washington, DC: The Urban Land Institute).

Covello, V. T., and J. Mumpower. 1985. "Risk Analysis and Risk Management: An Historical Perspective," *Journal of Risk Analysis* 5(2):103–120.

Council on Environmental Quality. 1980. *11th Annual Report on Environmental Quality* (Washington, DC: U.S. Government Printing Office).

Council on Environmental Quality. 1983. *13th Annual Report on Environmental Quality* (Washington, DC: U.S. Government Printing Office).

Council on Environmental Quality. 1986. *16th Annual Report on Environmental Quality* (Washington, DC: U.S. Government Printing Office).

Council on Environmental Quality. 1987. *17th Annual Report on Environmental Quality* (Washington, DC: U.S. Government Printing Office).

Creekmore, A. T., Jr. 1989. "Emission Trades for Sources of Volatile Organic Compounds," paper presented at the 82nd Annual Meeting of the Air & Waste Management Association, June 25–30.

Crozier, M. 1964. *The Bureaucratic Phenomenon* (Chicago: University of Chicago Press).

Cummings-Saxton, J., S. J. Ratick, F. W. Talcott, C. P. Dougherty, A. V. Vliet, A. J. Barad, and A. E. Cook. 1988. "Accidental Chemical Releases and Local Emergency Response: Analysis Using the Acute Hazardous Events Data Base," *Industrial Crisis Quarterly* 2(3–4): 139–170.

Curtin, N. P. 1988. "Comparison of Amtrak Employee Injury Settlement Costs Under the Federal Employers' Liability Act and State Workers' Compensation Programs," testimony before the Subcommittee on Surface Transportation, U.S. Senate (Washington, DC: U.S. Government Printing Office, June 22).

Cyert, R., and J. G. March. 1963. *A Behavioral Theory of the Firm* (Englewood Cliffs, NJ: Prentice-Hall, Inc.).

Dahl, R., and C. E. Lindblom. 1953. *Politics, Economics, and Welfare* (New York: Harper & Row, Publishers, Inc.).

Davidson, R. H. 1981. "Subcommittee Government: New Channels for Policy Making," in *The New Congress*, T. E. Mann and N. J. Ornstein, Eds. (Washington, DC: American Enterprise Institute).

Davidson, R. H., and W. J. Olesznek. 1981. *Congress and Its Members* (Washington, DC: Congressional Quarterly, Inc.).

Davies, J. C. 1984. "Environmental Institutions and the Reagan Administration," in *Environmental Policy in the 1980s: Reagan's New Agenda,* N. J. Vig and M. E. Kraft, Eds. (Washington, DC: Congressional Quarterly, Inc.).

Davis, K. C. 1972. *Administrative Law Text* (St. Paul, MN: West Publishing).

deHaven-Smith, L. 1984. "Regulatory Theory and State Land-Use Regulation: Implications from Florida's Experience with Growth Management," *Public Administration Review* (September/October).

DeLong, J. V. 1983. "Federal Regulatory Statutes and Control of Risk," paper presented at the U.S.–Federal Republic of Germany Conference on Managing Risk, April 1.

deLucia, R. J. 1974. "An Evaluation of Marketable Effluent Permit Systems" (Washington, DC: U.S. EPA, September).

Demkovich, L. E. 1982. "Critics Fear the FDA Is Going Too Far in Cutting Industry's Regulatory Load," *National Journal* 14 (July 17).

DeMuth, C. C. 1983. "What is Regulation?" in *What Role for Government? Lessons from Policy Research*, R. J. Zeckhauser and D. Leebaert, Eds. (Durham, NC: Duke University Press).

Deutsch, K. W. 1966 *The Nerves of Government* (New York: The Free Press).

Diemer, J. S., and J. W. Eheart. 1988. "Transferable Discharge Permits for Control of SO_2 Emissions from Illinois Power Plants," *Journal of the Air Pollution Control Association* 38:997–1005.

Diver, C. 1980. "A Theory of Regulatory Enforcement," *Public Policy* 28:257–299.

Dodd, L. C., and B. I. Oppenheimer, Eds. 1985. *Congress Reconsidered*, 3rd ed. (Washington, DC: Congressional Quarterly, Inc.).

Dodge, C. H., and R. L. Civiak. 1981. "Risk Assessment and Regulatory Policy" (Washington, DC: Congressional Research Service).

Doherty, N., P. Kleindorfer, and H. Kunreuther. 1988. "Insurance Perspectives on an Integrated Hazardous Waste Management Strategy," in *Integrating Insurance and Risk Management for Hazardous Waste* (Philadelphia: University of Pennsylvania, Wharton School, May 18–19).

Doniger, D. D. 1978. *Law and Policy of Toxic Substances Control* (Baltimore: Johns Hopkins University Press).

Dore, M. 1988. *Law of Toxic Torts: Litigation/Defense/Insurance*. (New York: Clark Boardman, April 1).

Douglas, M. 1986. *Risk Acceptability According to the Social Sciences* (New York: Russell Sage Foundation).

Douglas, M., and A. Wildavsky. 1982. *Risk and Culture* (Berkeley: University of California Press).

Dowd, R. M. 1983. "FIFRA Update," *Environmental Science and Technology* 17:415A.

Downs, A. 1967. *Inside Bureaucracy* (New York: Little, Brown & Company).

Downs, A. 1972. "Up and Down with Ecology—The Issue-Attention Cycle," *The Public Interest* 28:38–50.

Drucker, P. F. 1974. *Management* (New York: Harper & Row, Publishers, Inc.).

Ehcart, J. W., E. J. Joeres, and M. H. David. 1980. "Distribution Methods for Transferable Discharge Permits," *Water Resources Research* 16.

Engelhardt, H. T., Jr., and A. L. Caplan, Eds. 1987. *Scientific Controversies* (Cambridge, U.K. and New York: Cambridge University Press).

Etzioni, A. 1961. *A Comparative Analysis of Complex Organizations* (New York: The Free Press).

Evan, W. M. 1966. "The Organization Set: Toward a Theory of Interorganizational Relations," in *Approaches to Organizational Design*, J. D. Thompson, Ed. (Pittsburgh: University of Pittsburgh Press), pp. 173–188.

Executive Office of the President. 1989. *Economic Report of the President* (Washington, DC: U.S. Government Printing Office, January).

Executive Office of the President, Office of Management and Budget. 1985. "Historical Budget of the United States Government," Fiscal Year 1986 (Washington, DC: U.S. Government Printing Office).

Executive Office of the President, Office of Management and Budget. 1988. *Budget of the United States Government*, Fiscal Year 1988 (Washington, DC: OMB).

Executive Office of the President, Tort Policy Working Group. 1986. "Report of the Tort Policy Working Group on the Causes, Extent and Policy Implications of the Current Crisis in Insurance Availability and Affordability" (Washington, DC: U.S. Government Printing Office, February).

Executive Office of the President, Tort Policy Working Group. 1987. "An Update on the Liability Crisis" (Washington, DC: U.S. Government Printing Office, March).

Fairley, W. B. 1982. "Market Risk Assessment of Catastrophic Risks," in *The Risk Analysis Controversy: An Institutional Perspective*, H. C. Kunreuther and E. V. Ley, Eds. (New York: Springer-Verlag New York, Inc.).

Farber, H. S. 1987. "The Recent Decline of Unionization in the United States," *Science* 239:915–920 (November 13, 1987).

Feder, B. J. 1988. "Manville's Novel Plan: Profits for Claimants," *New York Times* (July 30).

Federal Regulatory Council. 1979. "Regulation of Chemical Carcinogens" (Washington, DC: Regulatory Council).

Feldman, E. J., and J. Milch. 1982. *Technocracy versus Democracy: The Comparative Politics of International Airports* (Boston: Auburn House Publishing Co.).

Fensterstock, J. C., and R. K. Fankhauser. 1968. "Thanksgiving 1966 Air Pollution Episode in the Eastern United States" (Durham, NC: U.S. Department of Health, Education, and Welfare, National Air Pollution Control Administration, July).

Ferreira, J., Jr. 1982. "Promoting Safety Through Insurance," in *Social Regulation*, E. Bardach and R. Kagan, Eds. (San Francisco: Institute for Contemporary Studies).

Field, R. 1980. "Statutory and Institutional Trends in Governmental Risk Management: The Emergence of a New Structure," draft (Washington, DC: Committee on Risk and Decision Making, National Research Council).

Fiksel, J. 1984. "The Problem of Indeterminate Causation in Victim Compensation," paper presented at the Society for Risk Analysis conference, Knoxville, TN, October 1.

Fischhoff, B., S. Lichtenstein, P. Slovic, S. L. Derby, and R. L. Keeney. 1981. *Acceptable Risk* (Cambridge, U.K. and New York: Cambridge University Press).

Fise, M. E. R. 1987. "The CPSC: Guiding or Hiding from Product Safety?" (Washington, DC: Consumer Federation of America, May).

Fisher, L. 1988. *Constitutional Dialogues* (Princeton, NJ: Princeton University Press).

Foreman, C. H., Jr. 1984. "Congress and Social Regulation in the Reagan Era," in *The Reagan Regulatory Strategy*, G. C. Eads and M. Fix, Eds. (Washington, DC: The Urban Institute).

Freeman, A. M., III. 1989. "Old Cars and Taxation," Letter to the Editor. *Science* 244:127.

Freudenberg, W. R. 1988. "Perceived Risk, Real Risk: Social Science and the Art of Probabilistic Risk Assessment," *Science* 242:44–49.

Freudenheim, M. 1988a. "Dalkon Shield Trust Fund Organizes," *New York Times* (August 8), p. D1.

Freudenheim, M. 1988b. "Dalkon Shield Trust Fund Faces Appeal," *New York Times* (August 16).

Gale, R. P. 1987. "The Environmental Movement Comes to Town: A Case Study of an Urban Hazardous Waste Controversy," in *The Social and Cultural Construction of Risk*, B. B. Johnson and V. T. Covello, Eds. (New York: Plenum Publishing Corporation), pp. 233–250.

Gilbert, P. 1988. "No Liability Amnesty for Pharmaceuticals," Letter to the Editor, *New York Times* (June 15).

Gladwin, T. N. 1981. "Environmental Mediation and a Contingency Theory of Preferred Third Party Intervention" (New York University Graduate School of Business Working Paper, March).

Goldsmith, R. I., and W. C. Banks. 1983. "Environmental Values: Institutional Responsibility and the Supreme Court," *Harvard Environmental Law Review*, 7:1–40.

Gordon, G. J. 1986. "Administrative Discretion and the Intergovernmental System," in *Administrative Discretion and Public Policy Implementation*, D. H. Shumavon and H. K. Hibbeln, Eds. (New York: Praeger Publishers), pp. 159–173.

Gormley, W. T., Jr. 1987. "Institutional Policy Analysis: A Critical Review," *Journal of Policy Analysis and Management* 6(2):153–169.

Gough, M. "Laws for the Regulation of Carcinogens: Identifying and Estimating the Risks that the Laws Seek to Reduce," *Toxic Substances Journal* 4.

Graymer, L., and F. Thompson, Eds. 1982. *Reforming Social Regulation* (Newbury Park, CA: Sage Publications, Inc.).

Green, H. P. 1980. "The Role of Law in Determining Acceptability of Risk," in *Societal Risk Assessment: How Safe Is Safe Enough?* R. C. Schwing and W. A. Albers, Jr., Eds. (New York: Plenum Publishing Corporation).

Greenwood, T. 1984. *Knowledge and Discretion in Government Regulation* (New York: Praeger Publishers).

Griesmeyer, J. M., and D. Okrent. 1981. "Risk Management and Decision Rules for Light Water Reactors," *Journal of Risk Analysis* 1:121–136.

Grunbaum, W. F. 1988. "Judicial Enforcement of Hazardous Waste Liability Laws," in *Dimensions of Hazardous Waste Politics and Policy*, C. E. Davis and J. P. Lester, Eds. (Westport, CT: Greenwood Press).

Gulick, L., and L. Urwick, Eds. 1937. *Papers on the Science of Administration* (New York: Columbia University, Institute of Public Administration).

Hadden, S. G. 1984. "Introduction: Risk Policy in American Institutions," in *Risk Analysis, Institutions, and Public Policy* S. G. Hadden, Ed. (Port Washington, NY: Associated Faculty Press).

Hadden, S. 1986. *Read the Label* (Boulder, CO: Westview Press).

Hamilton, R. 1978. "The Role of Nongovernmental Standards in the Development of Mandatory Federal Standards Relating to Safety or Health," *Texas Law Review* 56:1329.

Hargrove, E. C. 1983. "The Search for Implementation Theory," in *What Role for Government?* R. J. Zeckhauser and D. Leebaert, Eds. (Durham, NC: Duke University Press).

Harrington, S., and R. E. Litan. 1988. "Causes of the Liability Insurance Crisis," *Science* 239:737–741.

Harris, D., Jr., and P. R. Portney. 1982. "Who Loses from Reform of Environmental Regulation," in *Reform of Environmental Regulation*, W. A. Magat, Ed. (Cambridge, MA: Ballinger Publishing Co.).

Harter, P. 1982. "Negotiating Regulations: A Cure for Malaise," *Georgetown Law Journal* 71(1):1–118.

Havender, W. R. 1982. "Assessing and Controlling Risks," in *Social Regulation*, E. Bardach and R. Kagan, Eds. (San Francisco: Institute for Contemporary Studies).

Hay, A. 1982. *The Chemical Scythe: Lessons of 2,4,5-T and Dioxin* (New York: Plenum Publishing Corporation).

Hazarika, S. 1986. "$3 Billion Sought on Leak in Bhopal," *New York Times* (November 23).

Heffron, F., and N. McFeeley. 1983. *The Administrative Regulatory Process* (New York: Longman, Inc.).

Heller, J. L. 1988. "Limits on Non-Economic Damages" (Washington, DC: American Association of Retired Persons).

Hensler, D., M. E. Vaiana, J. S. Kakalik, and M. A. Peterson. 1987. "Trends in Tort Litigation: The Story Behind the Statistics" (Santa Monica, CA: Rand Corporation). Report No. R-3583-ICJ.

Hinds, R. deC. 1982. "Liability under Federal Law for Hazardous Waste Injuries," *Harvard Environmental Law Review* 6(1).

Huber, P. 1983. "Exorcists vs. Gatekeepers in Risk Regulation," *Regulation* (November/December), pp. 23–32.

Huber, P. 1987. "Injury Litigation and Liability Insurance Dynamics," *Science* 238:31–36.

Huber, P. W. 1988. *Liability* (New York: Basic Books).

Huber, P., D. McCarthy, and M. Mills. 1985. "The Role of the Price-Anderson Act in the Contemporary Tort System" (Washington, DC: Science Concepts, Inc., July).

Hutt, P. B. 1978. "The Basis and Purpose of Government Regulation of Adulteration and Misbranding of Food," *Food, Drug, Cosmetic Law Journal* 33(10):505–592.

Hutt, P. B. 1984. "Use of Quantitative Risk Assessment in Regulatory Decisionmaking under Federal Health and Safety Statutes," in *Risk Management of Existing Chemicals* (Washington, DC: Chemical Manufacturers Association, June).

Ingram, H. M., and D. E. Mann. 1984. "Preserving the Clean Water Act: The Appearance of Environmental Victory," in *Environmental Policy in the 1980s: Reagan's New Agenda*, N. J. Vig and M. E. Kraft, Eds. (Washington, DC: Congressional Quarterly, Inc.).

Ingram, H., and D. Mann, Eds. 1980. *Why Policies Succeed or Fail* (Newbury Park, CA: Sage Publications, Inc.).

Inhaber, H., and S. Norman. 1982. "The Increase in Risk Interest," Letter to the Editor. *Journal of Risk Analysis* 2.

Insurance Information Institute. 1988. *Insurance Facts. 1988–9 Property/Casualty Fact Book* (New York: Insurance Information Institute).

JAPCA. 1988. "Twelve Corporations Initiate EHS Program, Vol. 38 (August).

Jasanoff, S. 1985. "The Misrule of Law at OSHA," in *The Language of Risk*, D. Nelkin, Ed. (Newbury Park, CA: Sage Publications, Inc.).

Jasanoff, S. 1986. *Risk Management and Political Culture* (New York: Russell Sage Foundation).

J. H. Wiggins Co. 1981. "Risk Analysis in the Legislative Process," Report to the National Science Foundation (Redondo Beach, CA: J. H. Wiggins Co., September).

Johnson, B. B. 1985. "Congress as Hazard Manager," in *Perilous Progress: Managing the Hazards of Technology*, R. W. Kates, C. Hohenemser, and J. X. Kasperson, Eds. (Boulder, CO: Westview Press), pp. 455–475.

Kagan, R. A. 1978. *Regulatory Justice* (New York: Russell Sage Foundation).

Kakalik, J. S., and N. M. Pace. 1986. "Costs and Compensation Paid in Tort Litigation" (Santa Monica, CA: The Institute for Civil Justice, The Rand Corporation).

Kakalik, J. S., et al. 1983. "Costs of Asbestos Litigation" (Santa Monica, CA: The Institute for Civil Justice, The Rand Corporation).

Kakalik, J. S., et al. 1984. "Variation in Asbestos Litigation Compensation and Expenses" (Santa Monica, CA: The Institute for Civil Justice, The Rand Corporation).

Kasperson, R. E., and T. Bick. 1985. "The Consumer Product Safety Commission," in *Perilous Progress: Managing the Hazards of Technology*, R. W. Kates, C. Hohenemser, and J. X. Kasperson, Eds. (Boulder, CO: Westview Press), pp. 371–394.

Katz, D., and R. L. Kahn. 1978. *The Social Psychology of Organizations*, 2nd ed. (New York: John Wiley & Sons, Inc.).

Katzman, M. T. 1985. *Chemical Catastrophes: Regulating Environmental Risk Through Pollution Liability Insurance* (Homewood, IL: Richard D. Irwin, Inc.).

Katzman, M. T. 1988. "Pollution Liability Insurance and Catastrophic Environmental Risk," *Journal of Risk and Insurance* 55:75–100.

Kaufman, H. 1976. *Are Government Organizations Immortal?* (Washington, DC: The Brookings Institution).

Kawamura, K. 1984. "Risk Management Practices in the U.S," in *Proceedings of the First U.S.–Japan Workshop on Risk Management, October 28–31, 1984*, S. Ikeda and K. Kawamura, Eds. (Nashville, TN: Vanderbilt University).

Kawamura, K., and M. Boroush. 1984. "Managing Technological Risks: A Comparative Study of U.S. and Japanese Approaches," A Scoping Paper. Nashville, TN, Vanderbilt University, January 5. Unpublished memorandum.

Kelman, S. 1980. "Occupational Safety and Health Administration," in *Politics of Regulation*, J. Q. Wilson, Ed. (New York: Basic Books).

Kelman, S. 1981. *What Price Incentives? Economists and the Environment* (Boston: Auburn House Publishing Company).

Kenski, H. C., and H. M. Ingram. 1986. "The Reagan Administration and Environmental Regulation: The Constraint of the Political Market," in *Controversies in Environmental Policy*, S. Kamieniecki, R. O'Brien, and M. Clarke, Eds. (Albany, NY: State University of New York Press).

Kenski, H. C., and M. C. Kenski, 1984. "Congress Against the President: The Struggle Over the Environment," in *Environmental Policy in the 1980s: Reagan's New Agenda*, N. J. Vig and M. E. Kraft, Eds. (Washington, DC: Congressional Quarterly, Inc.).

Kessler, D. A. 1984. "Food Safety: Revising the Statute," *Science* 223:1034–1040.

Kilmann, R. H. 1981. "Organization Design for Knowledge Utilization," *Knowledge* 3:211–231.

Klaus, A. 1987. "Practical Aspects of Environmental Impairment Liability," in *Insuring and Managing Hazardous Risks: From Seveso to Bhopal and Beyond*, P. R. Kleindorfer and H. C. Kunreuther, Eds. (New York: Springer-Verlag New York, Inc.), pp. 448–453.

Kleindorfer, P. R. 1986. "Environmental Liability Insurance: Perspectives on the U.S. Insurance Crisis" (Philadelphia: University of Pennsylvania, Wharton School, September 3).

Kleindorfer, P. R., and H. Kunreuther, Eds. 1987. *Insuring and Managing Hazardous Risks: From Seveso to Bhopal and Beyond* (Heidelberg: Springer-Verlag).

Kneese, A. V., and C. L. Schultze. 1975. *Pollution, Prices and Public Policy* (Washington, DC: The Brookings Institution).

Knight, F. H. 1921. *Risk, Uncertainty and Profit* (New York: Harper & Row, Publishers, Inc.).

Kraft, M. E. 1984. "Risk Analysis in Regulatory Decision-Making: Perspectives from the Agencies and Capitol Hill," in *Risk Analysis, Institutions, and Public Policy* S. G. Hadden, Ed. (Port Washington, NY: Associated Faculty Press).

Kraft, M. E. 1986. "The Political and Institutional Setting for Risk Analysis," in *Risk Evaluation and Management*, V. T. Covello, J. Menkes, and J. Mumpower, Eds. (New York: Plenum Publishing Corporation).

Kraft, M. E., and R. Kraut. 1988. "Citizen Participation and Hazardous Waste Policy Implementation," in *Dimensions of Hazardous Waste Politics and Policy*, C. E. Davis and J. P. Lester, Eds. (Westport, CT: Greenwood Press), pp. 62–80.

Krupnick, A., W. Oates, and E. Van De Verg. 1983. "On Marketable Air-Pollution Permits: The Case for a System of Pollution Offsets," *Journal of Environmental Economics and Management* 10:233–247.

Kuhn, T. 1970. *The Structure of Scientific Revolutions*, 2nd ed. (Chicago: University of Chicago Press).

Kunreuther, H., et al. 1978. *Disaster Insurance Protection: Public Policy Lessons* (New York: John Wiley & Sons, Inc.).

Kunreuther, H. 1986. "Executive Summary. Workshop on The Role of Insurance and Compensation in Environmental Pollution Problems" (Philadelphia: University of Pennsylvania, Wharton School, May 15).

Kunreuther, H. 1987. "Problems and Issues of Environmental Liability Insurance," *The Geneva Papers on Risk and Insurance* 12:180–197.

Kunreuther, H., and M. Willingham. 1986. "Environmental Liability Insurance," unpublished manuscript, June.

Kurtz, H. 1983. "Since Reagan Took Office, EPA Enforcement Actions Have Fallen," *Washington Post* (March 1).

Labaton, S. 1988. "Chapter 11 Shield Protects Insurers," *New York Times* (June 27).

LaGrega, M. D., et al. 1987. "Legal Responsibility of Industry for Superfund Sites," in Official Proceedings of the Conference on Solid Waste Management and Materials Policy, Volume IX, February 11–14 (Albany, NY: The New York State Legislative Commission on Solid Waste Management).

Landes, W., and R. Posner. 1987. *The Economic Structure of Tort Law* (Cambridge, MA: Harvard University Press).

Laporte, G. J., T. P. Gies, and R. D. Baum. 1975. "Consumer Product Safety Commission: An Agency Manual," in *George Washington Law Review*, pp. 1077–1172.

Latin, H. 1988. "Good Science, Bad Regulation and Toxic Risk Assessment," *Yale Journal on Regulation* 5:89–148.

Lave, L. B. 1981. *The Strategy of Social Regulation: Decision Frameworks for Policy* (Washington, DC: The Brookings Institution).

Lave, L. B. 1986. "Approaches to Risk Management: A Critique," in *Risk Evaluation and Management*, V. T. Covello, J. Menkes, and J. Mumpower, Eds. (New York: Plenum Publishing Corporation).

Lave, L. B., and G. S. Omenn. 1982. *Clearing the Air: Reforming the Clean Air Act* (Washington, DC: The Brookings Institution).

Lawrence, E. L., Jr., Ed. 1978. *Knowledge and Policy: The Uncertain Connection* (Washington, DC: National Academy of Sciences).

Lawrence, P. R., and J. W. Lorsch. 1969. *Organization and Environment: Managing Differentiation and Integration* (Homewood, IL: Richard D. Irwin, Inc).

Leventhal, H. 1974. "Environmental Decision-Making and the Role of the Courts," *University of Pennsylvania Law Review* 122.

Leventhal, H. 1976. "Appellate Procedures: Design, Patchwork, and Managed Flexibility," *UCLA Law Review* 23.

Levin, M. H. 1989. "Incentives, Resurgent: Project 88's Campaign," *Journal of the Air and Waste Management Association* 39(2).

Lewin, T. 1986. "Judge Skeptical on Bhopal Proposal," *New York Times* (March 29).

Lieberman, J. 1981. *The Litigious Society* (New York: Basic Books).

Lindblom, C. E. 1968. *The Policymaking Process* (Englewood Cliffs, NJ: Prentice-Hall, Inc.).

Lindblom, C. E., and D. K. Cohen. 1979. *Usable Knowledge* (New Haven, CT: Yale University Press).

Lindell, M. K., and T. C. Earle. 1983. "How Close is Close Enough: Public Perceptions of the Risks of Industrial Facilities," *Risk Analysis* 3:245–253.

Litan, R., and W. Nordhaus. 1983. *Reforming Federal Regulation* (New Haven, CT: Yale University Press).

Lowi, T. J. 1969. *The End of Liberalism* (New York: W. W. Norton & Co., Inc.).

Lowrance, W. 1976. *Of Acceptable Risk* (Los Altos, CA: William Kaufman).

Ludd, S. O. 1986. "The Essentiality of Judicial Review: Toward a More Balanced Understanding of Administrative Discretion in American Government," in *Administrative Discretion and Public Policy Implementation*, D. H. Schumavon and H. K. Hibbeln, Eds. (New York: Praeger Publishers).

Lynn, L. E., Jr. 1987. *Managing Public Policy* (Boston: Little, Brown & Company).

Lyon, R. M. 1982. "Actions and Alternative Procedures for Allocating Pollution Rights," *Land Economics* 58.

Maass, A. 1983. *Congress and the Common Good* (New York: Basic Books).

MacCarthy, M. 1987. "Closure in Occupational Safety and Health: The Benzene and Cotton Dust Decisions," in *Scientific Controversies*, H. T. Engelhardt, Jr. and A. L. Caplan, Eds. (Cambridge, U.K. and New York: Cambridge University Press), pp. 505–527.

MacIntyre, A. A. 1985. "A Court Quietly Rewrote the Federal Pesticide Statute: How Prevalent is Judicial Statutory Revision?" *Law and Policy* 7:249–279.

MacIntyre, A. A. 1986. "The Multiple Sources of Statutory Ambiguity: Tracing the Legislative Origins to Administrative Discretion" in *Administrative Discretion and Public Policy Implementation*, D. H. Shumavon and H. K. Hibbeln, Eds. (New York: Praeger Publishers), pp. 67–88.

Maeroff, G. I. 1986. "Trial Set in Suit Over 7 Cancer Cases," *New York Times* (March 9).

Magat, W. A. 1982. *Reform of Environmental Regulation* (Cambridge, MA: Ballinger).

Mann, T. E., and N. J. Ornstein, Eds. 1981. *The New Congress* (Washington, DC: American Enterprise Institute).

March, J. G., and H. P. Olsen. 1976. *Ambiguity and Choice in Organizations* (Bergen-Oslo-Tromso, Norway: Universitetsforlaget).

March, J. G., and H. A. Simon. 1958. *Organizations* (New York: John Wiley & Sons, Inc.).

Marcus, A. 1980. "Environmental Protection Agency," in *Politics of Regulation*, J. Q. Wilson, Ed. (New York: Basic Books).

Marcus, G. 1983. "A Review of Risk Assessment Methodologies," report prepared by the Congressional Research Service, Library of Congress (Washington, DC: U.S. Government Printing Office).

Marrett, C. B. 1987. "Closure and Controversy: Three Mile Island," in *Scientific Controversies*, H. T. Engelhardt, Jr. and A. L. Caplan, Eds. (Cambridge, U.K. and New York: Cambridge University Press), pp. 551–566.

Marsh and McLennan Co., Inc. 1980. *Risk in a Complex Society* (New York: Marsh & McLennan).

Mayhew, D. 1974. *Congress: The Electoral Connection* (New Haven, CT: Yale University Press).

Mayo, E. 1945. *The Social Problems of an Industrial Civilization* (Boston, MA: Harvard University, Graduate School of Business).

Mazmanian, D., and D. Morrell. 1988. "The Elusive Pursuit of Toxics Management," *The Public Interest* 90:81–98.

Mazmanian, D. A., and J. Nienaber. 1979. *Can Organizations Change?* (Washington, DC: The Brookings Institution).

Mazmanian, D. A., and P. A. Sabatier. 1983. *Implementation and Public Policy* (Glenview, IL: Scott, Foresman & Company).

McCaffrey, D. P. 1982. *OSHA and the Politics of Health Regulation* (New York: Plenum Publishing Corporation).

McFadden, R. D. 1988. "Court Upholds Malpractice Rate Cap," *New York Times* (December 16), p. B3.

McGarity, T. 1979a. "The Death and Transfiguration of Mirex: An Examination of the Integrity of Settlements under FIFRA," *Harvard Environmental Law Review* 3:112–135.

McGarity, T. 1979b. "Substantive and Procedural Discretion in Administrative Resolution of Science Policy Questions: Regulating Carcinogens in EPA and OSHA," *Georgetown Law Review* 67:729–810.

McGarity, T. 1984. "Judicial Review of Scientific Rulemaking," *Science Technology and Human Values* 9.

McGartland, A. 1988. "A Comparison of Two Marketable Discharge Permits Systems," *Journal of Environmental Economics and Management* 15:35–44.

Meier, K. J. 1985. *Regulation* (New York: St. Martin's Press).

Melnick, R. S. 1983. *Regulation and the Courts: The Case of the Clean Air Act* (Washington, DC: The Brookings Institution).

Melnick, R. S. 1985. "The Politics of Partnership," *Public Administration Review* 45:653–660.

Mendeloff, J. 1979. *Regulating Safety* (Cambridge, MA: The M.I.T. Press).

Mendeloff, J. 1986. "Regulatory Reform and OSHA Policy," *Journal of Policy Analysis and Management* 5(3):440–468.

Mendeloff, J. 1988. *The Dilemma of Toxic Substance Regulation* (Cambridge, MA: The M.I.T. Press).

Menza, W. P. 1983. Memorandum to the Consumer Product Safety Commission (August 5).

Merrill, R. A. 1977a. "Discussion of Priorities," U.S. FDA, Office of Planning and Evaluation (February).

Merrill, R. A. 1977b. "Risk-Benefit Decisionmaking by the Food and Drug Administration," *George Washington Law Review* 45:994–1012.

Merrill, R. A. 1988. "FDA's Implementation of the Delaney Clause: Repudiation of Congressional Choice or Reasoned Adaptation to Scientific Progress?" *Yale Journal on Regulation* 5:1–88.

Mintzberg, H. 1979. *The Structuring of Organizations* (Englewood Cliffs, NJ: Prentice-Hall, Inc.).

Mintzberg, H., D. Raisinghani, and A. Theoret. 1976. "The Structure of Unstructured Decision Processes," *Administrative Science Quarterly* 21:246–275.

Mitchell, R. C. 1979. "National Environmental Lobbies and the Apparent Illogic of Collective Action," in *Collective Decision-Making*, C. Russell, Ed. (Baltimore: Johns Hopkins University Press).

Mitchell, R. C. 1984. "Public Opinion and Environmental Politics in the 1970s and 1980s," in *Environmental Policy in the 1980s: Reagan's New Agenda*, N. J. Vig and M. E. Kraft, Eds. (Washington, DC: Congressional Quarterly, Inc.).

Mitnick, B. M. 1980. *The Political Economy of Regulation* (New York: Columbia University Press).

Mohr, C. 1989. "Environmental Groups Gain in Wake of Spill," *New York Times* (June 11).

Molton, L. S. and P. F. Ricci. 1985. "The Process of Risk Assessment: Administrative Law and its Effects on Science Policy," in *Principles of Health Risk Assessment*, P. Ricci, Ed. (Englewood Cliffs, NJ: Prentice-Hall, Inc.).

Morais, R. 1986. "Insurance," *Forbes* (January 13).

Morgan, G. 1986. *Images of Organization* (Newbury Park, CA: Sage Publications, Inc.).

Morgan, D. F., and J. A. Rohr. 1986. "Traditional Responses to American Administrative Discretion," in *Administrative Discretion and Public Policy Implementation*, D. H. Shumavon and H. K. Hibbeln, Eds. (New York: Praeger Publishers), pp. 211–232.

Morrall, J. F., III. 1986. "A Review of the Record," *Regulation* (Nov./Dec.).

Morris, F. A., and E. Duvernoy. 1984. "The Statutory Basis of Risk Assessment," in *Low Probability High Consequence Risk Analysis*, R. A. Waller and V. T. Covello, Eds. (New York: Plenum Publishing Corporation), pp. 455–480.

Mosher, F. C. 1979. *The GAO: The Quest for Accountability in American Government* (Boulder, CO: Westview Press).

Mosher, L. 1983. "Ruckelshaus's First Mark on EPA—Another $165.5 Million for Its Budget," *National Journal* (June 25).

National Research Council. 1977. *Decision-Making in the EPA* (Washington, DC: National Academy of Sciences).

National Research Council. 1980. *Regulating Pesticides* (Washington, DC: NRC).

National Research Council. 1983. *Risk Assessment in the Federal Government: Managing the Process* (Washington, DC: National Academy Press).

National Research Council. 1984. *55: A Decade of Experience* (Washington, DC: NRC).

Neely, R. 1981. *How Courts Govern America* (New Haven, CT: Yale University Press).

Nelkin, D. 1982. "The Role of the Expert at Three Mile Island," in *Accident at Three Mile Island*, D. L. Sills, C. P. Wolf, and V. B. Shelanski, Eds. (Boulder, CO: Westview Press).

Nelkin, D. 1989. "Communicating Technological Risk: The Social Construction of Risk Perception," *Annual Review of Public Health* 10:95–113.

Nelkin, D., and M. Pollak. 1980. "Consensus and Conflict Resolution: The Politics of Assessing Risk," in *Technological Risk*, M. Dierkes, S. Edwards, and R. Coppock, Eds. (Cambridge, MA: Oelgeschlager, Gunn & Hain).

New York Times. December 17, 1985. "U.S. Announces $900,000 Fine in Accident at Ohio Nuclear Plant."

New York Times. March 19, 1986. "Union Carbide's Fine is Cut to $4,400 for Chemical Leak."

New York Times. January 29, 1987. "Guilty Plea from W.R. Grace in a Case of Water Pollution."

New York Times. October 8, 1987. "$520,000 Awarded in an Asbestos-Related Illness."

New York Times. October 25, 1987. "Dioxin Case Juror Acted to Go Home."

New York Times. June 9, 1988. "Monsanto to Pay $1.5 Million in Poisoning Case."

New York Times. December 20, 1988. "Shell Loses Suit on Cleanup Costs."

Nichols, A. L., and R. Zeckhauser. 1977. "Government Comes to the Workplace: An Assessment of OSHA," *The Public Interest*, pp. 39–69.

Noble, J. H., J. S. Banta, and J. S. Rosenberg, Eds. 1977. *Groping Through the Maze* (Washington, DC: The Conservation Foundation).

Noll, R. G. 1971. *Reforming Regulation* (Washington, DC: The Brookings Institution).

O'Brien, D. M. 1982. "The Courts, Technology Assessment, and Science-Policy Disputes," in *The Politics of Technology Assessment*, D. M. O'Brien and D. A. Marchand, Eds. (Lexington, MA: Lexington Books).

O'Brien, D. M. 1985. "Managing the Business of the Supreme Court," *Public Administration Review* 45:667–678.

O'Brien, D. M. 1986a. "Administrative Discretion, Judicial Review, and Regulatory Politics," in *Administrative Discretion and in Public Policy Implementation*, D. H. Shumavon and H. K. Hibbeln, Eds. (New York: Praeger Publishers).

O'Brien, D. M. 1986b. *Storm Center: The Supreme Court in American Politics* (New York: W.W. Norton & Company).

O'Brien, D. M. 1988. *What Process is Due? Courts and Science-Policy Disputes* (New York: Russell Sage Foundation).

O'Brien, R. M., M. Clarke, and S. Kamieniecki. 1984. "Open and Closed Systems of Decision Making: The Case of Toxic Waste Management," *Public Administration Review* (July/August), pp. 334–340.

O'Connell, J. 1979. *The Lawsuit Lottery* (New York: The Free Press).

Office of the Federal Register. 1988. *United States Government Manual, 1988/89* (Washington, DC: U.S. Government Printing Office, June).

Office of Science and Technology Policy. 1984. "Chemical Carcinogens; Review of the Science and Its Associated Principles," 49 *Federal Register* (May 22), 21594–21661.

Ogul, M. S. 1981. "Congressional Oversight: Structures and Incentives," in *Congress*

Reconsidered, L. C. Dodd and B. I. Oppenheimer, Eds. (Washington, DC: Congressional Quarterly, Inc.).

O'Hare, M., L. Bacow, and D. Sanderson. 1983. *Facility Siting and Public Opposition* (New York: Van Nostrand Reinhold Company).

Okrent, D. 1986. "Alternative Risk Management Policies for State and Local Governments," in *Risk Evaluation and Management*, V. T. Covello, J. Menkes, and J. Mumpower, Eds. (New York: Plenum Publishing Corporation), pp. 359–380.

Oleinick, A., L. D. Disney, and K. S. East. 1986. "Institutional Mechanisms for Converting Sporadic Agency Decisions into Systematic Risk Management Strategies," in *Risk Evaluation and Management*, V. T. Covello, J. Menkes, and J. Mumpower, Eds. (New York: Plenum Publishing Corporation), pp. 381–412.

Omenn, G. S. 1986. "Values in the Debate over Workplace Safety and Health: The Rancorous Rhetoric about Regulation," in *Scientific Controversies*, H. T. Engelhardt, Jr. and A. L. Caplan, Eds. (Cambridge, U.K. and New York: Cambridge University Press).

O'Reilly, J. T., K. A. Gaynor, D. W. Carroll, and P. F. Cronin, Jr. 1987. *Federal Regulation of the Chemical Industry* (Colorado Springs: Shepard's–McGraw-Hill).

Page, T. 1978. "A Generic View of Toxic Chemicals and Similar Risks," *Ecology Law Quarterly* 7:207–243.

Parker, B. J. 1986. "Bearing the Risk. The Insurance Crisis and Environmental Protection," *Environment* 28:14–18.

Parsons, T. 1960. *Structure and Process in Modern Societies* (Glencoe, IL: Free Press).

Pechman, J. A. 1983. *Federal Tax Policy*, 4th ed. (Washington, DC: The Brookings Institution).

Pechman, J. A. 1987. *Federal Tax Policy*, 5th ed. (Washington, DC: The Brookings Institution).

Pederson, W. F. 1975. "Formal Records and Informal Rulemaking," *Yale Law Journal* 84.

Peltzman, S. 1976. "Toward a More General Theory of Regulation," *Journal of Law and Economics* 19(2):211–240.

Perrow, C. 1984. *Normal Accidents* (New York: Basic Books).

Peterson, J. M. 1977. "Estimating an Effluent Charge: The Reserve Mining Case," *Land Economics* (August), pp. 328–341.

Pfeffer, J., and G. R. Salancik. 1978. *The External Control of Organizations* (New York: Harper & Row, Publishers, Inc.).

Pressman, J. L., and A. Wildavsky. 1984. *Implementation*, 3rd ed. (Berkeley: University of California Press).

Prosser, W., W. Keeton, D. Dobbs, R. Keeton, and D. Owen. 1984. *Prosser and Keeton on Torts* (St. Paul, MN: West Publishing).

Pugliaresi, L. P., and D. T. Berliner. 1989. "Policy Analysis at the Department of State: The Policy Planning Staff," *Journal of Policy Analysis and Management* 8(6):379–394.

Quirk, P. J. 1980. "Food and Drug Administration," in *Politics of Regulation*, J. Q. Wilson, Ed. (New York: Basic Books).

Reagan, M. D., and J. G. Sanzone. 1981. *The New Federalism* (New York: Oxford University Press).

Reinhold, R. 1986. "Jurors Assess Monsanto $108 Million over Death," *New York Times* (December 13).

Ricci, P. F., and L. A. Cox, Jr. 1987. "De Minimis Considerations in Health Risk Assessment," *Journal of Hazardous Materials* 15:77–96.

Ricci, P. F., and L. S. Molton. 1981. "Risk Benefit in Environmental Law," *Science* 214:1096–1100.

Ricci, P. F., and L. S. Molton. 1984. "Risk Analysis in the United States Law," in *Low Probability High Consequence Risk Analysis*, R. A. Waller and V. T. Covello, Eds. (New York: Plenum Publishing Corporation), pp. 373–392.

Ricci, P. F., and L. S. Molton. 1985. "Regulating Cancer Risks," *Environmental Science and Technology* 19(6):473–479.

Ripley, R. B., and G. A. Franklin. 1987. *Congress, the Bureaucracy and Public Policy* (Homewood, IL: Dorsey Press).

Rodenhausen, G. A. 1989. "Linking Landfills to Health Effects," *Journal of the Air Pollution Control Association* 39(1).

Rodgers, W. H., Jr. 1979. "A Hard Look at Vermont Yankee: Environmental Law Under Close Scrutiny," *Georgetown Law Journal* 67:699–727.

Rodgers, W. H., Jr. 1981. "Judicial Review of Risk Assessments: The Role of Decision Theory in Unscrambling the Benzene Decision," *Environmental Law* 11:301–320.

Rose, J. G. 1975. *The Transfer of Development Rights: A New Technique of Land Use Regulation* (New Brunswick, NJ: Center for Urban Policy Research, Rutgers, The State University of New Jersey).

Rowe, W. 1976. *An Anatomy of Risk* (New York: John Wiley & Sons, Inc.).

Ruckelshaus, W. D. 1983. "Science, Risk, and Public Policy," *Science* 221:1026–1028.

Rushefsky, M. 1982. "Technical Disputes: Why Experts Disagree," *Policy Studies Review* 1:676–685.

Rushefsky, M. 1984. "Institutional Mechanisms for Resolving Risk Controversies," in *Risk Analysis, Institutions, and Public Policy*, S. G. Hadden, Ed. (Port Washington, NY: Associated Faculty Press).

Sabatier, P. 1978. "The Acquisition and Utilization of Technical Information by Administrative Agencies," *Administrative Science Quarterly* 23:396–417.

Sabatier, P., and D. Mazmanian. 1979. "The Conditions of Effective Implementation: A Guide to Accomplishing Policy Objectives," *Policy Analysis* (Fall), pp. 481–504.

Sage, A. P., and E. B. White. 1980. "Methodologies for Risk and Hazard Assessment: A Survey and Status Report," *IEEE Transactions on Systems, Man and Cybernetics*, SMC-10(8):425–445.

Schick, A. 1980. *Congress and Money: Budgeting, Spending and Taxing* (Washington, DC: The Urban Institute).

Schnaibert, A. 1982. "Who Should be Responsible for Public Safety?" in *Accident at TMI*, D. L. Sills, C. P. Wolf, and V. B. Shelanski, Eds. (Boulder, CO: Westview Press).

Schneiderman, M. A. 1980. "The Uncertain Risks We Run: Hazardous Materials," in *Societal Risk Assessment: How Safe is Safe Enough?* R. C. Schwing and W. A. Albers, Jr., Eds. (New York: Plenum Publishing Corporation).

Schuck, P. 1983. *Suing Government* (New Haven, CT: Yale University Press).

Schuck, P. 1986. *Agent Orange on Trial: Mass Toxic Disasters in the Courts* (Cambridge, MA: The Belknap Press of Harvard University Press).

Scott, W. R. 1981. *Organizations* (Englewood Cliffs, NJ: Prentice-Hall, Inc.).

Segerson, K. 1988. "Uncertainty and Incentives for Nonpoint Pollution Control," *Journal of Environmental Economics and Management* 15:87–98.

Selznick, P. 1957. *Leadership in Administration* (Evanston, IL and White Plains, NY: Row, Peterson & Co).

Shabecoff, P. 1987. "Union Carbide Agrees to Pay $408,500 Fine for Safety Violations," *New York Times* (July 25).

Shanley, M. G. and M. A. Peterson. 1987. "Posttrial Adjustments to Jury Awards" (Santa Monica, CA: Rand Corporation) Report No. R-3511-ICJ.

Shapiro, M. 1979. "Judicial Activism," in *The Third Century*, S. M. Lipset, Ed. (Stanford, CA: Hoover Institution Press).

Shapiro, M. 1984. "Administrative Discretion: The Next Stage," *Yale Law Journal*.

Shughart, W. F., II, and R. D. Tollison. 1985. "The Cyclical Character of Regulatory Activity," *Public Choice* 45:303–311.

Shumavon, D. H., and H. K. Hibbeln. 1986. "Administrative Discretion: Problems and Prospects," in *Administrative Discretion and Public Policy Implementation*, D. H. Shumavon and H. K. Hibbeln, Eds. (New York: Praeger Publishers).

Silver, L. D. 1986. "The Common Law of Environmental Risk and Some Recent Applications," *Harvard Environmental Law Review* 10:61–98.

Simon, H. A. 1955. "A Behavioral Model of Rational Choice," *Quarterly Journal of Economics* 69:99–118.

Simon, H. A. 1969. *The Science of the Artificial* (Cambridge, MA: The M.I.T. Press).

Simon, H. A. 1976. *Administrative Behavior*, 3rd ed. (New York: The Free Press).

Simon, H. A. 1985. "Human Nature in Politics: The Dialogue of Psychology with Political Science," *The American Political Science Review* 79:293–404.

Skaff, R. B. 1979. "The Emergency Powers in Environmental Protection Statutes," *Harvard Environmental Law Review* 3.

Slovic, P., B. Fischhoff, and S. Lichtenstein. 1980. "Facts and Fears: Understanding Perceived Risk," in *Societal Risk Assessment*, R. C. Schwing and W. A. Albers, Jr., Eds. (New York: Plenum Publishing Corporation).

Smart, C., and I. Vertinsky. 1977. "Designs for Crisis Decision Units," *Administrative Science Quarterly* 22:640–657.

Smets, H. 1987. "Compensation for Exceptional Environmental Damage Caused by Industrial Activities," in *Insuring and Managing Hazardous Risks: From Seveso to Bhopal and Beyond*, P. R. Kleindorfer and H. Kunreuther, Eds. (Heidelberg: Springer-Verlag), pp. 79–137.

Smith, J. M. 1983. "Role of Risk Assessment at Chemical Manufacturers Association," in *Risk Management in the Chemical Industry* (Washington, DC: Chemical Manufacturers Association).

Soble, S. 1977. "A Proposal for the Administrative Compensation of Victims of Toxic Substance Pollution," *Harvard Journal of Legislation* 14.

Soular, L. W. 1986. "A Study of Large Product Liability Claims Closed in 1985" (Schaumburg, IL: Alliance of American Insurers).

Spuehler, J. 1988. "Insurability Issues Associated with Managing Existing Hazardous Waste Facilities," paper presented at the Geneva Association and Risk and Decision Processes Center conference, Philadelphia, University of Pennsylvania, May 18–19.

Starr, C. 1969. "Social Benefit vs. Technological Risk," *Science* 165:1232–1238.

Starr, C. 1985. "Risk Management, Assessment, and Acceptability," *Journal of Risk Analysis* 5(2).

Steuber, D. W. 1987. "Insurance Coverage: How to Effectively Receive the Maximum Benefits from Insurance Converage," Official Proceedings of the Conference on Solid Waste Management and Materials Policy, February 11–14 (Albany: The New York State Legislative Commission on Solid Waste Management).

Stever, D. W., Jr. 1980. *Seabrook and the Nuclear Regulatory Commission* (Hanover, NH: University Press of New England).

Stever, D. W., Jr. 1988. *The Law of Chemical Regulation and Hazardous Wastes* (New York: Clark Boardman).

Stigler, G. J. 1971. "The Theory of Economic Regulation," *Bell Journal of Economics and Management Science* 2(1):3–21.

Stigler, G. J. 1974. "Free Riders and Collective Action: An Appendix to Theories of Economic Regulation," *Bell Journal of Economics and Management Science* 5(2):359–365.

Stone, A. 1982. *Regulation and its Alternatives* (Washington, DC: Congressional Quarterly, Inc.).

Stoner, J. A. F. 1981. *Management* (Englewood Cliffs, NJ: Prentice-Hall, Inc.).

Strickler, S. E., and M. L. Williams. 1986. "Congressional Oversight: The Legislative Veto," in *Administrative Discretion and Public Policy Implementation*, D. H. Shumavon and H. K. Hibbeln, Eds. (New York: Praeger Publishers).

Stewart, R. B. 1975. "The Reformation of American Administrative Law," *Harvard Law Review* 88.

Sundquist, J. L. 1978. "Research Brokerage: The Weak Link," in *Knowledge and Policy: The Uncertain Connection*, L.E. Lynn, Jr., Ed. (Washington, DC: National Academy of Sciences).

Sundquist, J. 1981. *The Decline and Resurgence of Congress* (Washington, DC: The Brookings Institution).

Susskind, L. E., L. S. Bacow, and M. Wheeler, Eds. 1983. *Resolving Environmental Regulatory Disputes* (Cambridge, MA: Schenkman Publishing Co., Inc).

Talbot, A. R. 1983. *Settling Things: Six Case Studies in Environmental Mediation* (Washington, DC: Conservation Foundation).

Tarr, J. A. 1984. "The Evolution of the Urban Infrastructure in the Nineteenth and Twentieth Centuries," in *Perspectives on Urban Infrastructure*, R. Hanson, Ed. (Washington, DC: National Academy Press), pp. 4–66.

Taylor, F. W. 1911. *Principles of Scientific Management* (New York: Harper & Row, Publishers, Inc.).

Taylor, S., Jr. 1988. "Justices, 5–4, Give U.S. Contractors Liability Immunity," *New York Times* (June 28).

Thompson, F. J. 1982. "Deregulation by the Bureaucracy: OSHA and the Augean Quest for Error Correction," *Public Administration Review* (May/June), pp. 202–212.

Thompson, J. D. 1967. *Organizations in Action* (New York: McGraw-Hill Book Company).

Thompson, J. D., and R. W. Hawkes. 1962. "Disaster, Community Organization, and Administrative Process," in *Man and Society in Disaster*, G. W. Baker and D. W. Chapman, Eds. (New York: Basic Books).

Tietenberg, T. H. 1980. "Transferable Discharge Permits and the Control of Stationary Source Air Pollution: A Survey and Synthesis," *Land Economics* 56.

Toffler, A. 1970. *Future Shock* (New York: Bantam Books).

Trauberman, J. 1981. "Compensating Victims of Toxic Substances: Existing Federal Mechanisms," *Harvard Environmental Law Review* 5.

Trauberman, J. 1983. "Statutory Reform of 'Toxic Torts': Relieving Legal, Scientific and Economic Burdens on the Chemical Victim," *Harvard Environmental Law Review*.

Travis, C. C., and H. A. Hattemer-Frey. 1988. "Determining an Acceptable Level of Risk," *Environmental Science and Technology* 22:873–876.

Travis, C. C., S. A. Richter, E. A. C. Crouch, R. Wilson, and E. D. Klema. 1987. "Cancer Risk Management," *Environmental Science & Technology* 21(5):415–420.

Trolin, B. 1986. "State Constitutional Provisions Prohibiting Enactment of Legislation Limiting the Amount of Damages to be Recovered for Causing the Death or Injury of Any Person" (Denver, CO: National Conference of State Legislatures, September 15).

Turner, B. A. 1976. "The Organizational and Interorganizational Development of Disasters," *Administrative Science Quarterly* 21:378–397.

U.S. Bureau of the Census. 1975. *Historical Statistics of the United States* (Washington, DC: U.S. Government Printing Office).

U.S. Code Annotated. 1987. (St. Paul, MN: West Publishing Co.).

U.S. Congress. 1988. "Harnessing Market Forces to Protect Our Environment: Initiatives for the New President," Project 88 Report (Washington, DC: U.S. Senate, December).

U.S. Congress, Energy and Commerce Committee, Subcommittee on Oversight and Investigations. 1988. "The Threat from Substandard Fasteners: Is America Losing Its Grip?" (Washington, DC: U.S. Government Printing Office).

U.S. Congress, Congressional Budget Office. 1983. "Public Works Infrastructure: Policy Considerations for the 1980s" (Washington, DC: Congressional Budget Office, April).

U.S. Congress, Congressional Research Service. 1983. "A Review of Risk Assessment Methodologies" (Washington, DC: U.S. Government Printing Office).

U.S. Congress, House of Representatives, Commission on Administrative Review. 1977. "Administrative Organization and Legislative Management," H. Doc. 95–232, 95th Congress, 1st session.

U.S. Congress, House of Representatives. 1986a. "Safe Drinking Water Act Amendments of 1985. Report Together with Additional Views" (Washington, DC: House of Representatives).

U.S. Congress, House of Representatives. 1986b. "Superfund Amendments and Reauthorization Act of 1986, Conference Report" (Washington, DC: House of Representatives).

U.S. Congress, Office of Technology Assessment. 1981. *Assessment of Technologies for Determining Cancer Risks from the Environment* (Washington, DC: OTA, June).

U.S. Congress, Office of Technology Assessment. 1984. *Protecting the Nation's Groundwater From Contamination* Volume 1 (Washington, DC: OTA, October).

U.S. Congress, Office of Technology Assessment. 1986. *Serious Reduction of Hazardous Waste* (Washington, DC: OTA, September).

U.S. Congress, Office of Technology Assessment. 1988. *Are We Cleaning Up? 10 Superfund Case Studies* (Washington, DC: OTA, June).

U.S. Congress, Superfund Study Group. 1982. "Injuries and Damages from Hazardous Wastes-Analysis and Improvement of Legal Remedies" (Washington, DC: U.S. Congress, July 1).

U.S. Consumer Product Safety Commission. 1978. "Interim Policy and Procedure for Classifying, Evaluating and Regulating Carcinogens in Consumer Products," *Federal Register* 43(114):25658–25665 (June 13).

U.S. Consumer Product Safety Commission. 1983. Memorandum from W. P. Menza to the Commission (August 5).

U.S. Consumer Product Safety Commission. 1986. Annual Report. Fiscal Year 1986 (Springfield, VA: National Technical Information Service).

U.S. Department of Commerce, Interagency Task Force on Product Liability. 1977. *Final Report* (Washington, DC: U.S. Government Printing Office).

U.S. Department of Health, Education and Welfare, Food and Drug Administration. 1979. "Chemical Compounds in Food Producing Animals: Criteria and Procedures for Evaluating Assays for Carcinogenic Residues," *Federal Register* 44(55):17070–17114 (March 20).

U.S. Department of Labor, OSHA. 1980. "Identification, Classification, and Regulation of Potential Occupational Carcinogens," *Federal Register* 45(15):5002–5296 (January 22).

U.S. EPA. May 1976. "Interim Procedures and Guidelines for Health Risk and Economic Impact Assessments of Suspected Carcinogens," *Federal Register* 41(102):21402 (May 25).

U.S. EPA. July 1976. *Quality Criteria for Water* (Washington, DC: U.S. Government Printing Office).

U.S. EPA. 1982a. "Emissions Trading Policy Statement," *Federal Register* 47:15076–15086 (April 7).

U.S. EPA. 1982b. *Federal Register* 47:9350 (March 4).

U.S. EPA. 1983. *Federal Register* 48:45502 (October 6).

U.S. EPA. February 1984. *Federal Register* 49:5857–5858 (February 15).

U.S. EPA. September 1984. "Criteria Document for Lead" (Washington, DC: U.S. EPA, September 6).

U.S. EPA. November 1984. *Federal Register* 49:46294 (November 23).

U.S. EPA. December 1984a. "CERCLA 301 Study Executive Summary," Report to Congress (Washington, DC: U.S. EPA).

U.S. EPA. December 1984b. "Impact of CERCLA Taxes on the U.S. Balance of Trade," Report to Congress (Washington, DC: U.S. EPA).

U.S. EPA. December 1984c. "Risk Assessment and Management: Framework for Decision Making" (Washington, DC: U.S. EPA).

U.S. EPA. 1984. *Federal Register* 49:23498 (June 6).

U.S. EPA. 1984. *Federal Register* 49:24330 (June 13).

U.S. EPA. 1986a. "Final Emissions Trading Policy," *Federal Register* 51:43814 (December 4).

U.S. EPA. 1986b. *Federal Register* 51(185):33992–34052 (September 24).

U.S. EPA. August 1988. "Solid Waste Disposal Facility Criteria; Proposed Rule," *Federal Register* 53(168):33314–333403 (August 30).

U.S. EPA. October 1988. "Guidance for Conducting Remedial Investigations and Feasibility Studies under CERCLA—Interim Final" (Washington, DC: U.S. EPA, Office of Emergency and Remedial Response, October).

U.S. EPA. February 1989. "Standards for the Disposal of Sewage Sludge; Proposed Rule," *Federal Register* 54(23):5746–5902 (February 6).

U.S. EPA. May 1989. "National Primary and Secondary Drinking Water Regulations; Proposed Rule," *Federal Register* 54(97):22062–22160 (May 22).

U.S. EPA, Office of Health and Environmental Assessment. 1989. "Chemical Assessments and Related Activities (Washington, DC: U.S. EPA, June 1).

U.S. EPA, Office of the General Counsel. 1984. "Tolerance Issues Relating to Ethylene Dibromide," Memorandum (Washington, DC: U.S. EPA, January 3).

U.S. EPA, Office of Water Regulations and Standards. 1984. "National Water Quality Inventory. 1982 Report to Congress" (Washington, DC: U.S. EPA, February).

U.S. General Accounting Office. 1980. "Delays and Unresolved Issues Plague New Pesticide Protection Programs" (Washington, DC: U.S. GAO, February 15).

U.S. General Accounting Office. 1982. "Attrition of Scientists and Engineers at Three Regulatory Agencies" (Washington, DC: U.S. Government Printing Office, December 27).

U.S. General Accounting Office. 1983. "Delays in EPA's Regulation of Hazardous Air Pollutants" (Washington, DC: U.S. GAO).

U.S. General Accounting Office. 1984. "Attrition of Scientists and Engineers at Seven Agencies" (Washington, DC: U.S. Government Printing Office, May 29).

U.S. General Accounting Office. January 1985. "Status of the Department of Energy's Implementation of the Nuclear Waste Policy Act of 1982 as of December 31, 1984" (Washington, DC: U.S. GAO, January 31).

U.S. General Accounting Office. February 1985. "Illegal Disposal of Hazardous Waste: Difficult to Detect or Deter" (Washington, DC: U.S. GAO, February 22).

U.S. General Accounting Office. December 1985. "EPA's Strategy to Control Emissions of Benzene and Gasoline Vapor" (Washington, DC: U.S. GAO, December 18).

U.S. General Accounting Office. April 1986. "Pesticides. EPA's Formidable Task to Assess and Regulate Their Risks" (Washington, DC: U.S. Government Printing Office).

U.S. General Accounting Office. November 1986. "Liability Insurance" (Washington, DC: U.S. Government Printing Office).

U.S. General Accounting Office. December 1986a. "EPA's Standard Setting Process Should Be More Timely and Better Planned" (Washington, DC: U.S. GAO).

U.S. General Accounting Office. December 1986b. "Water Quality. An Evaluation Method for the Construction Grants Program-Methodology" (Washington, DC: U.S. GAO).

U.S. General Accounting Office. December 1986c. "Water Quality. An Evaluation Method for the Construction Grants Program—Case Studies" (Washington, DC: U.S. GAO).

U.S. General Accounting Office. April 1987. "Consumer Product Safety Commission. Administrative Structure Could Benefit from Change" (Washington, DC: U.S. Government Printing Office).

U.S. General Accounting Office. June 1987. "Nuclear Regulation. A Perspective on Liability Protection for a Nuclear Plant Accident" (Washington, DC: U.S. Government Printing Office).

U.S. General Accounting Office. October 1987. "Hazardous Waste. Issues Surrounding Insurance Availability" (Washington, DC: U.S. Government Printing Office).

U.S. General Accounting Office. June 1988. "Hazardous Waste. Many Enforcement Actions Do Not Meet EPA Standards" (Washington, DC: U.S. GAO), p. 9.

U.S. General Accounting Office. November 1988a. Transition Series. "Department of Labor Issues" (Washington, DC: U.S. GAO).

U.S. General Accounting Office. November 1988b. Transition Series. "Environmental Protection Agency Issues" (Washington, DC: U.S. GAO).

U.S. General Accounting Office. November 1988c. Transition Series. "Program Evaluation Issues" (Washington, DC: U.S. GAO).

U.S. Nuclear Regulatory Commission. 1988. "NUREG-0325—NRC Functional Organization Charts" (Washington, DC: U.S. NRC, January 1).

Vertinsky, I., and P. Vertinsky. 1981. "Communicating Environmental Health Risk Assessment and Other Risk Information: Analysis of Strategies," in *Risk: A Seminar Series*, H. Kunreuther, Ed. (Laxenburg, Austria: International Institute for Applied Systems Analysis).

Vig, N. J. 1979. "Environmental Decision Making in the Lower Courts: The Reserve Mining Case," in *Energy and Environmental Issues*, M. Steinman, Ed. (Lexington, MA: Lexington Books).

Vig, N. J. 1984. "The Courts: Judicial Review and Risk Assessments," in *Risk Analysis, Institutions and Public Policy*, S. G. Hadden, Ed. (Port Washington, NY: Associated Faculty Press).

Vig, N. J., and M. E. Kraft, 1984. "Environmental Policy from the Seventies to the Eighties," in *Environmental Policy in the 1980s: Reagan's New Agenda*, N. J. Vig and M. E. Kraft, Eds. (Washington, DC: Congressional Quarterly, Inc).

Vlek, C., and P. Stallen. 1980. "Rational and Personal Aspects of Risk," *Acta Psychologica* 45:273–300.

Vogler, D. 1980. *The Politics of Congress* (Boston: Allyn and Bacon).

Wald, M. L. 1987. "Congress Debates Law Limiting Liability in Nuclear Accidents," *New York Times* (June 15).

Wald, M. L. 1989. "Liability for Exxon Oil Spill: Untested Waters," *New York Times* (April 7), p. B5.

Warren, R. L. 1967. "The Interorganizational Field as a Focus for Investigation," *Administrative Science Quarterly* 12:396–419.

Weber, M. 1946. *From Max Weber: Essays in Sociology*, H. H. Gerth and C. Wright Mills, translators (New York: Oxford University Press, 1906–1924; translated).

Weingast, B. R. 1978. "A Positive Model of Public Policy Formation: The Case of Regulatory Agency Behavior," Working Paper No. 25 (St. Louis: Washington University, Center for the Study of American Business, January).

Weiss, C. H. 1989. "Congressional Committees as Users of Analysis," *Journal of Policy Analysis and Management* 8(6):411–431.

Wenner, L. M. 1982. *The Environmental Decade in Court* (Bloomington: Indiana University Press).

Wenner, L. M. 1984. "Judicial Oversight of Environmental Deregulation," in *Environmental Policy in the 1980s: Reagan's New Agenda*, N. J. Vig and M. E. Kraft, Eds. (Washington, DC: Congressional Quarterly, Inc).

West, W. F. 1988. "The Growth of Internal Conflict in Administrative Regulation, *Public Administration Review* 48:773-782.

White, L. J. 1981. *Reforming Regulation. Processes and Problems* (Englewood Cliffs, NJ: Prentice-Hall, Inc.).

Whitney, S. C. 1973. "The Case for Creating a Special Environmental Court System," *William and Mary Law Review* 14.

Whittemore, A. S. 1983. "Facts and Values in Risk Analysis for Environmental Toxicants," *Journal of Risk Analysis* 3.

Wildavsky, A. 1988. *Searching for Safety* (New Brunswick, NJ: Transaction Books).

Wilson, J. Q., Ed. 1980. *The Politics of Regulation* (New York: Basic Books).

Windsor, D. A. 1988. "Disposal Tax," Letter to the Editor. *New York Times* (August 16).

Wood, W. C. 1983. *Nuclear Safety Risks and Regulation* (Washington, DC: American Enterprise Institute).

Wood, W. C. 1984. "Public Policy and the Imponderable Number: Insurance for Nuclear Accidents," in *Risk Analysis, Institutions and Public Policy*, S.G. Hadden, Ed. (Port Washington, NY: Associated Faculty Press).

Woodward, J. 1965. *Industrial Organization: Theory and Practice* (New York: Oxford University Press).

Wu, C., and P. F. Colwell. 1988. "Moral Hazard and Moral Imperative," *Journal of Risk and Insurance* 55:101–117.

Yellin, J. 1981. "High Technology and the Courts: Nuclear Power and the Need for Institutional Reform," *Harvard Law Review* 94.

Zimmerman, R. 1982. "Formation of New Organizations to Manage Risk," *Policy Studies Review* 1:736–748 (May).

Zimmerman, R. 1984. "Management Systems for Low-Probability/High-Consequence Events," in *Low Probability/High Consequence Risk Analysis*, R. A. Waller and V. T. Covello, Eds. (New York: Plenum Publishing Corporation), pp. 425–454.

Zimmerman, R. 1985a. "The Relationship of Emergency Management to Governmental Policies on Man-Made Technological Disasters," *Public Administration Review* 45:29–39 (January).

Zimmerman, R. 1985b. "Private Sector Response Patterns to Risks from Chemicals," in *Risk Analysis in the Private Sector*, C. Whipple and V. T. Covello (New York: Plenum Publishing Corporation), pp. 15–32.

Zimmerman, R. 1986. "The Management of Risk," in *Risk Evaluation and Management*, V. T. Covello, J. Menkes, and J. Mumpower, Eds. (New York: Plenum Publishing Corporation), pp. 435–460.

Zimmerman, R. 1987a. "A Process Framework for Risk Communication," *Science, Technology and Human Values* 12:131–137.

Zimmerman, R. 1987b. "Policy, Legal and Administrative Considerations for the Control of the Outdoor Environment," in *Public Health and the Environment*, M. R. Greenberg, Ed. (New York: Guilford Press), pp. 230–270.

Zimmerman, R. 1989. "Profiling Community Characteristics for the Application of Risk Communication Strategies to Hazardous Facility Siting and Remediation," project funded by the U.S. EPA. September 1, 1988–August 31, 1990 (New York: Graduate School of Public Administration, New York University).

Glossary of Abbreviations and Acronyms

ACGIH	American Conference of Governmental Industrial Hygienists
AEA	Atomic Energy Act
AEC	Atomic Energy Commission
AFL–CIO	American Federation of Labor–Congress of Industrial Organizations
AHERA	Asbestos Hazard Emergency Response Act
ALARA	As Low As Reasonably Achievable
ANPRM	Advance Notice of Proposed Rulemaking
AOUSC	Administrative Office of the U.S. Courts
APA	Administrative Procedure Act
ARAR	Applicable or Relevant and Appropriate Requirement
ATSDR	Agency for Toxic Substances and Disease Registry
BAT	Best Available Control Technology
BPT	Best Practical Control Technology
CAA	Clean Air Act
CAG	Carcinogen Assessment Group
CASAC	Clean Air Scientific Advisory Committee
CDC	Centers for Disease Control
CEQ	Council on Environmental Quality
CERCLA	Comprehensive Environmental Response, Compensation, and Liability Act
CFCs	chlorofluorocarbons
CFR	Code of Federal Regulations
CG	Coast Guard
CGLI	commercial (or comprehensive) general liability insurance
CMSA	Coal Mine Safety Act
COE	Army Corps of Engineers
CPSA	Consumer Product Safety Act
CPSC	Consumer Product Safety Commission
CWA	Clean Water Acts/Amendments
CZMA	Coastal Zone Management Act
DCA	Dangerous Cargo Act
DES	diethylstilbestrol
DHEW	Department of Health, Education, and Welfare
DHHS	Department of Health and Human Services
DOE	Department of Energy
DOL	Department of Labor

ECAO	Environmental Criteria and Assessment Office
EIA	Energy Information Administration
EIL	Environmental Impairment Liability
EIS	Environmental Impact Statement
EPA	Environmental Protection Agency
EPCRA	Emergency Planning and Community Right-to-Know Act
EPIC	Environmental Protection Insurance Corporation
ERDDAA	Environmental Research, Development, and Demonstration Authorization Act
FAA	Federal Aviation Act/Administration
FACA	Federal Advisory Committee Act
FDA	Food and Drug Administration
FDCA	Food, Drug, and Cosmetic Act
FEMA	Federal Emergency Management Agency
FEPCA	Federal Environmental Pesticide Control Act
FERC	Federal Energy Regulatory Commission
FHWA	Federal Highway Administration
FIFRA	Federal Insecticide, Fungicide, and Rodenticide Act
FMIA	Federal Meat Inspection Act
FRA	Federal Railroad Administration
FSIS	Food Safety and Inspection Service
FWPCA	Federal Water Pollution Control Act
FWPCAA	Federal Water Pollution Control Act Amendments
FWS	Fish and Wildlife Service
GAO	General Accounting Office
GRAS	Generally Recognized As Safe
HHS	Health and Human Services
HLPSA	Hazardous Liquid Pipeline Safety Act
HMTA	Hazardous Materials Transportation Act
HSA	Highway Safety Act
HSA	Hazardous Substances Act
HSWA	Hazardous and Solid Waste Amendments
HUD	Housing and Urban Development
IRLG	Interagency Regulatory Liaison Group
ITC	Interagency Testing Committee
LCCA	Lead Contamination Control Act
LLRWPA	Low Level Radioactive Waste Policy Act
LOEL	Lowest Observed Effect Level
LPPPA	Lead-based Paint Poisoning Prevention Act

MCLs	Maximum Contaminant Levels
MOCA	4,4'-methylene(bis)-2-chloroaniline
MPRSA	Marine Protection, Research, and Sanctuaries Act
MSHA	Mine Safety and Health Act/Administration
MVCSA	Motor Vehicle Carrier Safety Act
NAAQS	National Ambient Air Quality Standards
NAS	National Academy of Sciences
NBIP	National Bridge Inspection Program
NCA	National Cancer Act
NCA	Noise Control Act
NCSL	National Conference of State Legislatures
NDIA	National Dam Inspection Act
NEPA	National Environmental Policy Act
NESHAP	National Emission Standards for Hazardous Air Pollutants
NHTSA	National Highway Traffic Safety Administration
NIH	National Institutes of Health
NIOSH	National Institute for Occupational Safety and Health
NOEL	No Observed Effect Level
NPDES	National Pollutant Discharge Elimination System
NRC	Nuclear Regulatory Commission
NRDC	National Resources Defense Council
NSPS	New Source Performance Standards
NTMVSA	National Traffic and Motor Vehicle Safety Act
NTSB	National Transportation Safety Board
NWPA	Nuclear Waste Policy Act
OAQPS	Office of Air Quality Planning and Standards
ODBA	Ocean Dumping Ban Act
OHEA	Office of Health and Environmental Assessment
OSHA	Occupational Safety and Health Act/Administration
OSHRC	Occupational Safety and Health Review Commission
OSTP	Office of Science and Technology Policy
OTA	Office of Technology Assessment
PCBs	polychlorinated biphenyls
PLIA	Pollution Liability Insurance Association
PMN	Premanufacturing Notice
PPIA	Poultry Products Inspection Act
PPPA	Poison Prevention Packing Act
PRPs	potentially responsible parties
PSA	Pipeline Safety Act
PSD	Prevention of Significant Deterioration

PTSA	Port and Tanker Safety Act
PWSA	Ports and Waterways Safety Act
RCHSA	Radiation Control for Health and Safety Act
RCRA	Resource Conservation and Recovery Act
RHA	Rivers and Harbors Act
RMCLs	Recommended Maximum Contaminant Levels
RPAR	Rebuttable Presumption Against Registration
RSA	Railroad Safety Act
RSIA	Rail Safety Improvement Act
SAA	Safety Appliance Acts
SAB	Science Advisory Board
SAP	Science Advisory Panel
SARA	Superfund Amendments and Reauthorization Act
SBA	Seat Belts Act
SDWA	Safe Drinking Water Act
SDWAA	Safe Drinking Water Act Amendments
SIP	State Implementation Plan
SMCRA	Surface Mining Control and Reclamation Act
SNARL	Suggested No Adverse Response Level
STIC	System for Tracking the Inventory of Chemicals
SWDA	Solid Waste Disposal Act
TDP	Transferable Discharge Permit
TDRs	Transfers of Development Rights
TSCA	Toxic Substances Control Act
TSDF	treatment, storage, and disposal facility
TTHMs	total trihalomethanes
UMTA	Urban Mass Transportation Administration
UMTRCA	Uranium Mill Tailings Radiation Control Act
VDRESCA	Veterans' Dioxin and Radiation Exposure Compensation Standards Act
WILL	Waste Insurance Liability Limited
WLAs	wasteload allocations

Index